43
Topics in Heterocyclic Chemistry

Aims and Scope

The series Topics in Heterocyclic Chemistry presents critical reviews on present and future trends in the research of heterocyclic compounds. Overall the scope is to cover topics dealing with all areas within heterocyclic chemistry, both experimental and theoretical, of interest to the general heterocyclic chemistry community.

The series consists of topic related volumes edited by renowned editors with contributions of experts in the field.

More information about this series at http://www.springer.com/series/7081

Jean-Christophe M. Monbaliu
Editor

The Chemistry of Benzotriazole Derivatives

A Tribute to Alan Roy Katritzky

With contributions by

T. Albers · K. Bajaj · J.K. Beagle · O.I. Bolshakov · A.F. Gameiro · R. Gérardy · D.N. Haase · I.O. Lebedyeva · J.-C.M. Monbaliu · M.V. Povstyanoy · V. Povstyanoy · R. Sakhuja · D.L. Watkins

Springer

Editor
Jean-Christophe M. Monbaliu
Center for Integrated Technology and Organic Synthesis – CiTOS
Department of Chemistry
University of Liège
Liège
Belgium

ISSN 1861-9282 ISSN 1861-9290 (electronic)
Topics in Heterocyclic Chemistry
ISBN 978-3-319-31552-2 ISBN 978-3-319-31554-6 (eBook)
DOI 10.1007/978-3-319-31554-6

Library of Congress Control Number: 2016938784

Printed on acid-free paper

This Springer imprint is published by Springer Nature
The registered company is Springer International Publishing AG Switzerland

Dedicated to Alan Roy Katritzky

Preface

Each chapter within this volume critically surveys the applications of benzotriazole derivatives in a variety of important synthetic applications ranging from heterocyclic chemistry to peptide constructs. The most significant developments and concepts in benzotriazole methodology are presented using selected examples. Each chapter is designed to provide the non-specialist reader with all the important concepts and methodology that led to the development of a general benzotriazole methodology. Beyond the concepts and important developments, these contributions also offer an outlook on potential future developments in benzotriazole chemistry.

Chapter "Preparation, Reactivity, and Synthetic Utility of Simple Benzotriazole Derivatives" introduces the reader with the specific reactivity of benzotriazole and its simple derivatives. The preparation, synthetic properties and applications of a representative set of important and versatile benzotriazole derivatives are illustrated. Chapter "Acylbenzotriazoles: New Allies for Short Linear and Cyclic Peptide Constructs" gathers the most important advances in the preparation and use of acylbenzotriazole derivatives for the preparation of oligopeptide constructs. Chapter "Benzotriazole-Based Strategies Towards Peptidomimetics, Conjugates and Other Peptide Derivatives" extends the discussion on the versatility of acylbenzotriazole derivatives towards the preparation of peptidomimetics, conjugates and other peptide derivatives. Chapter "Benzotriazole-Mediated Synthesis of Oxygen Containing Heterocycles" discusses benzotriazole-mediated strategies towards oxygen-containing heterocycles, and Chapter "Benzotriazole-Mediated Synthesis of Nitrogen Containing Heterocycles" reviews benzotriazole-mediated strategies towards nitrogen-containing heterocycles. Last but not least, Chapter "Benzotriazole: Much More Than Just Synthetic Heterocyclic Chemistry" regroups various applications of benzotriazole and its derivatives to medicinal chemistry and materials sciences.

Dedication

Every domain in society has its own icons who are well known for their contributions and their leadership in the field. Certainly in sports, the role of icons is enormous and stimulates many youngsters to engage in certain sports.

Also, sciences have their icons.

When talking about heterocyclic synthesis, I believe that no one will doubt that Alan Katritzky is and has been an icon during his career and will stay one for a long time. Together with his good friend Charles Rees, they had a tremendous impact on the field and studied heterocycles in a very systematic way. The numerous books that Alan edited on heterocyclic chemistry organised the field in different classes of heterocycles. Alan studied extensively their specific reactivity, their conformational behaviour, their physicochemical properties and their aromaticity and heterocyclic rearrangements.

Picture: Alan R. Katritzky receives the honorary doctorate at Ghent University in 2001.

Alan Katritzky (born August 18, 1928) was raised and educated in England and prepared his first heterocyclic compound at the age of 15. From 1948 to 1958, he spent time at Oxford, obtaining his degree in 1952 and publishing his first benzotriazole paper in 1953. He performed doctoral work with Sir Robert Robinson and received a PhD in 2 years. In 1957, Alan moved to Cambridge and became the founding fellow of Churchill College, of which Sir John Cockcroft, Nobel Prize winner in 1951 for his work on atom splitting, was the first master. In 1962, Alan

Katritzky moved to the new University of East Anglia in Norwich as founding professor of chemistry where he met Sir Christopher Ingold, member of the Academic Planning Board. Quickly, he became dean of the faculty in East Anglia and succeeded in convincing the authorities to construct a new chemistry building which was opened by her Majesty the Queen. In 1980, Prof. Katritzky moved to Florida as Kenan professor of chemistry where he has been running since then a big international group of postdocs and PhDs studying several aspects of heterocyclic chemistry. Alan passed away on February 10, 2014, in Gainesville after a fully filled life of chemistry.

The contributions of Alan Katritzky to the international scientific literature are elaborate with over 2500 international peer-reviewed papers and numerous book series (*Advances in Heterocyclic Chemistry*, *Comprehensive Heterocyclic Chemistry*, etc.). His scientific drive was exceptional. Up to the last moments, Alan was taking initiatives and leading a big multicultural research group. His knowledge and drive have led to the creation of the Center for Heterocyclic Compounds at the Department of Chemistry in Gainesville, Florida, with a research group of around 50 postdocs and PhDs working on heterocyclic chemistry over the years. Organised as ever, he was able to get the best out of his co-workers and built an impressive network covering all continents. His work was also internationally recognised with more than ten honorary doctorate titles and numerous prestigious awards. His good friend Prof. Al Padwa categorised Alan as a super achiever.

Alan Katritzky also spread the word. He travelled all over the globe and presented his views with a clear voice. As we all remember, Alan did not need a microphone even lecturing in front of hundreds of chemists. He also loved to lecture in other languages since he was eager to learn foreign languages. Not only in academic circles was Alan well known, he was also very well recognised in the chemical industry and has been consulting for all major chemical and pharmaceutical companies in the USA and Europe.

Alan Katritzky was also a person with a great humanitarian spirit. He founded Arkivoc, an electronic scientific journal, in order to give opportunities to developing countries to publish their work for free and also download all other Arkivoc articles for free. He strongly believed that scientific information is key to human and social development and therefore he created Arkivoc. Arkivoc is functioning on a personal donation of Alan and his wife Linde, is supported by the FLOHET conference in Gainesville (Florida, USA) and is run through the efforts of a team of scientists who perform all the editorial work for free. Arkivoc was very close to his heart since he wanted really to help change the world for the better. He wanted to get things done and "bochra" (tomorrow) was not high on the list of his vocabulary.

He was a charismatic mentor of his team of collaborators. With an amazing working power and with firm leadership, diplomacy and British humour, he paved the way for a tremendous scientific career and helped many collaborators throughout their careers. Members and ex-group members were always warmly welcomed by Linde and Alan. Many of us have enjoyed the dinners and the selected excellent wines "at Prof's place".

Alan also enjoyed windsurfing and was an outstanding wine expert, with a huge interest in languages and travelling, but above all, he had an enormous scientific drive and passion for all kinds of science. We will always remember Alan for his incredible scientific memory and amazing personality.

Ghent, Belgium
August 2015

Christian Stevens

Contents

Top Heterocycl Chem (2016) 43: 1–66
DOI: 10.1007/7081_2015_179

Published online: 26 November 2015

Preparation, Reactivity, and Synthetic Utility of Simple Benzotriazole Derivatives

Romaric Gérardy and Jean-Christophe M. Monbaliu

Abstract The benzotriazole fragment is known to behave as (1) an excellent leaving group, (2) an electron-donating or an electron-withdrawing group, (3) an anion precursor, and (4) a radical precursor. It confers unique physicochemical properties to its immediate vicinity on various molecular scaffolds. This review covers the preparation and synthetic utility of versatile benzotriazole derivatives. The selected compounds are conveniently prepared from 1*H*-benzotriazole and are characterized by a huge synthetic potential. Their specific reactivity is discussed and illustrated with various examples ranging from methodology in organic chemistry to the total synthesis of complex structures.

Keywords Benzotriazole derivatives · Benzotriazole methodology · Molecular diversity · Versatile synthons

Contents

R. Gérardy and J.-C.M. Monbaliu (✉)
Center for Integrated Technology and Organic Synthesis, Department of Chemistry, University of Liège, 4000 Liège, Belgium
e-mail: jc.monbaliu@ulg.ac.be

1 Introduction

1*H*-Benzotriazole is a versatile synthetic auxiliary endowed with a unique set of physicochemical properties. Alan R. Katritzky spent most of his research career demonstrating that benzotriazole might be an ideal synthetic companion: easy on, easy off, stable under a variety of conditions, and recyclable [1–10]. Besides, benzotriazole is readily available in large quantities and is, most importantly, inexpensive. Anchored on molecular scaffolds, the benzotriazole moiety acts as an enabling group conveying its unique electronic, steric, and stereoelectronic properties to the surroundings. Four major properties of the benzotriazole fragment interplay and are responsible for the synthetic versatility of its derivatives: (1) excellent leaving group ability, (2) electron-donating or electron-withdrawing character, (3) stabilization of α-negative charges, and (4) stabilization of radicals. Most of benzotriazole derivatives are characterized by a long shelf-life, and their preparations are amenable to large scales. A variety of halogenated synthons have been advantageously replaced by benzotriazole surrogates, with benefits in terms of increased stability and reduced toxicity [1–10]. Most of benzotriazole derivatives are prepared as a mixture of two isomers – the 1*H*- and 2*H*-benzotriazole isomers. In some instances, both isomers display a similar reactivity while in some specific cases they display distinct reactivity profiles. A simple chromatography on silica gel usually suffices for their separation.

This review covers the preparation and synthetic utility of a representative set of simple, yet versatile, 1*H*-benzotriazole derivatives. Two important criteria were retained for selecting these benzotriazole derivatives: (1) the practical aspect of their preparation, with preferably no more than two steps from 1*H*-benzotriazole, and (2) the scope of their synthetic utility either in terms of molecular versatility and achievable diversity or in terms of enabling difficult synthetic transformations. The different benzotriazole reagents discussed below are organized into three main classes according to their structural features rather than to their tremendous variety of reactivity (Fig. 1). Class A includes benzotriazole reagents where the benzotriazol-1-yl fragment is directly connected to an activating heteroatom or group of atoms, such as 1*H*-chlorobenzotriazole (**1a**) and its bromo- and iodo-analogs, 1*H*-trimethylsilylbenzotriazole (**1b**), 1*H*-aminobenzotriazole (**1c**), 1-(methylsulfonyl)-1*H*-benzotriazole (**1d**) and other *N*-sulfonylbenzotriazole

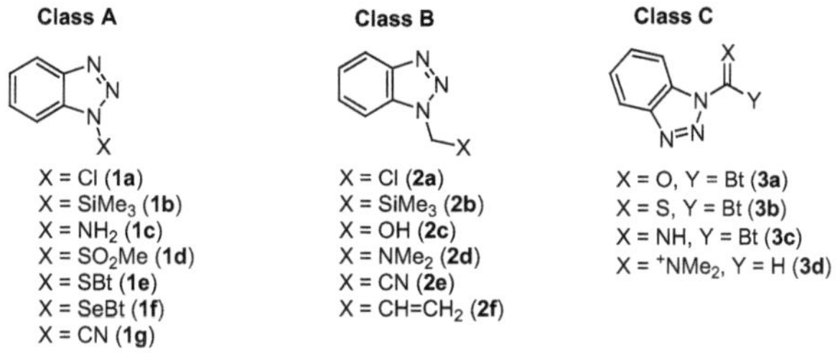

Fig. 1 Classes A–C emphasizing the structural features of selected benzotriazole derivatives

derivatives, bis(1*H*-benzotriazol-1-yl)sulfide (**1e**) and bis(1*H*-benzotriazol-1-yl) selenide (**1f**), and 1*H*-cyanobenzotriazole (**1g**). Class B regroups active methylene derivatives bearing one benzotriazol-1-yl fragment and another activating heteroatom or group of atoms such as 1-(chloromethyl)-1*H*-benzotriazole (**2a**), 1-(trimethylsilylmethyl)-benzotriazole (**2b**), 1*H*-benzotriazole-1-methanol (**2c**) and its alkoxy/aryloxymethyl-1*H*-benzotriazole derivatives, *N*,*N*-dimethylaminomethylbenzotriazole (**2d**) and its 1-(α-aminoalkyl/aryl)benzotriazole derivatives, 2-(1*H*-benzotriazol-1-yl)acetonitrile (**2e**), and 1-allylbenzotriazole (**2f**). Class C gathers compounds with sp^2 carbon centers bearing at least one benzotriazol-1-yl fragment such as bis-(1*H*-benzotriazol-1-yl)-methanone (**3a**), bis-(1*H*-benzotriazol-1-yl)-methanethione (**3b**), di(1*H*-benzotriazol-1-yl)methanimine (**3c**) and its derivatives, and the cation *N*,*N*-dimethylaminobenzotriazol-1-ylmethyleniminium (**3d**).

The different methods for preparing compounds **1a–g**, **2a–f**, and **3a–d** and some of their relevant derivatives, as well as their reactivity profile and synthetic utility, are thoroughly reviewed. Obviously, while the Katritzky group has tremendously contributed to spread and generalize this benzotriazole methodology, the contribution of other groups is also included in this review. Some benzotriazole derivatives are not considered in the following pages despite their widespread utilization in synthetic chemistry, such as 1-hydroxybenzotriazole and other benzotriazole-based peptide coupling reagents (pyBOP, HBTU, etc.), since they have been extensively reviewed over the last few years [11–13]. Acylbenzotriazole derivatives will also be discussed in this volume.

2 Class A Benzotriazole Derivatives 1a–g

2.1 1-Halogenobenzotriazoles (1-Chloro-, 1-Bromo-, and 1-Iodobenzotriazole)

1-Halogenobenzotriazoles **1a** and **4,5** are conveniently prepared from benzotriazole. While 1-bromo- and 1-iodobenzotriazoles were scarcely studied, 1-chlorobenzotriazole appeared as a convenient chloronium cation donor and as a versatile oxidant.

Scheme 1 Preparation of 1-halogenobenzotriazoles

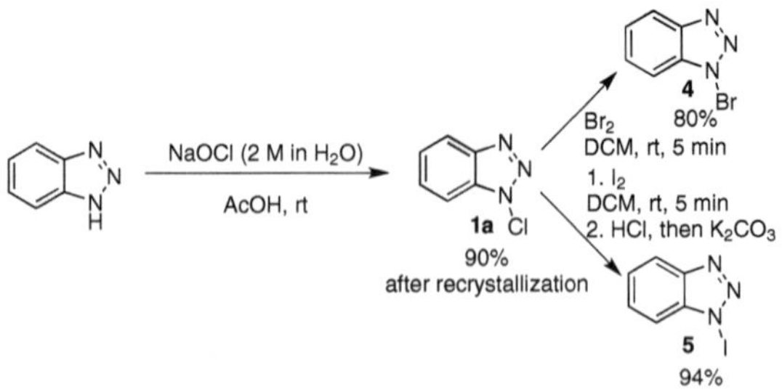

2.1.1 Preparation of 1-Halogenobenzotriazoles

The first report on 1-chlorobenzotriazole (**1a**) dates back to 1969, with Storr's study of its preparation and oxidizing properties [14]. In the original procedure, **1a** was conveniently prepared from benzotriazole and sodium hypochlorite in high yield (Scheme 1). Commercial bleach or *tert*-butyl hypochlorite can also be used [15]. The procedure is amenable on large scale (>50 g), and a pure sample of 1-chlorobenzotriazole can be obtained by recrystallization in dichloromethane/ petroleum ether [14] or hexanes [15]. The white crystals obtained after recrystallization are extremely shock sensitive. Thermogravimetric analysis showed that **1a** starts to decompose at 90°C [15]. A violent reaction occurs when **1a** is dissolved in DMSO [14].

Little is reported regarding the preparation of 1-bromobenzotriazole (**4**) and 1-iodobenzotriazole (**5**). **4** is prepared by the addition of **1a** to a solution of bromine in dichloromethane (Scheme 1) [16]. Similarly, 1-iodobenzotriazole (**5**) is prepared by the addition of **1a** to a solution of iodine in dichloromethane (Scheme 1) [16]. In contrast to 1-chlorobenzotriazole, 1-bromo and 1-iodobenzotriazole are barely soluble in organic solvents, due to an increased polarity of the *N*-X bond [16].

2.1.2 Reactivity and Synthetic Utility of 1-Halogenobenzotriazoles

1-Halogenobenzotriazoles are oxidants, but **4** and **5** appeared to be less convenient than **1a** due to their low solubility in organic solvents and despite a higher reactivity [14, 16].

The reactivity of 1-chlorobenzotriazole (**1a**) was widely studied [14–60]. By contrast to other *N*-chloro derivatives, **1a** displays a moderate global electrophilicity and behaves as a strong electropositive chlorine atom donor [15]. Computations at the B3LYP/6-31+G* level of theory revealed that **1a** will rather undergo homolytic (44.7 kcal mol^{-1} in dichloromethane) than heterolytic cleavage (203.7 kcal mol^{-1} in dichloromethane) [15]. Rees and Storr demonstrated that the addition of **1a** to olefins is ionic, while its reaction as an oxidant proceeds through radical addition [60].

The original study of the reactivity of **1a** was reported by Storr in 1969 [60]. 1-Chlorobenzotriazole reacted with olefins to give both 1*H*- and 2*H*-(2-chloroethyl)

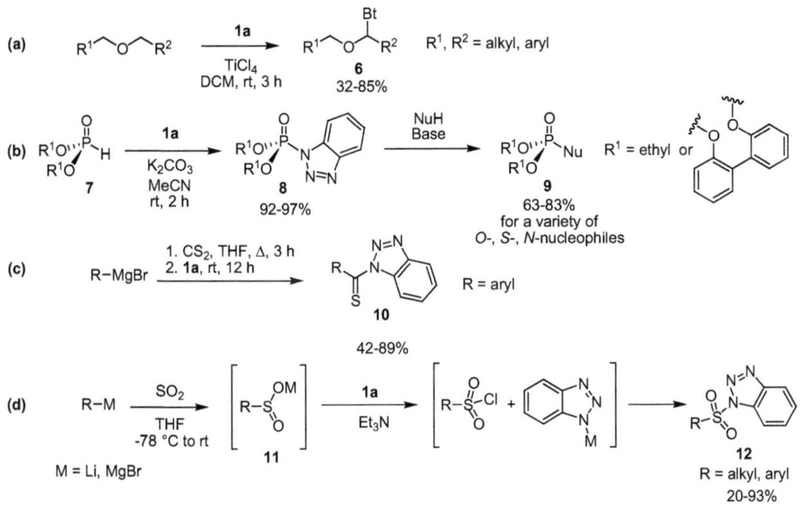

Scheme 2 (**a**) Preparation of α-(benzotriazol-1-yl)alkyl ethers; (**b**) preparation of 1*H*-benzotriazol-1-yl-1-phosphonate derivatives; (**c**) synthesis of thiocarbonylbenzotriazoles; (**d**) preparation of *N*-substituted sulfonylbenzotriazoles

benzotriazole derivatives in good yields and gave exclusively the *trans*-Markovnikov adduct [60]. The reaction proceeded through the initial addition of the chloronium electrophile and then the addition of the benzotriazolate anion. Katritzky reported the reaction of **1a** with ethers in the presence of a Lewis acid ($TiCl_4$) to afford α-(benzotriazol-1-yl)alkyl ethers **6** (Scheme 2a) [41]. It was postulated that the presence of a Lewis acid would increase the ionic nature of the reaction and hence improves the yield compared to the non-catalyzed reaction.

Similarly to its reaction with aldehydes [38], 1-chlorobenzotriazole (**1a**) reacts with dialkyl and diarylphosphites **7** in the presence of a base, yielding 1*H*-benzotriazol-1-yl-1-phosphonate derivatives **8** of synthetic value for the phosphonylation of various *N*-, *S*-, and *O*-nucleophiles [18]. 1*H*-Benzotriazol-1-yl-1-phosphonate derivatives **8** displayed higher stabilities than their corresponding chloro-analogs (Scheme 2b).

1-Chlorobenzotriazole (**1a**) was also used as a mediating reagent for the synthesis of thiocarbonylbenzotriazoles **10** from Grignard reagents and carbon disulfide. The reaction proceeds with moderate to good yields with a range of aromatic substrates (Scheme 2c) [19, 31]. Similarly, Katritzky developed a method for the preparation of *N*-substituted sulfonylbenzotriazoles **12** by the reaction of sulfinic salts **11** and **1a**. Sulfinic salts **11** were generated in situ from the reaction of organometallic reagents and sulfur dioxide (Scheme 2d) [28, 34]. *N*-substituted sulfonylbenzotriazoles **12** are convenient sulfonylating agents and are important building blocks for the synthesis of a variety of sulfur-containing molecules such as sulfonamides, α-cyanoalkylsulfones,

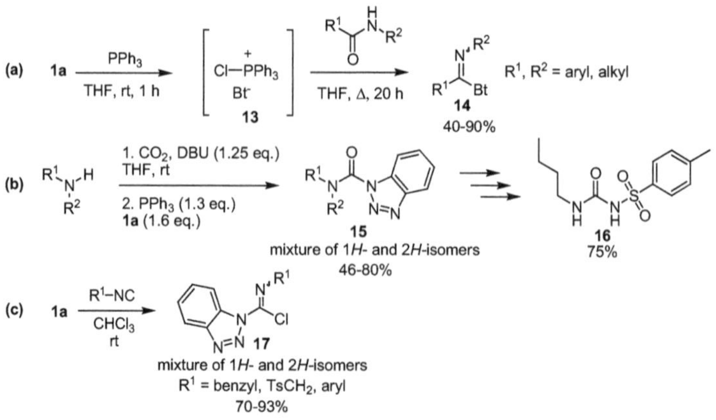

Scheme 3 (**a**) Preparation of imidoylbenzotriazoles from amides and chlorotriphenylphosphonium benzotriazolate; (**b**) one-pot synthesis of carbamoylbenzotriazole derivatives from CO_2 and synthesis of Tolbutamide; (**c**) reaction of 1-chlorobenzotriazole and isonitriles

sulfonyl-heteroaromatics, α-sulfonylalkylheterocycles, α-sulfonylalkyl sulfones, esters of α-sulfonyl acids, and alkyl/arylsulfonyl azides (see Sect. 2.4) [28, 32].

Secondary amides react with 1-chlorobenzotriazole (**1a**) in the presence of triphenylphosphine to afford imidoylbenzotriazoles **14**, which are important building blocks for accessing enaminones [33]. The mechanism proceeds through the formation of an oxophilic chlorotriphenylphosphonium **13**, which reacts with the amide similarly to a Vilsmeier-Haack reaction. The method developed by Katritzky using **1a** afforded a more efficient and versatile entry towards imidoylbenzotriazoles **14** than the existing strategies and is compatible with both aliphatic and aromatic substituents (Scheme 3a) [33]. The oxophilic chlorotriphenylphosphonium **13** was also utilized by Hunter et al. for the synthesis of carbamoylbenzotriazole derivatives **15** directly from CO_2 [23]. CO_2 gas was trapped with a primary or a secondary amine in the presence of DBU to give carbamate salts, which reacted with chlorotriphenylphosphonium benzotriazolate **13** to give carbamoylbenzotriazoles **15**. The strategy was then applied for the synthesis of Tolbutamide **16** (Scheme 3b) [23].

Benzotriazole-1-carboximidoyl chlorides **17**, which are nontoxic and stable synthetic equivalents for isocyanide dichlorides, were obtained from the reaction of 1-chlorobenzotriazole (**1a**) with isonitriles in chloroform at room temperature (Scheme 3c) [39]. *N*-Functionalized benzotriazole-1-carboximidoyl chlorides **17** were then reacted with a variety of nucleophiles to give polysubstituted guanidines, *S*-aryl isothioureas, and 2-aminoquinazolin-4-thiones. **1a** also reacts with cyanide to form 1-cyanobenzotriazole (**1g**) through the intermediate formation of cyanogen chloride (see Sect. 2.6) [42].

More recently, Hunter et al. reported a high-yielding and convenient procedure for the synthesis of heterodisulfides using 1-chlorobenzotriazole (**1a**) as an oxidant (Scheme 4) [17, 24, 25, 30]. The scope of this methodology is quite impressive and

Scheme 4 Hunter's procedure utilizing 1-chlorobenzotriazole for the preparation of unsymmetrical disulfides

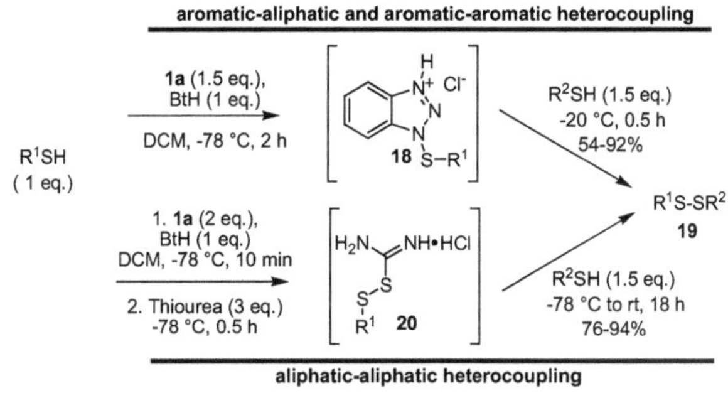

is applicable to all types of thiols (aliphatic, aromatic, heteroaromatic). The key feature of this one-pot strategy is the in situ formation of benzotriazol-1-yl sulfenyl derivative **18**, which prevents homocoupling. Intermediate **18** is then further reacted with a second thiol to effect heterocoupling towards the desired heterodisulfide **19**. Aromatic-aliphatic and aromatic-aromatic disulfide synthesis required a slight excess of **1a** as an oxidant, 1 eq. of benzotriazole as a carrier and 1.5 eq. of the second thiol. The procedure for aliphatic-aliphatic heterocoupling required some adjustments to avoid contamination with primary and secondary homodimers. Primary homocoupling was suppressed with an increased excess of **1a**. The addition of a large excess of thiourea suppressed secondary homocoupling by the intermediate of compound **20** and annihilated the excess of **1a**. Cysteine hetereosulfides were also prepared in excellent yield. The method was successfully applied to sensitive glycosyl substrates [25], for the preparation of redox probes [22] and cationic lipophosphoramidates [20]. Sulfur activation with **1a** was also studied for the cleavage of benzyl thiols [35].

1-Chlorobenzotriazole (**1a**) also oxidizes a variety of substrates, including sulfinic acids [27, 34], sulfinamides [21, 55], sulfides [48], triarylformazans [46], and alcohols [52]. It was recently demonstrated by Monbaliu and Katritzky that **1a** oxidizes quantitatively a variety of oximes into the corresponding α-chloronitroso derivatives **21** (Scheme 5) [15], while its reaction with imines leads to the benzotriazole analogs of α-haloenamines [40]. Similarly, the benzotriazole analogs of α-haloenols were obtained from enol ethers [44].

1-Chlorobenzotriazole (**1a**) was utilized for the radical chlorination of various indole scaffolds [57, 59] and other aromatic substrates such as carbazole [36, 37, 43, 51] and proton sponge derivative **22** (Scheme 6) [26].

The reactivity of 1-bromobenzotriazole (**4**) and 1-iodobenzotriazole (**5**) was scarcely studied in the literature. Due to an increased *N*-X polarization, they are much more reactive than **1a** for the electrophilic addition to alkenes. For instance, the reaction of 1-bromobenzotriazole (**4**) with cyclohexene was instantaneous at room temperature, while it took 4 h with **1a** [14, 16].

Scheme 5 Preparation of α-chloronitroso derivatives

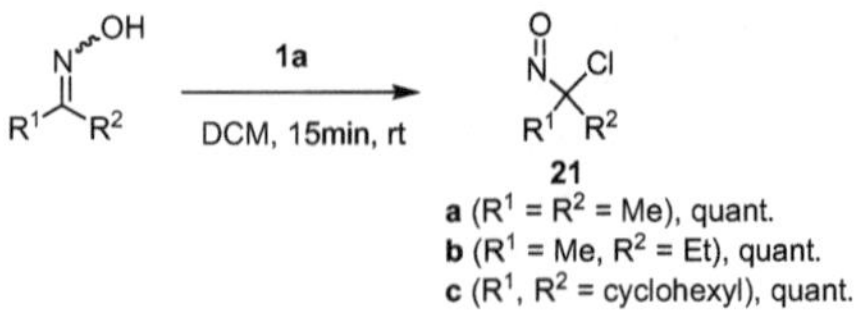

Scheme 6 Chlorination of a proton sponge derivative using 1-chlorobenzotriazole

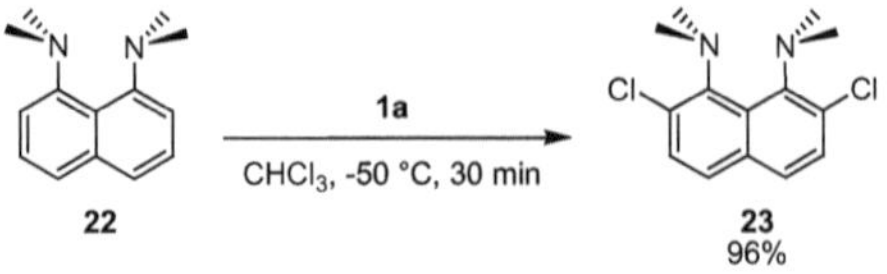

2.2 1-(Trimethylsilyl)benzotriazole

1-(Trimethylsilyl)benzotriazole (**1b**) is an activated source of the nucleophilic benzotriazolate anion [61, 62]. The trimethylsilyl counterpart intervenes as a Lewis acid [63].

2.2.1 Preparation of 1-(Trimethylsilyl)benzotriazole

1-(Trimethylsilyl)benzotriazole (**1b**) is usually prepared from the reaction of benzotriazole and a silylation derivatization reagent, such as hexamethyldisilazane (Birkofer's procedure) or *N*,*O*-bis(trimethylsilyl)acetamide (Scheme 7) [64–66]. In a typical procedure, benzotriazole is treated with an excess of *N*,*O*-bis (trimethylsilyl)acetamide under an inert atmosphere and refluxed for 1 h. The volatile compounds are removed under vacuum, and the entire operation is repeated twice, affording **1b** in quantitative yield [64–66]. Alternatively, **1b** can be prepared from benzotriazole and trimethylsilylchloride in the presence of a base, although this method is less convenient [66].

2.2.2 Reactivity and Synthetic Utility of 1-(Trimethylsilyl)benzotriazole

1-(Trimethylsilyl)benzotriazole (**1b**) is an activated source of the nucleophile benzotriazolate anion [61, 62] and is a convenient starting material for a variety of benzotriazole-based reagents. In some examples, the trimethylsilyl counterpart was believed to intervene as a Lewis acid [63]. 1-(Trimethylsilyl)benzotriazole (**1b**) reacts with thiophosgene to afford bis(benzotriazol-1-yl)methanethione (**3b**) [67–70], which is a nontoxic equivalent of thiophosgene and a common thioacylating reagent (see Sect. 4.1). **1b** was also used as a starting material for the synthesis of 1,1′-(sulfonyl)bisbenzotriazole (**24**) by reaction with sulfuryl chloride in toluene (Scheme 8) [71]. Similarly, the reaction of **1b** with sulfonyl chloride in tetrahydrofuran (THF) at room temperature affords 1,1′-(sulfinyl)bisbenzotriazole (**25**)

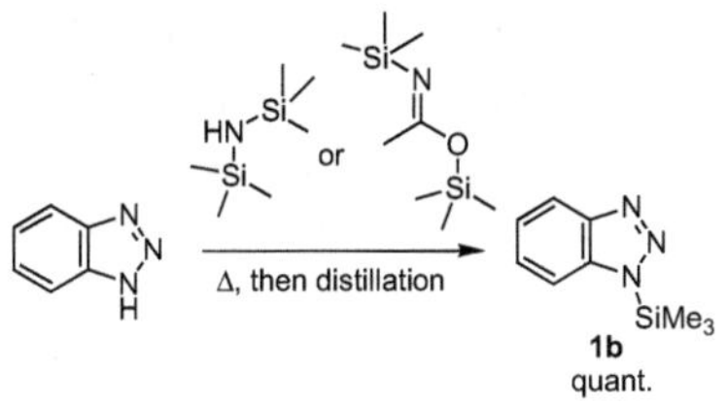

Scheme 7 Preparation of 1-(trimethylsilyl)benzotriazole

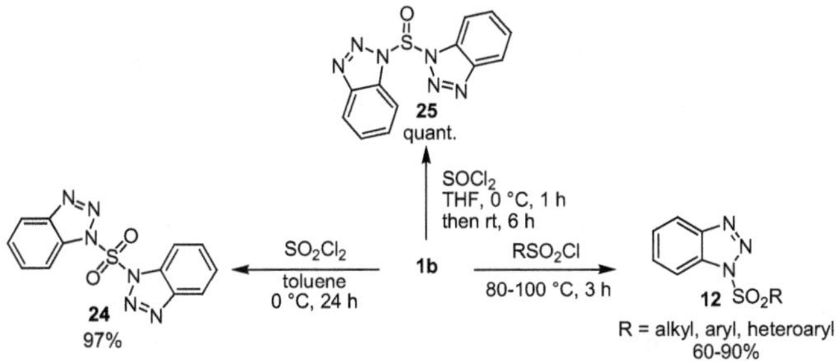

Scheme 8 Reactions of 1-(trimethylsilyl)benzotriazole with sulfuryl chloride, sulfonyl chloride and aryl/alkylsulfonyl chlorides

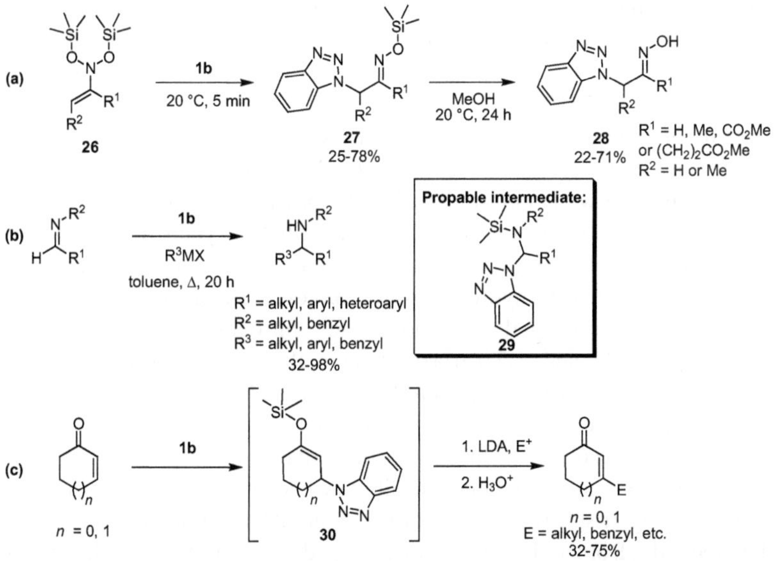

Scheme 9 (**a**) Preparation of α-benzotriazolyl-substituted oximes from *N*,*N*-bis(siloxy)enamines and 1-(trimethylsilyl)benzotriazole; (**b**) reaction of 1-(trimethylsilyl)benzotriazole with various imines in the presence of organometallic reagents; (**c**) reaction of 1-(trimethylsilyl)benzotriazole with enones

[72, 73]. In the presence of *N*,*N*-dimethylformamide (DMF), the same mixture of reagents yields *N*,*N*-dimethylaminobenzotriazolylmethyleniminium (**3d**) chloride, which is a carbene precursor (see corresponding Sect. 4.3) and a synthetic equivalent of Vilsmeier's salt [74]. 1-(Methanesulfonyl)benzotriazole (**1d**) and other alkyl/aryl sulfonylbenzotriazole derivatives **12**, which are useful reagents for converting carboxylic acids into their 1-acylbenzotriazole derivatives, are obtained in moderate to high yields by reacting **1b** and the corresponding aryl/alkylsulfonyl chlorides (Scheme 8, see also Sect. 2.4) [62].

1-(Trimethylsilyl)benzotriazole also reacts with enamine derivatives, imines, enones, aldehydes, and ketones [65, 75–79]. For example, Ioffe et al. developed a procedure for the synthesis of α-azolyl-substituted oximes **27** from *N*,*N*-bis(siloxy) enamines **26** and *N*-silylated azoles such as **1b** (Scheme 9a). Silylated α-azolyl-substituted oximes **27** were desilylated to yield the corresponding α-azolyl-substituted oximes **28** [65].

Katritzky et al. reported a general procedure towards secondary amines from the addition of organometallic reagents to various imines in the presence of **1b** [77]: equimolar amounts of **1b** and the appropriate imine were dissolved in dry toluene in the presence of alkyl/aryl Grignard or Reformatsky reagents. The reaction proceeded smoothly and afforded the corresponding amines in 32–98% yield. The presence of **1b** was crucial since in its absence; the yield for the addition of organometallic reagents to imines remained very low. It was believed to proceed through the reversible addition of **1b** to the imine, giving silylated intermediate **29** (Scheme 9b). Katritzky also studied the reaction of **1b** with enones (Scheme 9c) [76]. The reaction proceeded through the formation of a 1,4-silylated benzotriazole-adduct **30**. The latter was successively treated with LDA and an electrophile, to give the corresponding 2-cycloalkenones after acidic treatment (one-pot procedure) in moderate yields (32–75%).

Some reports emphasized that 1-(trimethylsilyl)benzotriazole (**1b**) has a markedly different reactivity than other silylated *N*-containing heterocycles [63, 79, 80]. For instance, Howell et al. studied the ring-opening reactions of 1,5-dioxaspiro [3.2]hexanes **31** with various nitrogen-containing heterocycles [63]. Depending on the nature of the nucleophilic *N*-containing heterocycle, two major products were observed: α-substituted-β′-hydroxyketones **32** or 2,2-disubstituted oxetanes **33** (Scheme 10). As a rule of thumb, more acidic nitrogen-containing heterocycles (imidazole, pyrazole and 1,2,4-triazole) gave the corresponding α-substituted-β′-hydroxyketones **32** as the major product, while 1,2,3-triazole and benzotriazole gave the corresponding 2,2-disubstituted oxetane derivatives **33**. Silylated derivatives gave similar results, and the TMS counterpart was supposed to intervene as a Lewis acid. The same reaction with lithium benzotriazolate unexpectedly gave the corresponding α-substituted-β′-hydroxyketone **32** in poor yield [63].

More recently, Boto and Hernández developed a new strategy for the preparation of highly functionalized iminosugar-based nucleosides from readily available proline derivatives [64, 81]. The procedure combines a tandem radical decarboxylation-oxidation-β-iodination with the addition of activated nitrogen bases such as 1-(trimethylsilyl)benzotriazole (**1b**). Proline derivative **34** was treated with iodine

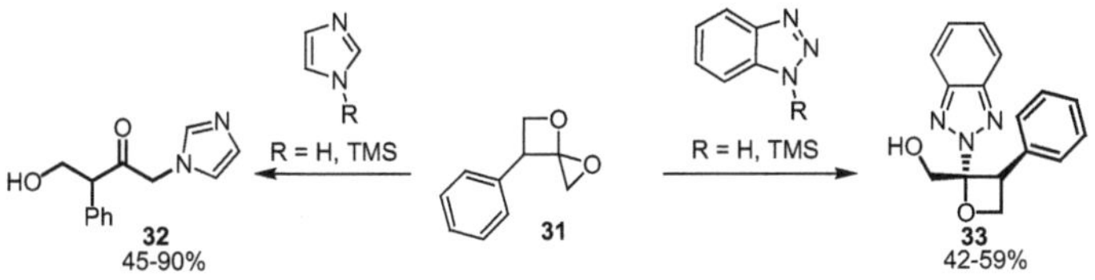

Scheme 10 Ring-opening reactions of 1,5-dioxaspiro[3.2]hexanes with various nitrogen-containing heterocycles

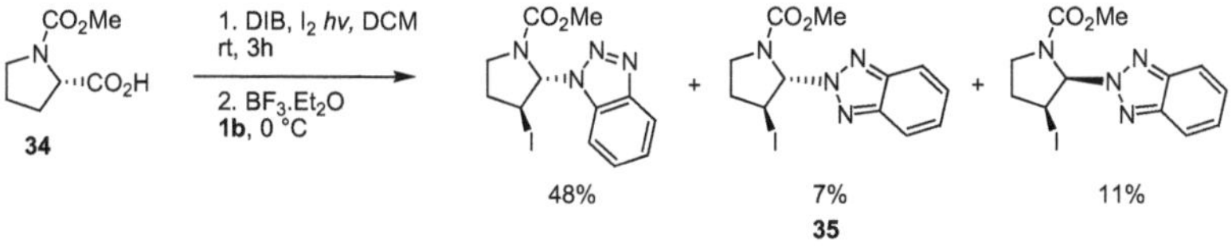

Scheme 11 Preparation of highly functionalized iminosugar-based nucleosides from proline derivatives

in the presence of (diacetoxyiodo)benzene (DIB) in dichloromethane under visible light irradiation to trigger a radical decarboxylation-oxidation-β-iodination sequence. Then, **1b** was added in the presence of $BF_3.Et_2O$ to effect the addition of the activated benzotriazole. The one-pot process afforded the desired iodinated iminosugar-based nucleoside analog **35** as a mixture of several isomers (Scheme 11). The procedure was successfully applied to other substrates [71]. **1b** also reacts with less conventional electrophiles such as benzeneselenyl chloride to give unconventional selenylating reagents [82], or with bromodimethylborane to afford triazabole analogs [83].

2.3 *1*H*-Aminobenzotriazole*

1*H*-Aminobenzotriazole (**1c**) is conveniently prepared according to a one-step procedure from benzotriazole. The synthetic utility of **1c** revolves around the generation of benzyne under oxidative conditions and its use as a reactive nitrogen-containing nucleophile.

2.3.1 Preparation of 1*H*-Aminobenzotriazole

The original procedure reported by Rees and Campbell in 1969 [60, 84] started from 2-nitroaniline and required four steps to get 1-aminobenzotriazole (**1c**) with an overall 20% yield. An updated procedure was disclosed in 2000 by Knight and Little, requiring a single step from benzotriazole [85]. **1c** was prepared accordingly by the reaction of benzotriazole with hydroxylamine-*O*-sulfonic acid in the

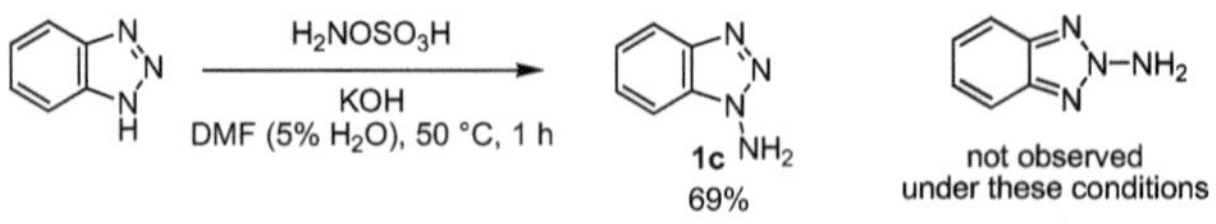

Scheme 12 Chemoselective preparation of 1*H*-aminobenzotriazole

presence of KOH in DMF containing 5% water at 50°C (Scheme 12). Contamination with significant amounts of the 2*H*-isomer was observed when the reaction was carried out in other solvents and at higher reaction temperatures. The procedure was amenable to large scale with high yield and selectivity (38.8 g, up to 69% yield).

2.3.2 Reactivity and Synthetic Utility of 1*H*-Aminobenzotriazole

1*H*-Aminobenzotriazole (**1c**) is primarily utilized as an alternative to conventional benzyne sources such as benzene-diazonium-2-carboxylate, benzothiadiazole dioxide, diphenyliodonium-2-carboxylate, dihalogenobenzenes, and aryltriflates [86–88] for the generation of benzyne under non-basic conditions [60, 84, 85, 89–100]. Since the original report by Rees [60], lead tetraacetate is preferentially used for the clean oxidation of **1c**, while other oxidants such as *N*-bromosuccinimide [60], iodobenzene diacetate [60, 100], activated MnO_2 [60], selenium dioxide [60], mercuric oxide [60], and potassium permanganate were also reported, yet often yielding complex mixtures [60, 100]. Cytochrome P-450 is suspected to oxidize **1c** to benzyne [101–103]. By contrast, the 2*H*-isomer of aminobenzotriazole is not a benzyne precursor.

The oxidation of 1*H*-aminobenzotriazole (**1c**) with lead tetraacetate is a unique method for the generation of benzyne since it can be carried out over a large range of temperatures, from −78°C to room temperature [93, 95]. The oxidation, if not carried out in the presence of a proper quench, is known to produce large amounts of the benzyne dimer (**36**), probably due to the formation of a metal-benzyne complex (Scheme 13) and phenyl acetate because of the presence of acetic acid in $Pb(OAc)_4$ [95]. The mechanism probably goes through the formation of a nitrene that decomposes to form 2 eq. of nitrogen gas and benzyne [95].

The oxidation of **1c** was reported in the presence of a variety of substrates and enabled the preparation of complex structures such as germathiiranes [98] and various azulenes [97]. In 1989, Rigby et al. reported the oxidation of 1-aminobenzotriazole in the presence of vinyl isocyanates to afford the corresponding [4+2] cycloadducts **37** in moderate yields (Scheme 14a) [94]. The oxidation of **1c** was studied in the presence of oxazoles **38** (Scheme 14b) by Rickborn [93, 96]. The order of addition is important; **1c** and lead acetate should be added simultaneously to avoid building-up high concentrations of benzyne and hence avoid the formation of the corresponding dimer (**36**) [93]. The benzyne-oxazole cycloadduct **39** underwent a thermal retro Diels-Alder process, releasing benzonitrile and isobenzofuran (**40**), which could be trapped in the presence of

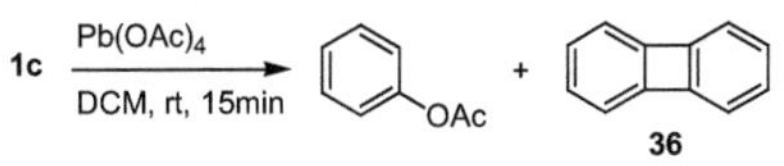

Scheme 13 Oxidation of 1*H*-aminobenzotriazole in the presence of lead tetraacetate

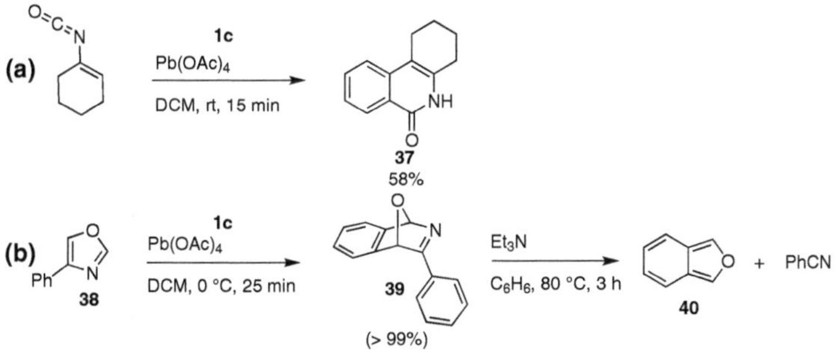

Scheme 14 (**a**) Oxidation of 1*H*-aminobenzotriazole in the presence of vinyl isocyanates; (**b**) Oxidation of 1*H*-aminobenzotriazole in the presence of oxazoles

other dienophiles [93]. Other less conventional dienes were also reported for trapping benzyne generated from **1c** [89].

In 2000, Knight and coworkers developed an elegant route towards chromanes **45** and chromenes **47** starting from **1c** (Scheme 15) [85, 91]. The sequence started with the Boc-protection of **1c**. Since the enhanced nucleophilicity (see below) of the amino group of **1c** precluded direct mono-Boc protection, its reaction with 2 eq. of Boc_2O gave the corresponding *bis*-Boc derivative, which was then selectively hydrolyzed with aqueous NaOH. *N*-Boc-1-amino-benzotriazole **41** was then treated with 2.2 eq. of *n*-BuLi to afford the corresponding *ortho*-bis anion. The latter was then quenched with diodoethane in the presence of $CeCl_3$ to give selectively the *ortho*-iodided derivative **42**, which was next used in a Sonogashira coupling to furnish **43**. Complete alkyne reduction of **43** gave benzotriazole derivative **44**, and alkene **46** was obtained by partial reduction of **43**. After reduction and Boc-deprotection, the corresponding 1*H*-benzotriazole derivatives were oxidized with *N*-iodosuccinimide (NIS) to generate benzyne transient species, which were quenched via an intramolecular reaction with the hydroxyl group leading to chromanes **45** or chromenes **47**, respectively, in high yields.

Besides its use as a benzyne precursor, 1*H*-aminobenzotriazole (**1c**) behaves also as an enhanced nitrogen-containing nucleophile due to a pronounced α-effect, [85] and reacts with a variety of electrophiles, such as aldehydes and ketones, to give the corresponding *N*-benzotriazol-1-ylimines [104–106]. El Kaim et al. reported that *N*-benzotriazol-1-ylimines are suitable precursors of iminyl radicals in the presence of Bu_3SnH and azobisisobutyronitrile (AIBN). These iminyl radicals are capable of interesting synthetic behaviors. For instance, nitrile **48** was formed after the radical fragmentation of the parent *N*-benzotriazol-1-ylimine, while 3,4-dehydropyrrole **49**

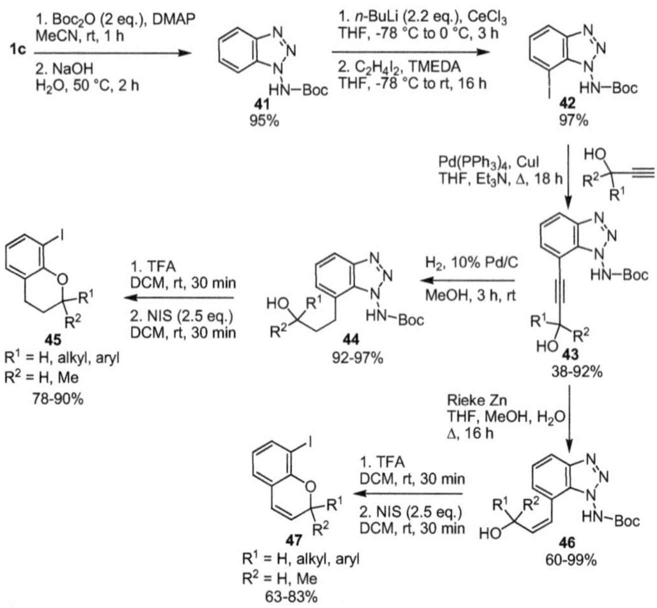

Scheme 15 Preparation of chromanes and chromenes from 1*H*-aminobenzotriazole

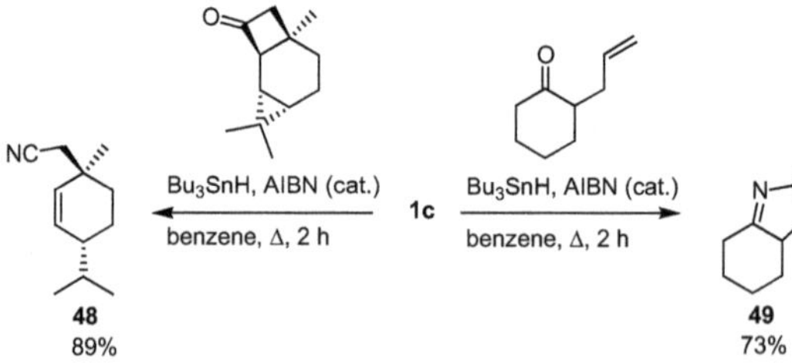

Scheme 16 Preparation of *N*-benzotriazol-1-ylimines and their subsequent radical fragmentation (*left*) or cyclization (*right*)

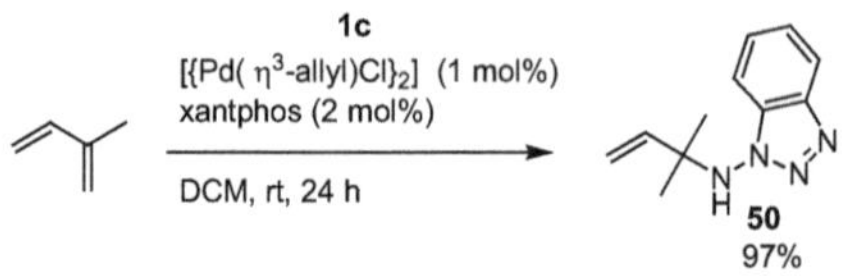

Scheme 17 Pd-catalyzed hydroamination of 1,3-dienes with 1*H*-aminobenzotriazole

was formed after the intramolecular cyclization of the parent *N*-benzotriazol-1-ylimine (Scheme 16) [105].

1*H*-Aminobenzotriazole (**1c**) underwent addition to 1,3-dienes in the presence of $[\{Pd(\eta^3\text{-allyl})Cl\}_2]$ to afford the branched addition product **50** in high yield (Scheme 17) [107].

2.4 N-Sulfonylbenzotriazoles

N-Sulfonylbenzotriazoles **12** were widely utilized by the Katritzky group as versatile reagents for the preparation of acylbenzotriazoles, and this topic will be thoroughly covered in the present volume. *N*-Sulfonylbenzotriazoles **12** were also employed for the preparation of sulfonamides and for the *C*-sulfonylation of various substrates towards sulfones. The preparation and reactivity of 1,1′-(sulfinyl)bisbenzotriazole (**25**) will be illustrated as well.

2.4.1 Preparation of *N*-Sulfonylbenzotriazoles

1-(Methylsulfonyl)-1*H*-benzotriazole (**1d**) is prepared according to a convenient one-pot procedure from benzotriazole and was generalized to a variety of alkyl/aryl *N*-sulfonylbenzotriazole derivatives [32, 108–113]. In a typical procedure, benzotriazole is directly reacted with methanesulfonyl chloride [114] in the presence of a base, generally pyridine, affording **1d** in 89% yield (Scheme 18a). Alkyl/aryl *N*-sulfonylbenzotriazole derivatives typically have a long shelf-life and are usually obtained as the benzotriazol-1-yl isomer [113]. An earlier paper by Katritzky also reported the preparation of **1d** in 60% yield by the reaction of 1-(trimethylsilyl)benzotriazole (**1b**) and methanesulfonyl chloride (see Scheme 8) [62]. Alternatively, *N*-sulfonylbenzotriazoles **12** were also prepared according to a two-step process, involving 1*H*-chlorobenzotriazole (**1a**, see Scheme 2d) [34]. This method was reported as quite general and especially useful for the preparation of *N*-sulfonylbenzotriazoles **12** derived from sulfonyl halides difficult to access or unstable. This procedure, however, failed with allylic and acetylenic organometallic starting materials [34]. Perfluoroalkyl *N*-sulfonylbenzotriazoles, such as benzotriazole trifluoromethanesulfonate (**51**), were obtained by the reaction of benzotriazole and the corresponding perfluoroalkanesulfonic anhydride [109] (Scheme 18b) or by the reaction of lithium benzotriazolate and perfluoroalkanesulfonyl fluoride [115].

Benzotriazol-1-yl-sulfonyl azide (**52**) is prepared according to a one-pot two-step procedure from benzotriazole (Scheme 19) [116]. The first step involves the reaction of sodium azide and sulfuryl chloride to produce chlorosulfonyl azide, which is then converted into **52** by addition of benzotriazole and pyridine. Benzotriazol-1-yl-sulfonyl azide (**52**) has a long shelf-life but is shock and friction-sensitive and must be carefully handled and stored.

2.4.2 Reactivity and Synthetic Utility of *N*-Sulfonylbenzotriazoles

1-(Methylsulfonyl)-1*H*-benzotriazole (**1d**) is one of the most common *N*-sulfonylbenzotriazole derivatives [32, 34, 62, 111, 114, 117–145] and was widely utilized as an activating agent for carboxylic acids. Carboxylic acids are converted in the

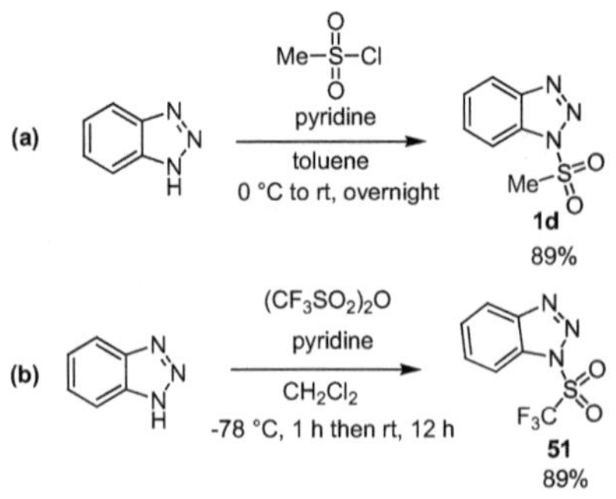

Scheme 18 (**a**) One-pot procedure towards 1-(methylsulfonyl)-1*H*-benzotriazole; (**b**) Preparation of 1-trifluoromethylsulfonylbenzotriazole

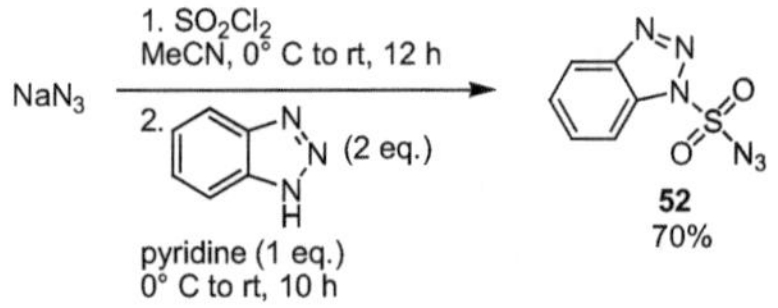

Scheme 19 One-pot two-step procedure from benzotriazole towards benzotriazol-1-yl-sulfonyl azide

presence of **1d** and a base into the corresponding acylbenzotriazoles. Acylbenzotriazoles are powerful and versatile acylation reagents, and their synthetic applications have been extensively reviewed over the last few decades [1, 2, 6–8, 10].

For instance, Katritzky and Lebedyeva developed a straightforward one-pot decarboxylative acylation of various nucleophiles using malonic acid derivatives activated with **1d** (Scheme 20a) [119]. Selective monocarbonyl activation of 2,2-disubstituted malonic acid derivates with **1d** in the presence of triethylamine under mild microwave heating led to the mono decarboxylation of the substrates at the free carboxylic acid. The corresponding acylbenzotriazoles **53** were then reacted in situ with various nucleophiles, leading to various amides, alcohols, and thioesters **54** in high yields with retention of chirality. **1d** was utilized by Stevens and coworkers for the preparation of ferulic acid derivatives [118] and as an activation reagent of various key intermediates bearing a carboxylic acid moiety for the total synthesis of complex structures such as (+)-Roseophilin [122], Rolloamide B [123], and clomiphene metabolites [132]. Marchand-Brynaert and coworkers reported the activation of ibuprofen by reaction with **1d** en route towards FAAH inhibitors [131]. Interestingly, Zhu et al. developed a procedure for the preparation of α,β-unsaturated acylbenzotriazoles **55**, which were further reacted with *o*-phenylenediamine or *o*-aminothiophenol to give 1,3,4,5-tetrahydro-4-substituted-1,5-benzodiazepine-2-ones **56** or 2,3-dihydrobenzo[b][1,4]thiazepin-4(5*H*)-ones **57**, respectively (Scheme 20b) [134].

Katritzky pioneered the *C*-sulfonylation of various substrates with *N*-sulfonylbenzotriazoles **12** (Scheme 21) [32]. Active methylene compounds such as various nitriles, sulfones, and esters were lithiated and then reacted with a variety of

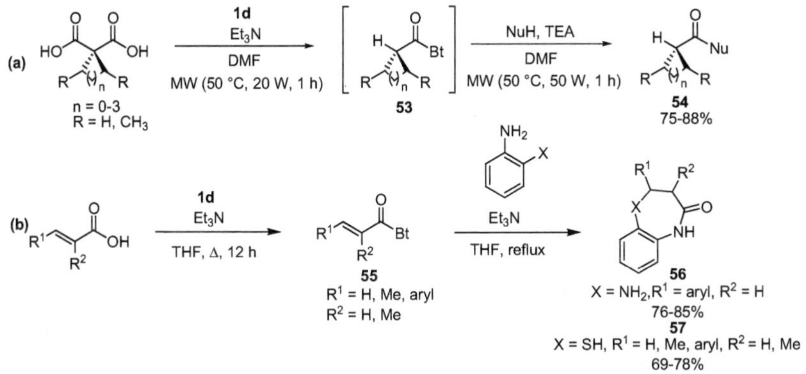

Scheme 20 (**a**) One-pot decarboxylative acylation of various nucleophiles using malonic acid derivatives; (**b**) preparation of 1,3,4,5-tetrahydro-4-substituted-1,5-benzodiazepine-2-ones or 2,3-dihydrobenzo[b][1,4]thiazepin-4(5*H*)-ones from α,β-unsaturated acylbenzotriazoles

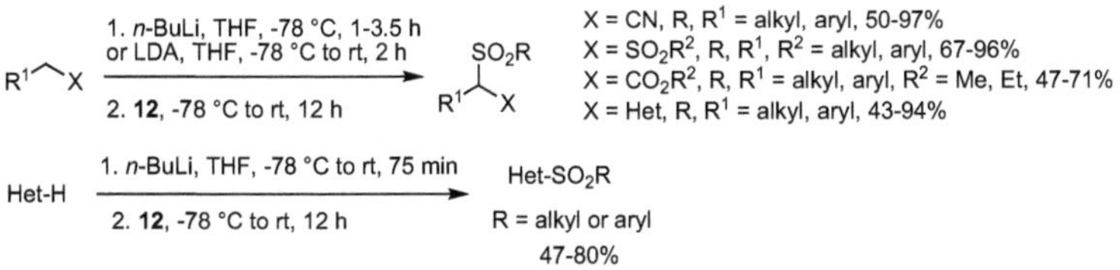

Scheme 21 *C*-sulfonylation of various substrates with *N*-sulfonylbenzotriazoles

alkyl/aryl *N*-sulfonylbenzotriazoles **12** to give the corresponding sulfones in high yields. The method was further extended with the sulfonation of heteroaromatics such as furane, thiophene, and pyrrole derivatives [32]. A similar strategy was employed by Ruano and coworkers for the preparation of a library of α-cyano-α-arylsulfones [110].

N-Sulfonylbenzotriazole derivatives **12** were also reported for the preparation of various sulfonamides by reaction with nitrogen-containing nucleophiles [112, 146–148]. Typically, sulfonamides were prepared in high yields (64–100%) by the reaction of alkyl/aryl *N*-sulfonylbenzotriazole derivatives **12** and primary or secondary aliphatic/aromatic amines (Scheme 22) [34, 112, 148]. It turned out that sulfonylbenzotriazoles **12** are less reactive than the corresponding sulfonyl chlorides and enable the selective sulfonylation of aliphatic amines over aromatic amines [148]. Katritzky also studied the preparation of *N*-acylsulfonamides and devised two different strategies [138]. The first strategy involved the activation of carboxylic acids by reaction with alkyl/aryl *N*-sulfonylbenzotriazole derivatives **12**, followed by coupling with various sulfamides. Alternatively, amides were reacted with alkyl/aryl *N*-sulfonylbenzotriazoles **12** in the presence of NaH. In both cases, *N*-acylsulfonamides were obtained in moderate to excellent yields. Another example involving the preparation of *N*-acylsulfonamides from *N*-sulfonylbenzotriazole derivatives **12** was reported by Mohanakrishnan and coworkers for the total synthesis of Calothrixin B and analogs [146].

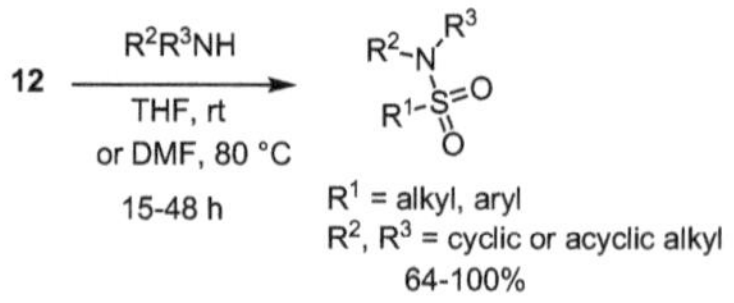

Scheme 22 Preparation of sulfonamides from alkyl/aryl *N*-sulfonylbenzotriazole derivatives and primary or secondary aliphatic or aromatic amines

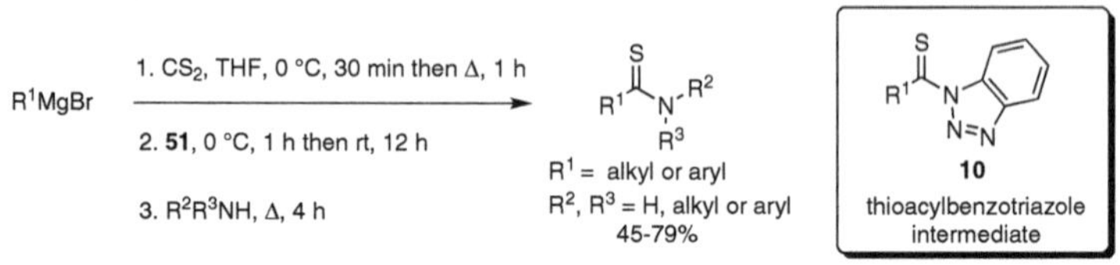

Scheme 23 One-pot procedure for the preparation of thioamides

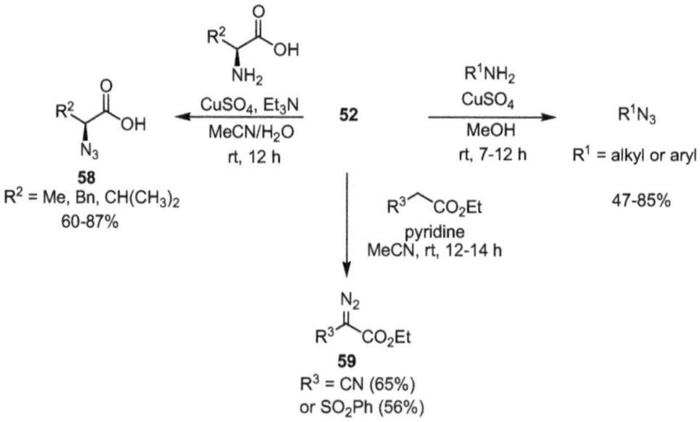

Scheme 24 Applications of benzotriazol-1-yl-sulfonyl azide

Katritzky reported an interesting one-pot procedure for the preparation of (thio) amides [109, 114]. Thioamides were prepared as follows: Grignard reagents were successively treated with carbon disulfide, 1-trifluoromethylsulfonylbenzotriazole (**51**), and with an amine (Scheme 23) [109]. Katritzky assumed that the reaction proceeded through the formation of a thioacylbenzotriazole intermediate **10**.

Benzotriazol-1-yl-sulfonyl azide (**52**) was developed by Katritzky and coworkers and reported in 2010 as a convenient diazotransfer reagent [116]. It reacted with primary amines in the presence of a catalytic amount of copper sulfate to give the corresponding alkyl or arylazides. Similarly, amino acids were converted in the presence of copper sulfate and triethylamine into α-azido acids **58**, with complete retention of the stereochemistry [116]. Substrates containing an activated methylene group reacted with **52** in the presence of pyridine at room temperature to afford the corresponding diazo compounds **59** in moderate yields (Scheme 24) [116].

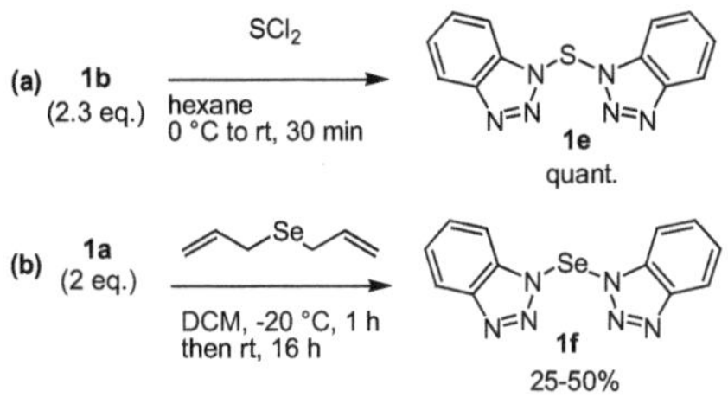

Scheme 25 (**a**) Preparation of bis(1*H*-benzotriazol-1-yl)sulfide; (**b**) preparation of bis(1*H*-benzotriazol-1-yl)selenide

2.5 *Bis(1*H*-benzotriazol-1-yl)sulfide and Bis(1*H*-benzotriazol-1-yl)selenide*

Bis(1*H*-benzotriazol-1-yl)sulfide (**1e**) and bis(1*H*-benzotriazol-1-yl)selenide (**1f**) were scarcely studied in the literature, despite their promising properties as sulfur and selenium transfer reagents.

2.5.1 Preparation of Bis(1*H*-benzotriazol-1-yl)sulfide and Bis(1*H*-benzotriazol-1-yl)selenide

Bis(1*H*-benzotriazol-1-yl)sulfide (**1e**) was prepared by the reaction of 1-(trimethylsilyl)benzotriazole (**1b**) and sulfur dichloride in dry hexane at 0°C (Scheme 25a) [149]. Bis(1*H*-benzotriazol-1-yl)selenide (**1f**) was obtained from the reaction of 1-chlorobenzotriazole (**1a**) and diallyl selenide in dichloromethane at −20°C (Scheme 25b) [150, 151]. By contrast to its sulfide analog, bis(1*H*-benzotriazol-1-yl)selenide is very moisture sensitive and must be stored under controlled atmosphere.

2.5.2 Reactivity and Synthetic Utility of Bis(1*H*-benzotriazol-1-yl)sulfide and Bis(1*H*-benzotriazol-1-yl)selenide

Bis(1*H*-benzotriazol-1-yl)sulfide (**1e**) was recently reported by Wyatt for the preparation of 10,11-dibenzo[*b*,*f*]thiepin derivatives such as **60** and was used as an alternative to other conventional sulfur transfer reagents (Scheme 26a) [149]. Bis(1*H*-benzotriazol-1-yl)selenide (**1f**) was first described in 1992 by Harpp as a unique selenium transfer reagent [151]. It was successfully utilized for the preparation of symmetrical dithioselenides **61** in high yields (75–98%) (Scheme 26b) [150].

Scheme 26 (**a**) Preparation of chiral 10,11-dibenzo[*b*,*f*] thiepin derivatives; (**b**) preparation of symmetrical dithioselenides

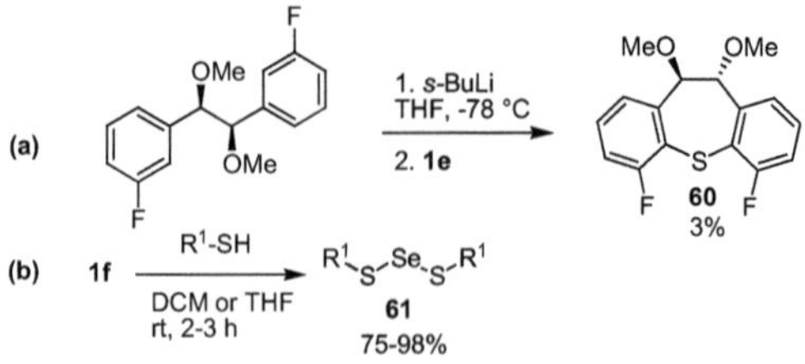

2.6 *1-Cyanobenzotriazole*

1-Cyanobenzotriazole, also known as 1*H*-benzotriazole-1-carbonitrile (**1g**), is a convenient and stable source of the CN^+ cation and is therefore frequently used as an electrophilic cyanation reagent. 1-Cyanobenzotriazole was also utilized for the preparation of substituted ureas and thioureas.

2.6.1 Preparation of 1-Cyanobenzotriazole

The original procedure reported by Marsh and Hermes in 1967 involved the reaction of sodium benzotriazolate and cyanogen chloride [152]. Katritzky optimized the original procedure and obtained 1-cyanobenzotriazole (**1g**) in higher yield (92%) by reacting sodium benzotriazolate and cyanogen bromide in THF (Scheme 27a) [153]. Cava developed a convenient alternative from 1-chlorobenzotriazole (**1a**) and sodium cyanide (Scheme 27b) [42]. Treatment of **1a** with sodium cyanide in acetonitrile gave **1g** as a crystalline solid in 70% yield after sublimation. The main side product of the reaction is di(1*H*-benzotriazol-1-yl) methanimine (**3c**), which arises from the reaction of the intermediate benzotriazolate anion and **1g**.

2.6.2 Reactivity and Synthetic Utility of 1-Cyanobenzotriazole

1-Cyanobenzotriazole (**1g**) is primarily used as an electrophilic cyanation reagent [154–157]. Cava was one of the first to report the electrophilic cyanation of organic substrates with **1g** [42] and studied quite extensively the electrophilic cyanation of arylacetonitrile anions. The corresponding arylmalonitriles **62** were obtained in moderate yields (30–66%, Scheme 28a). He then studied the cyanation of various substrates such as arenes, heteroarenes, and alkynes. High yields were obtained for the cyanation of alkynes (83%), while the cyanation of heteroarenes gave the

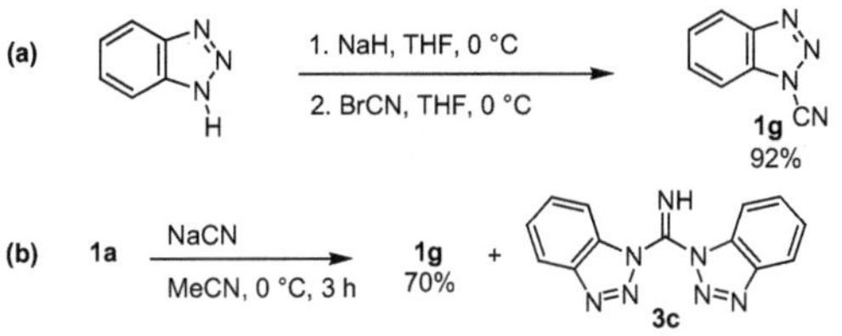

Scheme 27 (**a**) Katritzky's procedure for the preparation of 1-cyanobenzotriazole; (**b**) Cava's procedure for the preparation of 1-cyanobenzotriazole

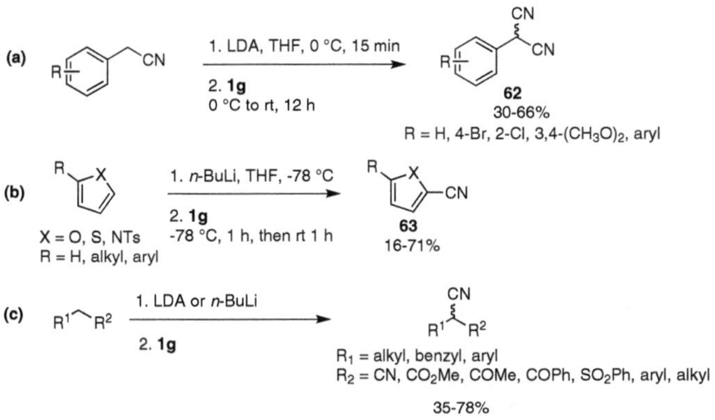

Scheme 28 (**a**) Electrophilic cyanation of arylacetonitrile anions; (**b**) cyanation of substituted thiophenes and pyrroles; (**c**) cyanation of activated methylene groups

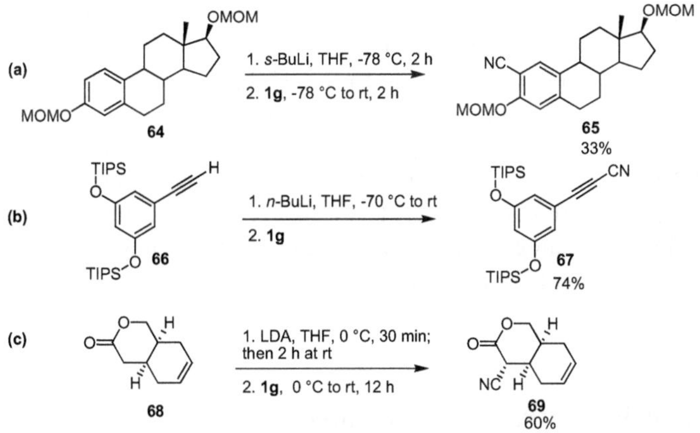

Scheme 29 (**a**) *Ortho* regioselective cyanation of a steroid derivative; (**b**) cyanation of an alkyne derivative with 1-cyanobenzotriazole; (**c**) intermediate cyanation en route towards (+)-Scholarisine A

corresponding heteroarylnitriles **63** in poor to moderate yields (16–71%) (Scheme 28b) [153, 158]. Katritzky demonstrated the general utility of **1g** as a cyanation reagent for a variety of substrates [153] and performed the *C*-cyanation of

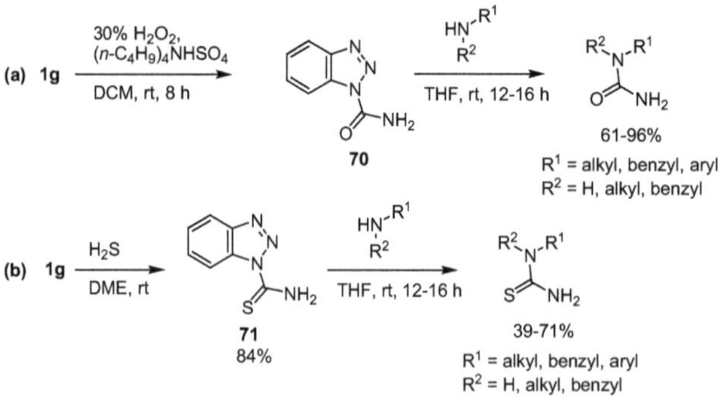

Scheme 30 (**a**) Preparation of ureas from 1-cyanobenzotriazole; (**b**) preparation of thioureas from 1-cyanobenzotriazole

activated methylene groups bearing sulfones, ketones, esters, nitriles, heterocycles, and diarylmethanes (Scheme 28c).

The cyanation of more complex substrates also succeeded with **1g**. For instance, an *ortho* regioselective cyanation of steroid derivative **64** was reported by Cushman (Scheme 29a) [159], and alkyne **66** was cyanated in good yield as well (Scheme 29b) [160, 161]. More recently, **1g** was employed for the preparation of a key intermediate **69** towards the synthesis of (+)-Scholarisine A [155, 156]. Lactone **68** was treated with 2.2 eq. of lithium diisopropylamide (LDA) in THF and was then treated with **1g** at 0°C, furnishing nitrile **69** in 60% yield (Scheme 29c) [155, 156].

Besides its frequent use as a cyanation reagent, **1g** was also utilized for the preparation of substituted ureas and thioureas. 1-Cyanobenzotriazole (**1g**) was reacted with hydrogen peroxide in the presence of a phase transfer catalyst in dichloromethane at room temperature to furnish benzotriazol-1-carboxamide (**70**) (Scheme 30a). The latter is an important synthon for the preparation of mono- and *N*,*N*-disubstituted ureas [162]. Katritzky then extended his strategy towards thioureas [163]: **1g** was reacted with H_2S in dimethoxyethane (DME) to give 1-benzotriazole-1-carbothioamide (**71**) in 84% yield. The reactivity of **71** with various amines was sluggish and only gave the corresponding thioureas in poor to moderate yields (39–71%) (Scheme 30b).

3 Class B Benzotriazole Derivatives 2a–f

*3.1 1-(Chloromethyl)-1*H*-benzotriazole*

1-(Chloromethyl)-1*H*-benzotriazole (**2a**) is an ambivalent electrophile. **2a** is commonly used to introduce an *N*-(benzotriazol-1-ylmethyl) moiety on a variety of nucleophilic substrates. The benzotriazole fragment also stabilizes α-negative

Scheme 31 Katritzky's preparation of 1-chloromethylbenzotriazole

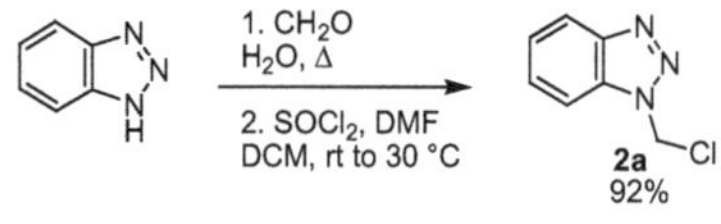

charges. In the presence of a Lewis acid, **2a** behaves as a source of the highly reactive methylene-benzotriazolium cation.

3.1.1 Preparation of 1-(Chloromethyl)-1*H*-benzotriazole

One of the earliest reports on the preparation of 1-(chloromethyl)-1*H*-benzotriazole (**2a**) dates back to 1952 [164]: it was prepared from 1-hydroxymethylbenzotriazole (**2c**) and thionyl chloride. Another procedure was reported in 1969, with Storr's study of 1-chlorobenzotriazole (**1a**). The latter reacted with diazomethane to afford **2a** in low yield (14% after purification) [58]. Nowadays, **2a** is conveniently prepared according to Katritzky's two-step procedure [165]. It involves the reaction of benzotriazole with formaldehyde followed by chlorination with thionyl chloride (Scheme 31) [166–168]. Alternatively, **2a** was prepared from sodium benzotriazolate and dichloromethane in refluxing DMF for 2 h [169].

3.1.2 Reactivity and Synthetic Utility of 1-(Chloromethyl)-1*H*-benzotriazole

1-(Chloromethyl)-1*H*-benzotriazole (**2a**) possesses a unique reactivity profile that comes from the presence of two different leaving/activating groups on a methylene fragment. **2a** is a strong electrophile due to stereoelectronic effects: the chlorine group is easily displaced with a variety of *C*-, *N*-, *S*-, *P*-, *O*-containing nucleophiles [165–191]. The corresponding *N*-benzotriazol-1-ylmethylated substrates can be further functionalized by displacement of the benzotriazole moiety [181, 183, 192].

For example, various (benzotriazol-1-yl)alkoxy/aryloxymethane derivatives were obtained by the reaction of 1-chloromethylbenzotriazole (**2a**) and the corresponding sodium alkoxides or aryloxides (see Sect. 3.3) [165, 166, 188]. Displacement of the chlorine atom with triethylphosphite or triphenylphosphine furnished, upon treatment with a base, the corresponding Wadsworth-Emmons-Horner or Wittig reagents. These reagents were used for the olefination of various aldehydes [167]. Reaction of **2a** with various *N*-containing nucleophiles – including indole, carbazole, pyrrole, imidazole, benzimidazole, and 1,2,4-triazole derivatives – afforded the corresponding *N*-(benzotriazol-1-ylmethyl) derivatives [190]. Recently, chiral benzotriazole-based phase transfer catalysts **73** derived from *Cinchona* alkaloids were prepared by reacting *Cinchona* alkaloids and **2a** [193]. Chiral phase transfer catalysts **73** were utilized for the enantioselective preparation of protected amino acids **74** from iminoglycyl derivative **72** (up to

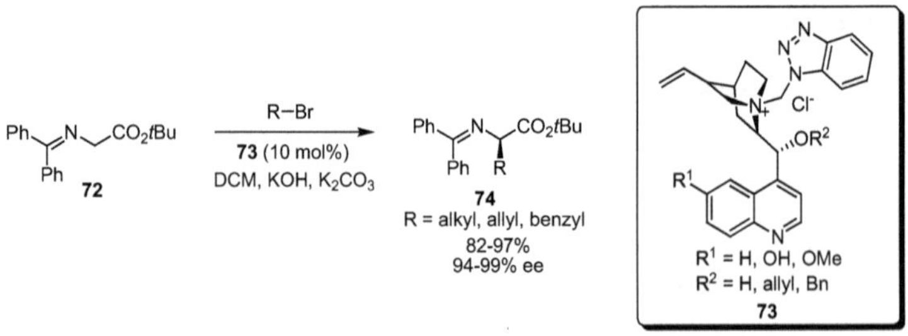

Scheme 32 Enantioselective preparation of amino acids

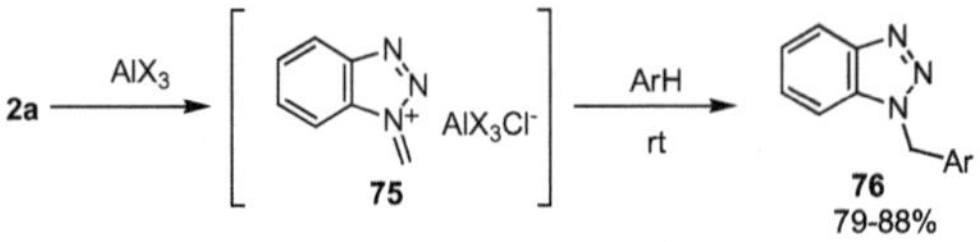

Scheme 33 Mannich/Friedel-Crafts preparation of 1-(arylmethyl)benzotriazoles

99% ee, Scheme 32) [193], nitro-Mannich reactions [194], and aza-Henry reactions [195].

Katritzky demonstrated that 1-chloromethylbenzotriazole (**2a**) can be used for Mannich/Friedel-Crafts alkylations of non-activated aromatic substrates (Scheme 33) [192]. The methylenebenzotriazolium salt **75** was generated in situ from **2a** and a Lewis acid. The methylenebenzotriazolium cation **75** is highly reactive and largely exceeds the reactivity of common Mannich intermediates such as methylene and dichloromethylene-iminium salts. The high electrophilicity of **75** is most likely related to the π-deficient nature of the benzotriazole ring, but also to the extra activation caused by coordination of the excess Lewis acid to the benzotriazole ring. Compounds **76** were obtained in excellent yields by the alkylation of various aromatic substrates, although the regioselectivity of the reaction was low [192].

In addition to its electron-donor properties, the benzotriazole fragment also possesses electron-acceptor properties with respect to the group attached to it, and stabilizes a negative charge on the α-carbon [188, 191, 196–199]. The combination of these properties led to the design of strategies targeting a variety of poly(hetero)cyclic compounds. For instance, Katritzky reported a convenient one-pot procedure for the preparation of furan and thiophene libraries [183]. 1-Chloromethylbenzotriazole (**2a**) was reacted with polysubstituted phenols (X = O, Scheme 34a), and the corresponding 1-(arylmethyl)benzotriazoles were then lithiated with *n*-BuLi and further quenched with various alkyl or benzyl halides to give alkylated benzotriazole derivatives. The latter were treated with an additional equivalent of *n*-BuLi and the corresponding anions were reacted with various aldehydes to afford intermediate alcohols. The last step required the addition of zinc bromide at 140°C to trigger a pinalcol rearrangement, and then treatment at 180°C led to the formation of the desired benzo[*b*]furans **77** (X = O). A similar procedure was applied for the preparation of benzo[*b*]thiophenes (X = S) in 49–80%. The

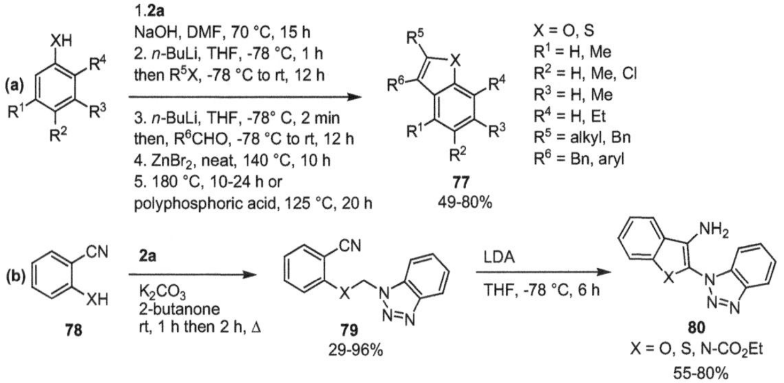

Scheme 34 (**a**) Preparation of benzo[*b*]furans and benzo[*b*]thiophenes from 1-chloromethylbenzotriazole; (**b**) 3-amino-2-(1*H*-benzotriazol-1-yl) substituted benzofurans, benzothiophenes and indoles from 1-chloromethylbenzotriazole

method was next extended to naphto[1,2-*b*]furans and naphto[1,2-*b*]thiophenes (71–83%) and then to other substrates in 2003 [179, 197, 200].

Rádl reported a modified Gewald reaction for the preparation of 3-amino-2-(1*H*-benzotriazol-1-yl)-substituted benzofurans, benzothiophenes, and indoles **80** from **2a** (Scheme 34b). The reaction of **2a** and various *ortho O*-, *S*-, *N*-substituted-benzonitriles **78** and the subsequent treatment of the *N*-benzotriazol-1-ylmethylated intermediates **79** with LDA afforded the corresponding 3-amino-2-(1*H*-benzotriazol-1-yl)-substituted benzofurans, benzothiophenes, and indoles **80** in moderate to good yields (55–80%) [177].

Katritzky prepared imidazolo[1,2-*a*]pyridines **82** and imidazolo[2,1-*a*] isoquinolines **84** from **2a** (Scheme 35a) [181]. 2-Aminopyridines and 1-aminoisoquinoline reacted with **2a** to give the corresponding 2-amino-1-[-α-benzotriazol-1-ylmethyl]pyridinium chlorides **81** and 1-amino-2-[-α-benzotriazol-1-ylmethyl]isoquinolinium chloride **83**, respectively. In the next step, intermediates **81** and **83** were reacted with non-enolizable aldehydes in the presence of a base to afford imidazolo[1,2-*a*]pyridines **82** and imidazolo[2,1-*a*] isoquinolines **84** in good yields. In 1999, Katritzky developed new procedures for the preparation of some 1,3-diarylaziridines **86** and substituted pyrroles **88** from **2a** (Scheme 35b). The sequence started with the deprotonation of **2a** with LiHMDS in THF/HMPA or DME/HMPA, which gave a new carbenoid species **85**. As a result of the electron-donating ability of the benzotriazole fragment, the carbenoid **85** was expected to behave as a transient species and was quenched in situ with diaryl imines. The corresponding 1,3-diarylaziridines **86** were obtained in high yields (85–90%). Treatment of aziridines **86** at 100°C in the presence of dialkyl acetylene-dicarboxylates led to the formation of various substituted pyrroles **88** in high yields. This reaction most likely proceeded through a formal [2+3] cyclization of azomethine ylides **87** [199].

Katritzky then studied the reactivity of 1-benzotriazolylchloromethyllithium, generated from **2a** and LDA. It reacted with aliphatic and aromatic enolizable

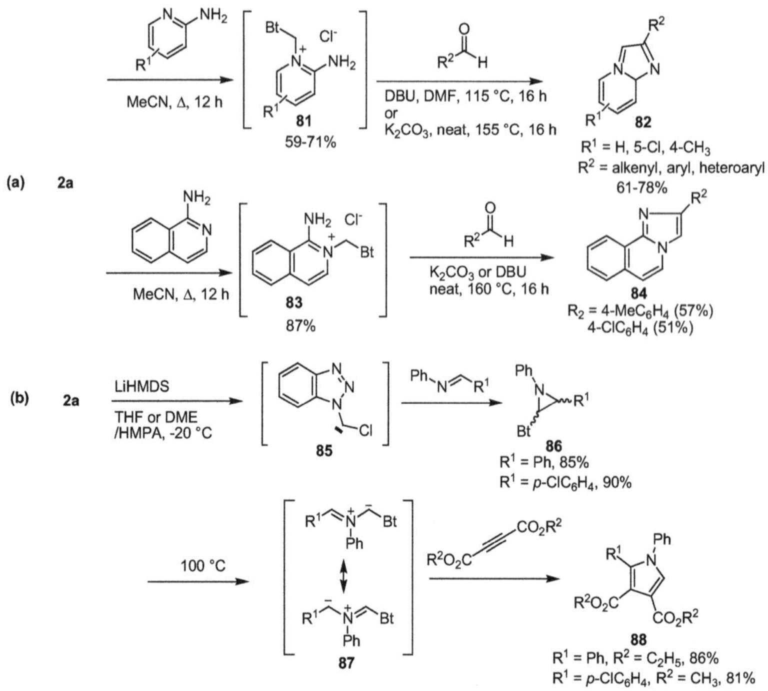

Scheme 35 (**a**) Preparation of imidazolo[1,2-*a*]pyridines and –[2,1-*a*]isoquinolines from 1-chloromethylbenzotriazole; (**b**) preparation of some 1,3-diarylaziridines and substituted pyrroles from 1-chloromethylbenzotriazole

and non-enolizable ketones to afford benzotriazolyloxiranes **89** in good yields (62–75%) [196], while the same reaction with aldehydes failed. Treatment of benzotriazolyloxiranes **89** with *n*-BuLi led to the formation of oxizanyl anions, which could be further quenched with a variety of electrophiles in good to excellent yields (Scheme 36a).

Finally, Katritzky disclosed an original synthetic strategy for the preparation of substituted cyclopropanes based on (benzotriazol-1-yl)methyldiphenylphosphine oxide **91** [186]. **91** was obtained by the lithiation of diphenylphosphine oxide with *n*-BuLi at −78°C followed by treatment with **2a**. When (benzotriazol-1-yl) methyldiphenylphosphine oxide **91** was treated with *n*-BuLi and various epoxides, the corresponding cyclopropane derivatives **92** were obtained in low to good yields (Scheme 36b).

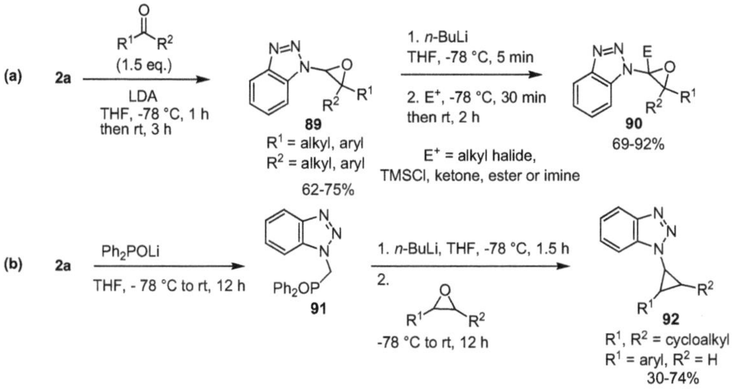

Scheme 36 (**a**) Preparation of substituted benzotriazolyloxiranes from 1-chloromethylbenzotriazole; (**b**) cyclopropanation of oxiranes with (benzotriazol-1-yl)methyldiphenylphosphine oxide

3.2 1-(Trimethylsilylmethyl)benzotriazole

1-(Trimethylsilylmethyl)benzotriazole (**2b**) is a source of the methylenebenzotriazolium anion and can be used for one-carbon homologations of carboxylic acid derivatives.

3.2.1 Preparation of 1-(Trimethylsilylmethyl)benzotriazole

Katritzky studied the preparation of 1-(trimethylsilylmethyl)benzotriazole (**2b**) from benzotriazole [201]. Two methods were assessed: (a) the silylation of 1-(lithiomethyl)benzotriazole and (b) the reaction of sodium benzotriazolate with chloromethyltrimethylsilane. The first method was abandoned since it afforded mainly the bis-silylated product, while the second method gave the desired compound together with the 2*H*-isomer (20%) [201, 202]. Recently, Stevens et al. revisited Katritzky's procedure, although the yield did not exceed 40% (Scheme 37) [203]. The rather modest yield was explained by the formation of the 2*H*-benzotriazole isomer.

3.2.2 Reactivity and Synthetic Utility of 1-(Trimethylsilylmethyl) benzotriazole

1-(Trimethylsilylmethyl)benzotriazole (**2b**) is a safe alternative to the Arndt-Eistert reaction for one-carbon homologations [198, 201–211]. It complements the classical strategies for one-carbon homologation of carbonyl compounds.

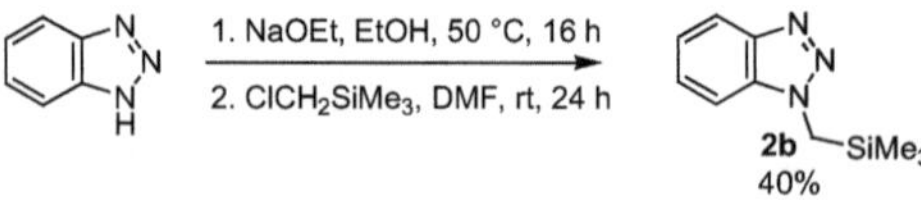

Scheme 37 Preparation of 1-(trimethylsilylmethyl)benzotriazole

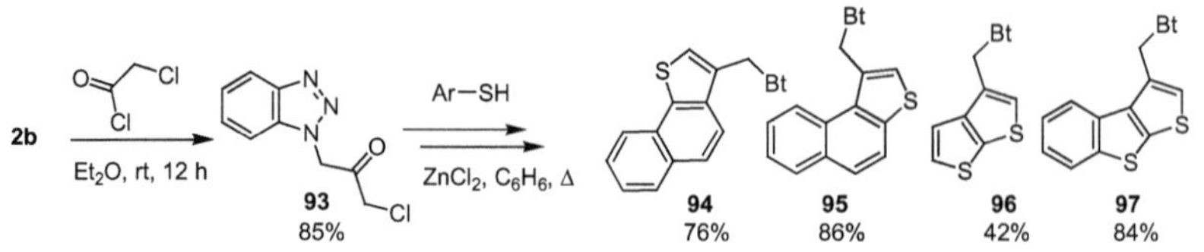

Scheme 38 One-carbon homologation of chloroacetylchloride in the presence of 1-(trimethylsilylmethyl)benzotriazole towards the preparation of various thiophene derivatives

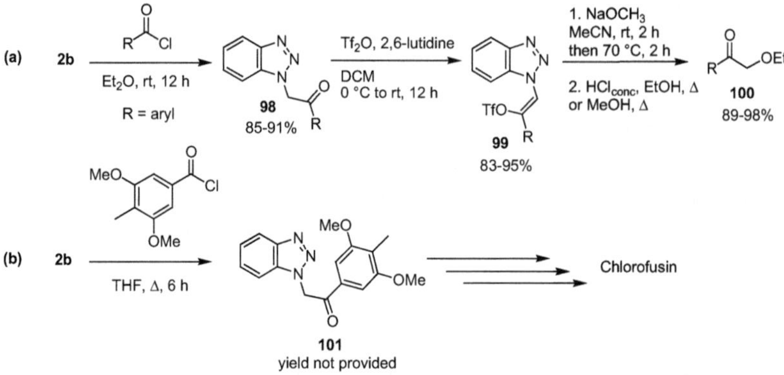

Scheme 39 (**a**) Katritzky's general one-carbon homologation of acyl chlorides; (**b**) 1-(trimethylsilylmethyl)benzotriazole one-carbon homologation towards Chlorofusin

Katritzky pioneered the use of 1-(trimethylsilylmethyl)benzotriazole (**2b**) for one-carbon homologations of carboxylic acids [207, 209, 210]. One-carbon homologation of chloroacetylchloride in the presence of **2b** led to the formation of 1-(1*H*-benzotriazol-1-yl)-3-chloroacetone **93**, which was further used as a synthon for heterocyclizations towards various heterocycles such as substituted pyrido[1,2-*a*] indoles [209], indolizines [209], pyridines [209], and benzothiazoles [209]. Scheme 38 illustrates the preparation of various thiophenes **94–97** from **93** [207].

Katritzky then extended this homologation procedure for a variety of aryl/alkyl carboxylic acids [208, 210]. 1-(Trimethylsilylmethyl)benzotriazole (**2b**) reacted with various acyl chlorides to furnish the corresponding *N*-acylmethylbenzotriazoles **98** in high yields (85–91%). Their reaction with triflic anhydride in the presence of 2,6-lutidine at 0–20°C overnight afforded the corresponding enol triflates **99** in high yields (83–95%). Aromatic enol triflates were then transformed into the corresponding one-carbon homologated esters **100** with $NaOCH_3$ followed by concentrated HCl in overall yields of 89–98% (Scheme 39a). For aliphatic derivatives, the treatment of the corresponding enol triflates led to the formation of alkynyl benzotriazoles, which were further transformed into the corresponding carboxylic acids by treatment with *p*-TsOH followed by tetrabutylammonium fluoride in THF (24–68%). More recently, **2b** was utilized for the development of new binapthyl ligands [206] and for the synthesis of key intermediate **101** for the total synthesis of Chlorofusin and related analogs (Scheme 39b) [204, 205].

*3.3 1*H*-Benzotriazole-1-methanol and Alkoxy/aryloxymethyl-1*H*-benzotriazole Derivatives*

Despite their structural analogy, 1*H*-benzotriazole-1-methanol and alkoxy/aryloxymethyl-1*H*-benzotriazole derivatives have distinct reactivity profiles. 1*H*-Benzotriazole-1-methanol is commonly used as a source of anhydrous formaldehyde. Besides, numerous other synthetic applications were reported, such as the benzotriazolylalkylation of various substrates, the preparation of *N*-acylaminals, the *N*-methylation of amines, the *para*- or *ortho*-benzotriazolylmethylation of aniline, phenol, methoxybenzene, and naphthalene derivatives. It was also used as a ligand, as is or after functionalization, for organometallic coupling reactions [212–216]. Alkoxy/aryloxymethyl-1*H*-benzotriazole derivatives are primarily used as synthetic equivalents of acetal anions for one-carbon homologations of aldehydes and for the preparation of ethers. Their unique set of properties makes them useful synthons for the preparation of ketones, alkynes, (thio)acylsilanes, and acyltrifluoroborates. Their anions are considered as carbenoid species [217].

3.3.1 Preparation of 1*H*-Benzotriazole-1-methanol

The preparation of 1*H*-benzotriazole-1-methanol (**2c**), also known as 1-hydroxymethylbenzotriazole, is straightforward and was originally reported in 1952 by Hall and coworkers (Scheme 40) [164]. Since then, the original procedure has remained unchanged [166, 168]. A mixture of benzotriazole, formalin, and acetic acid in water was stirred at room temperature for 2 h. **2c** was recrystallized from hot water or ethyl acetate. 1-(1-Hydroxyalkyl)benzotriazole derivatives were similarly prepared by the condensation of benzotriazole and aldehydes [218].

3.3.2 Reactivity and Synthetic Utility of 1*H*-Benzotriazole-1-methanol

Formaldehyde is a ubiquitous reagent for performing one-carbon homologations of various substrates. It is, however, quite challenging to get pure and anhydrous yet reactive sources of formaldehyde. 1*H*-Benzotriazole-1-methanol (**2c**) is a convenient, anhydrous, and reactive source of formaldehyde under basic conditions. It

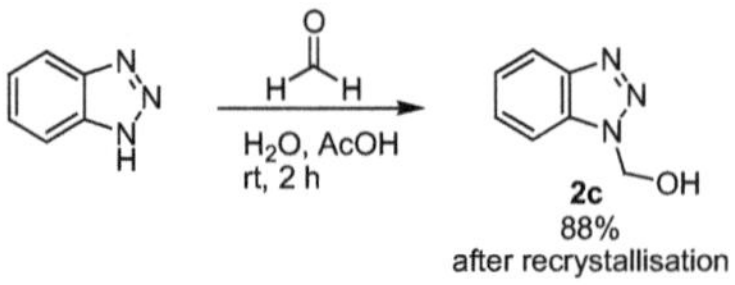

Scheme 40 Preparation of 1*H*-benzotriazole-1-methanol

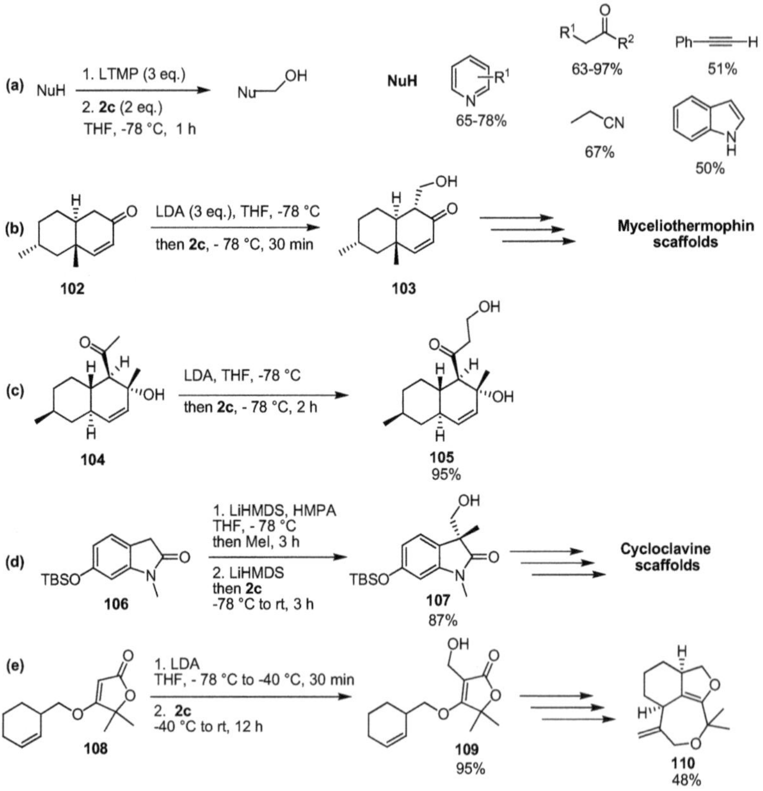

Scheme 41 (**a**) Bischoff's original procedure for the hydroxymethylation of a variety of nucleophiles with 1*H*-benzotriazole-1-methanol; (**b**) Nicolaou's total synthesis of Myceliothermophin scaffolds; (**c**) final hydroxymethylation performed with 1*H*-benzotriazole-1-methanol towards Aspermytin A; (**d**) preparation of cycloclavine scaffolds; (**e**) construction of various terpenoid tridecane skeletons using a key hydroxymethylation with 1*H*-benzotriazole-1-methanol

was successfully employed for the hydroxymethylation of a plethora of substrates [219–233].

The original report of the use of **2c** as a stable, convenient, and anhydrous source of formaldehyde in organic solvents is actually quite recent [233]. The procedure, disclosed by Bischoff and coworkers in 2007, is suitable for the hydroxymethylation of a variety of anions. It proceeded under basic conditions with an excess of LDA or lithium tetramethylpiperidide (LTMP) for the simultaneous generation of nucleophilic species and formaldehyde (Scheme 41a) and

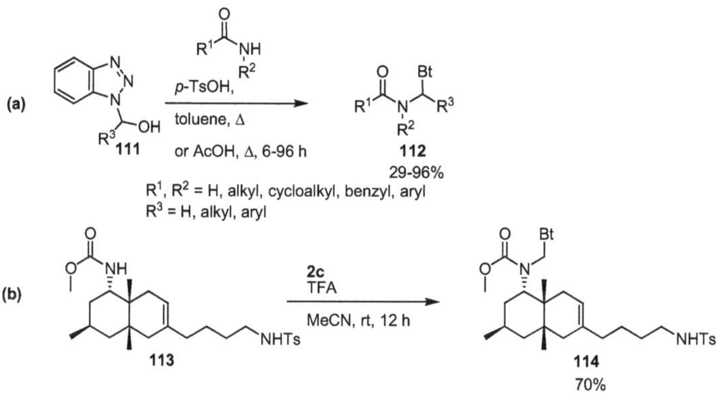

Scheme 42 (**a**) Katritzky's original procedure for the benzotriazol-1-ylmethyl/alkylation of primary and secondary amides; (**b**) benzotriazol-1-ylmethylation of a key intermediate for the total synthesis of Nankakurine A and B

enabled the hydroxymethylation of organometallic derivatives and enolates in high yields (50–97%) [233]. Bischoff's procedure was then utilized by various research groups including Nicolaou [219], Shishido [220], Marcaurelle [221], and many others [222–232] for the hydroxymethylation of complex substrates. For instance, a key hydroxymethylation step with **2c** was utilized for the total syntheses of Myceliothermophins C, D, and E (Scheme 41b) [219], Aspermytin A **105** (Scheme 41c) [220], Cycloclavine scaffolds (Scheme 41d) [226], and various terpenoid tridecane skeletons **110** (Scheme 41e) [227].

The other important synthetic facet of **2c** is its use as a Mannich electrophile, generally under acidic conditions, which triggers the release of a reactive iminium species. The latter is an efficient reagent for the benzotriazol-1-ylmethylation of various substrates, such as amides and derivatives [234–244], amines and derivatives [245–257], and (hetero)aromatics [258–265]. Other 1-(1-hydroxyalkyl) benzotriazole derivatives were also successfully employed as benzotriazol-1-ylalkylation reagents. The benzotriazol-1-ylalkylated substrates could be further functionalized by nucleophilic displacement of the benzotriazole moiety, either with sodium/lithium borohydride, organometallic reagents [239–242, 244, 266, 267], or other nucleophiles [236].

One of the earliest reports on this aspect was published in 1988 by Katritzky. He reported that 1-hydroxyalkylbenzotriazoles **111** reacted with amides to afford the corresponding *N*-monoalkylated products **112** (42–78%) in refluxing toluene with azeotropic removal of water (Scheme 42a) [266]. The primary adduct was further functionalized by displacement of the benzotriazole moiety with sodium borohydride (94–99%), while treatment with lithium borohydride afforded the corresponding secondary amines (58–83%). A few years later, the method was generalized to secondary amides (Scheme 42a) [239, 240]. **2c** was used for the benzotriazol-1-ylmethylation of key intermediate **113** for the total synthesis of Nankakurine scaffolds (Scheme 42b), and the corresponding benzotriazol-1-ylmethylated derivative **114** was obtained in 70% yield [235].

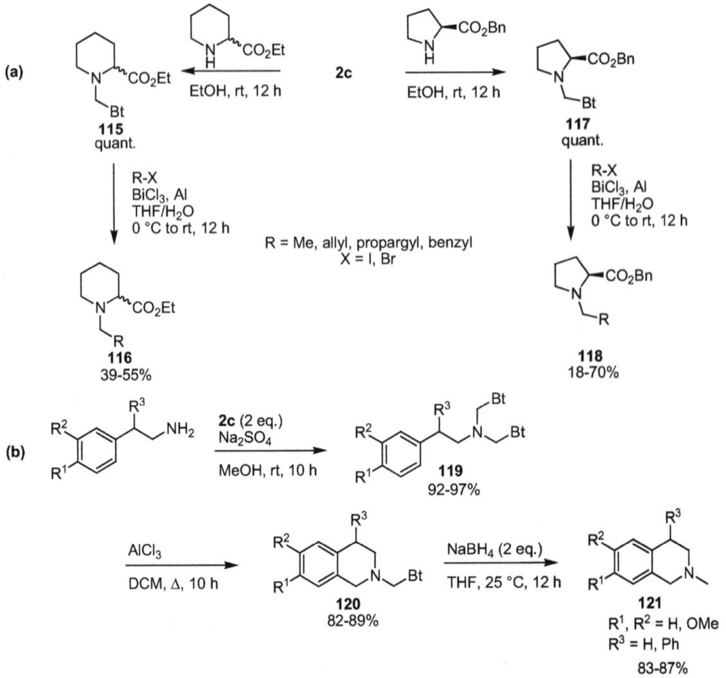

Scheme 43 (a) 1-Hydroxymethylbenzotriazole derivatization of proline and pipecolinic esters; (b) preparation of 2-substituted-1,2,3,4-tetrahydroisoquinolones utilizing 1-hydroxymethylbenzotriazole

Thioamides reacted in a similar way, affording *N*-(benzotriazole-1-ylmethyl/alkyl)thioamides [237]. Thioureas reacted with mixtures of benzotriazole and aldehydes under acidic conditions to give the corresponding *N*-substituted ureas through a Mannich-type condensation, as well. Katritzky assumed that the reaction proceeded through the formation of 1-hydroxyalkylbenzotriazole derivatives **111**. The benzotriazole moiety could then be displaced with $NaBH_4$ or Grignard reagents, affording *N,N*-disubstituted thioureas or carbodiimides [241]. Similarly, 1-acyl, 1-acyl-2-arylhydrazines, and hydroxamic acids reacted with **2c**, affording the corresponding *N*- or 2,2-bis-*N*-[(benzotriazol-1-yl)methyl] derivatives in high yields [243, 244].

Amines were also successfully benzotriazol-1-ylmethylated with **2c**. For instance, Katritzky described the *N*-derivatization of secondary amines with **2c** [257]. Pipecolinic and proline esters were reacted with **2c** in ethanol at room temperature for 12 h, affording the corresponding *N*-(benzotriazol-1-yl) adducts **115** and **117**, respectively. Next, the primary *N*-(benzotriazol-1-yl) adducts were reacted with various alkyl, allyl, propargyl, and aryl halides in aqueous medium in the presence of $BiCl_3$ and aluminum powder to furnish the corresponding *N*-substituted pipecolinic esters **116** and proline esters **118** in low to good yields (Scheme 43a). Katritzky also reported the preparation of isoquinolone derivatives from **2c** (Scheme 43b) [267]. Katritzky's procedure involved the reaction of sterically

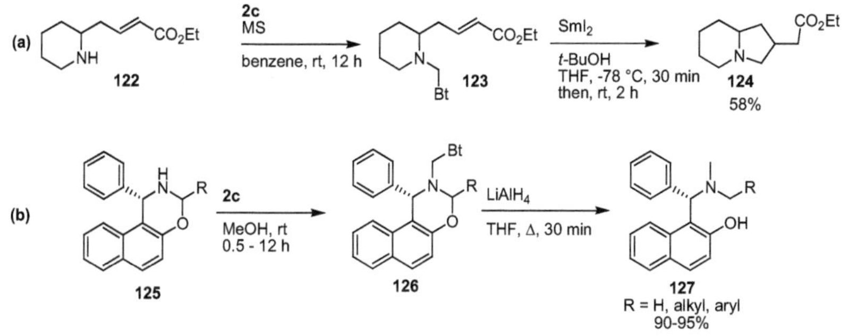

Scheme 44 (**a**) Two-step sequence cyclization towards polysubstituted pyrrolidines; (**b**) reductive *N*-methylation

unhindered primary amines with an excess of **2c**, leading to the formation of the corresponding bis(benzotriazol-1-ylmethyl) derivatives **119**. Upon treatment with $AlCl_3$, compounds **119** underwent intramolecular cyclization to furnish 2-benzotriazol-1-ylmethyl-1,2,3,4-tetrahydroisoquinolines **120**. Reduction with 2 eq. of sodium borohydride in THF replaced the remaining benzotriazole moiety with hydrogen to give *N*-methyl-1,2,3,4-tetrahydroisoquinolines **121** in high yields (83–87%). Other nucleophiles were also successfully used to afford a variety of substituted 1,2,3,4-tetrahydroisoquinolines. Abonia studied the preparation of 1,2,5,6-tetrahydro-4*H*-pyrrolo[3,2,1-*ij*]quinolones from indolines and **2c** [251, 268].

Similar strategies were developed for the preparation of 2,3,4,5-tetrahydro-1*H*-pyrrolo[1,2-*a*][1,4]diazepines and substituted 2,3,4,5-tetrahydro-1*H*-[1,4]diazepino [1,2-*a*]indoles from 3-(pyrrol-1-yl)-1-propylamine or 3-(3-methyl-indol-1-yl)-propylamine, respectively, and **2c** in chloroform in the presence of *p*-TsOH [249]. Aurrecoechea and coworkers reported an interesting two-step sequence involving the benzotriazol-1-ylmethylation of a secondary amine followed by a SmI_2-promoted radical cyclization towards polysubstituted pyrrolidine scaffolds (Scheme 44a) [250]. For instance, ethyl (*E*)-4-(piperidin-2-yl)prop-2-enoate (**122**) was condensed with **2c** in the presence of MS 4Å to give the corresponding *N*-(benzotriazol-1-ylmethyl)amine derivative **123**, which was next used in an intramolecular SmI_2-promoted radical cyclization step, affording the desired 2-(ethoxycarbonylmethyl)indolizidine (**124**). This method was further extended to access piperidine derivatives later on [245, 246]. Hu and coworkers studied new protocols for the preparation of non-racemic Betti base derivatives **127** [247]. The one-pot sequence involved the *N*-benzotriazol-1-ylmethylation of chiral hemiaminal ethers **125** with **2c**, followed by its reductive cleavage in the presence of $LiAlH_4$ (Scheme 44b), formally resulting in a reductive *N*-methylation.

Katritzky studied the preparation of symmetrical and unsymmetrical methylenebisanilines and methylenebis(*N*,*N*-dialkylanilines) **129** by the reaction of **2c** and anilines or *N*,*N*-dialkylanilines under acidic conditions (Scheme 45a) [265]. Anilines reacted with **2c** to afford the corresponding *N*-benzotriazol-1-ylmethylated products, before undergoing the Hofmann-Martius rearrangement [261, 265] towards *para*-benzotriazol-1-ylmethylated anilines **128**. *N*,*N*-dialkylanilines cannot

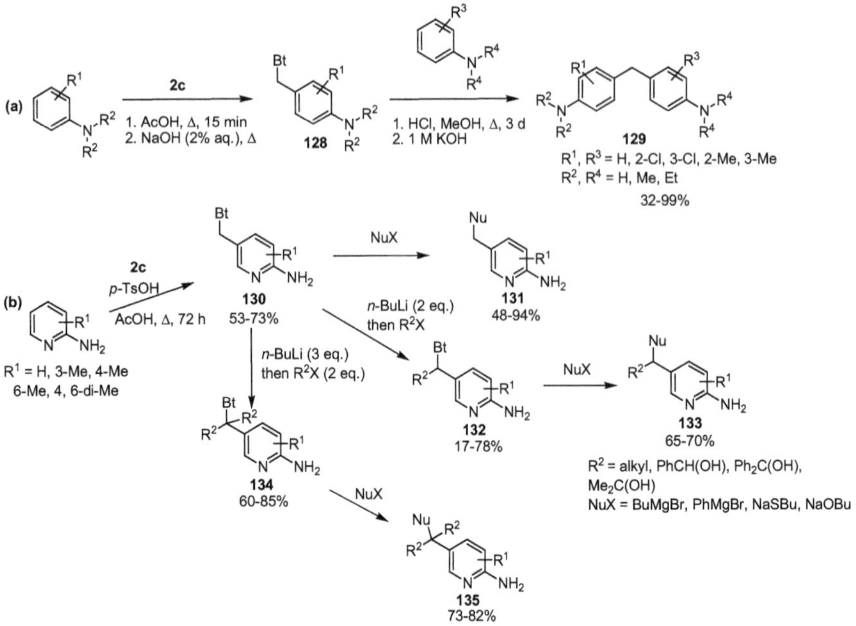

Scheme 45 (**a**) Preparation of methylenebisanilines from anilines and 1-hydroxymethyl-benzotriazole; (**b**) preparation of 2-amino-5-(benzotriazol-1-ylmethyl)pyridines from 1*H*-benzotriazole-1-methanol and 2-aminopyridine derivatives and functionalization thereof

undergo the Hofmann-Martius rearrangement, and in this case, the reaction must proceed through direct substitution. *p*-Benzotriazol-1-ylmethylated anilines **128** acted as iminium species in the presence of another aniline derivative under acidic conditions, ultimately affording the corresponding methylenebisanilines **129** in low to excellent yields. A similar procedure was applied for the preparation of HIV-1 integrase inhibitors by Neamati [258]. Katritzky extended the scope of this benzotriazol-1-ylmethylation procedure to the preparation of 5-substituted 2-aminopyridines (Scheme 45b) [261]. The corresponding 2-amino-5-(benzotriazol-1-ylmethyl)pyridines **130** were mono- or bislithiated and further functionalized by reaction with various electrophiles. The benzotriazole group was displaced by various nucleophiles such as alkyl or aryl Grignard reagents, alkoxides, and sulfides (Scheme 45b). Katritzky also reported the 3-benzotriazol-1-ylmethylation of indoles [262] and thiophenes [259].

Other aromatic compounds were successfully benzotriazol-1-ylmethylated with **2c**. For instance, **2c** was used for the direct *p*-benzotriazol-1-ylmethylation of methoxybenzenes and naphtalenes in acetic acid or in toluene in the presence of *p*-TsOH [263]. *Ortho*-substitution occurred in low yields when the *para* position was not available. Further molecular diversification was obtained either by lithiation and substitution with electrophiles or by displacement of the benzotriazole group with various Grignard reagents (Scheme 46a). Contrastingly, the reaction of **2c** and phenols or naphthols in refluxing acetic acid led to the exclusive formation of the *ortho*-benzotriazol-1-ylmethylation products **137**

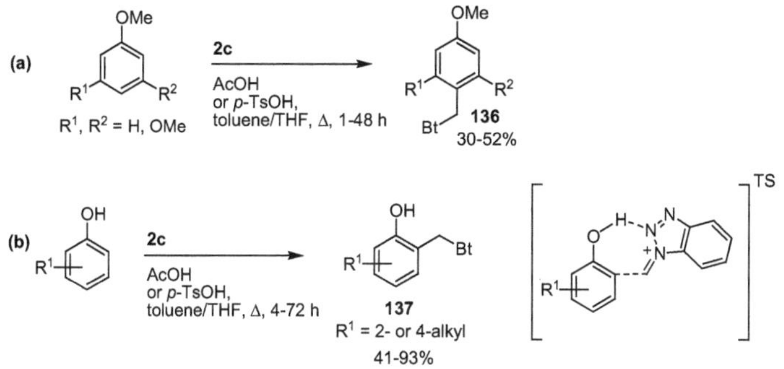

Scheme 46 (**a**) *p*-Benzotriazol-1-ylmethylation of methoxybenzenes and naphtalenes with 1*H*-benzotriazole-1-methanol; (**b**) *o*-benzotriazol-1-ylmethylation of phenols or naphthols

(Scheme 46b) [264]. *Para*-substitution was reported if both *ortho* positions were occupied. Katritzky postulated that the reaction most likely proceeded through the formation of a reactive benzotriazolyl iminium species, and the *ortho*-regioselectivity arose from its chelation with the phenolic or naphtholic hydrogen (Scheme 46b). As usual, the primary *N*-benzotriazol-1-ylmethylated adducts could be functionalized either by lithiation or direct nucleophilic displacement of the benzotriazole moiety.

3.3.3 Preparation of Alkoxy/Aryloxymethyl-1*H*-benzotriazole Derivatives

Several procedures were reported for the preparation of 1-alkoxy/aryloxymethyl-1*H*-benzotriazoles. Burckhalter's method was the first reported and involved the nucleophilic addition of alkoxide or aryloxide anions on 1-chloromethylbenzotriazole (**2a**) [164]. This method was particularly suitable for the preparation of benzotriazolyl derivatives of aromatic ethers. Katritzky developed two main strategies for the preparation of 1-alkoxy/aryloxymethyl-1*H*-benzotriazole derivatives [165, 218]: (a) the condensation of benzotriazole, a ketone or an aldehyde, an alcohol and a catalytic amount of sulfuric acid in carbon tetrachloride and (b) the reaction of benzotriazole and dialkyl ketals. These methods enabled the preparation of 1-alkoxy/aryloxymethylbenzotriazoles in low to excellent yields (20–97%) [165, 218]. Bode reported a slight modification of Katritzky's method in 2012 [166]: *N*,*O*-acetals **138** were obtained from a mixture of benzotriazole, an aldehyde, ethanol and triethylorthoformate in the presence of sulfuric acid in THF (Scheme 47). Juliusz prepared 1-(alkoxymethyl)benzotriazoles **139** by the DMAP-catalyzed *N*-alkylation of benzotriazole under basic conditions (Scheme 48, left) [269]: a variety of chloromethylalkyl ethers were utilized, giving the corresponding 1-(alkoxymethyl) benzotriazoles **139** in good yields. More recently, Wang et al. reported an alternative route towards 1*H*-ethoxymethyl benzotriazole (**139**) [270]. The procedure involved the *N*-alkoxymethylation of benzotriazole by the reaction of diethyl phosphite and

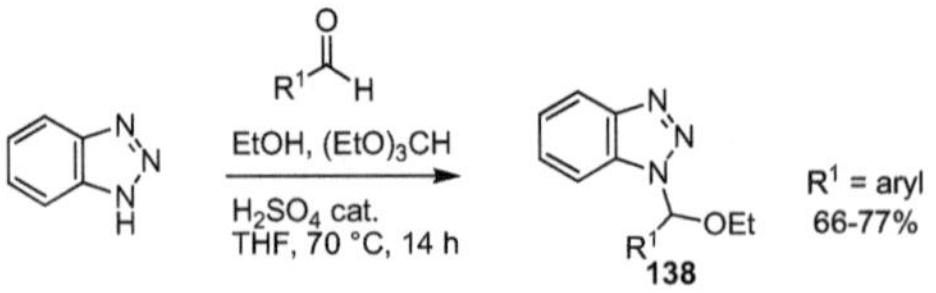

Scheme 47 Modification of Katritzky's procedure towards *N*,*O*-acetals

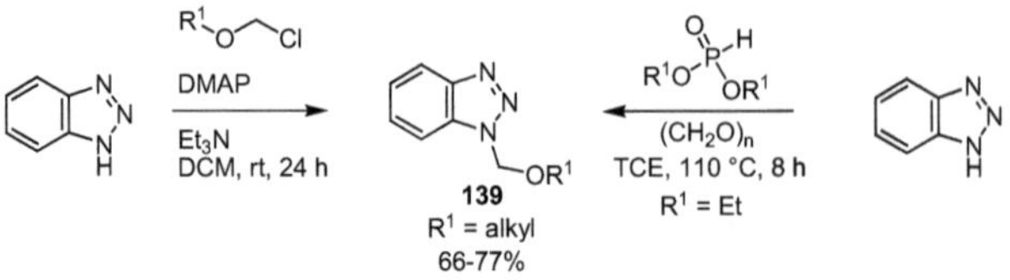

Scheme 48 Preparation of 1-(alkoxymethyl)benzotriazoles from chloromethylalkyl ethers or diethyl phosphite

paraformaldehyde at high temperature in tetrachloroethylene (Scheme 48, right). A significant amount of the 2*H*-isomer (16%) was formed.

3.3.4 Reactivity and Synthetic Utility of Alkoxy/aryloxymethyl-1*H*-benzotriazoles

The most obvious and direct application of benzotriazole *N*,*O*-acetals **139** is the preparation of ethers by direct nucleophilic benzotriazole displacement [165, 271]. For example, Katritzky demonstrated that the lithiation of 1-alkyloxymethylbenzotriazoles **139** at the methylene group, and its subsequent reaction with various electrophiles, provided the corresponding α-benzotriazol-1-ylalkyl alkylethers **141**. Displacement of the benzotriazole moiety afforded the corresponding alkylethers **142** in moderate to good yields (Scheme 49) [271]. He then studied benzotriazole-alkoxy/aryloxy *N*,*O*-acetals as potential acyl or acetal anions synthetic equivalents [188, 272–278]. α-Benzotriazolylalkyl alkylethers **141** were refluxed in methanol in the presence of *p*-TsOH and gave the corresponding α-functionalized dimethyl acetals **143** (Scheme 49) [188]. A few years later, Katritzky reported a one-carbon homologation of aldehydes and ketones towards α-alkoxy/aryloxy ketones **145** using a similar strategy [273, 276, 278]. The procedure involved trapping of the anion **140** with aldehydes and ketones to give oxyanion intermediates **144** (Scheme 49). Further treatment with an excess of zinc bromide triggered a rearrangement towards the formation of α-alkoxy/aryloxy ketones **145** in fair to good yields with a remarkable regioselectivity.

1,2-Diketones were obtained by quenching the anions from 1-(phenoxymethyl) benzotriazole derivatives **146** with benzoyl chlorides (Scheme 50) [275]. Two years later, Katritzky drastically extended the scope of this method for the preparation of alkyl, aryl, alkenyl, and alkynyl 1,6-diketones [272]. The lithiated anion of

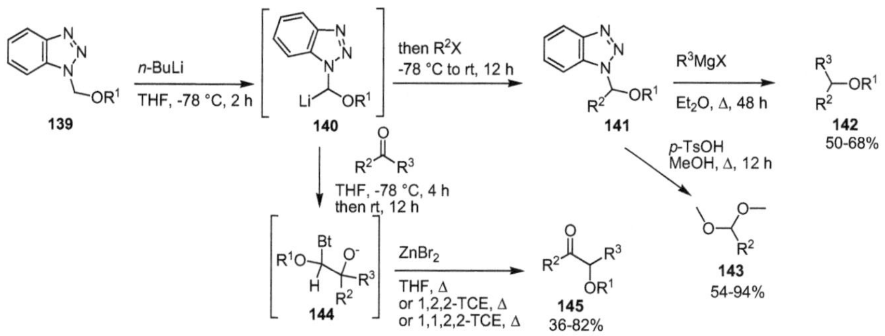

Scheme 49 Towards ethers, acetals, and α-alkoxy/aryloxy ketones from compounds **139**

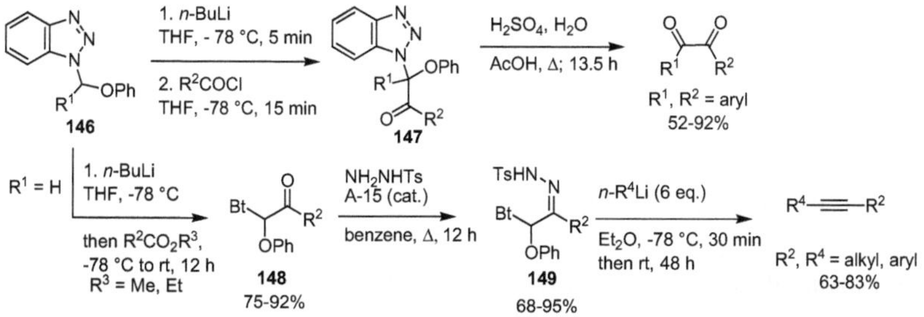

Scheme 50 Preparation of 1,2-diketones and alkynes from 1-(phenoxymethyl)benzotriazole derivatives

1-(phenoxymethyl)-1*H*-benzotriazole (**146**) also reacted with aromatic or aliphatic esters, affording α-benzotriazolyl ketones **148** in high yields (75–92%), which were converted into *p*-tosylhydrazones **149** upon treatment with (*p*-toluenesulfonyl)-hydrazine. Compounds **149** were then treated with an excess of organolithium reagents to afford alkynes in high yields (Scheme 50) [274].

Katritzky pioneered the preparation of acylsilanes from alkoxy/aryloxymethyl-1*H*-benzotriazole derivatives [277] and also succeeded to prepare thioacylsilanes (Scheme 51) [279]. In a first step, 1-(phenoxymethyl)-1*H*-benzotriazole (**146**) was successively alkylated and silylated towards the corresponding trimethylsilyl-benzotriazole **150**. Compound **150** was then reacted with hexamethyldisilathiane (HMDST) in the presence of trimethylsilyl trifluoromethanesulfonate (TMSOTf) to give the corresponding transient thioacylsilane **151**, which was trapped with 2,3-dimethylbuta-1,3-diene. More recently, Portella prepared acylsilane derivatives **154** from compound **146** (Scheme 51) [280]. 1-(phenoxymethyl)-1*H*-benzotriazole (**146**) was lithiated, and the corresponding anion quenched with chloroallyldi-methylsilane or chlorovinyldimethylsilane. The intermediate silane derivatives were lithiated again and treated with alkyl or allyl bromide to give the corresponding alkyl(1-benzotriazolyl-1-phenoxyalkyl)-dimethylsilane or allyl (1-benzotriazolyl-1-phenoxyalkyl)-dimethylsilane derivatives **153**. Further treatment with $FeCl_3$ afforded acylsilanes **154** in high yields. The unique reactivity of

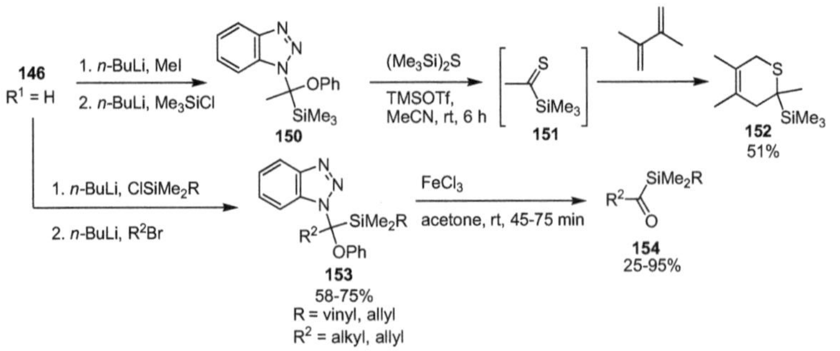

Scheme 51 Preparation of thioacylsilanes and acylsilane derivatives from 1-phenoxymethylbenzotriazole

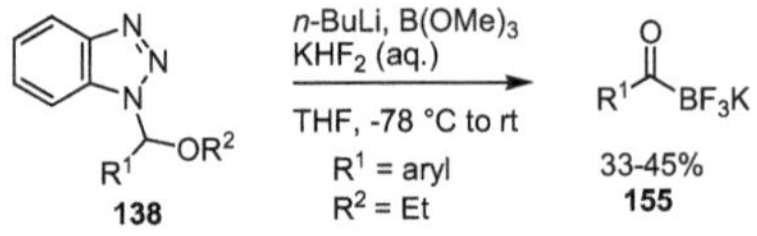

Scheme 52 Preparation of acyltrifluoroborates

benzotriazole-alkoxy/aryloxy *N*,*O*-acetals **138** also enabled the preparation of acyltrifluoroborates **155** (Scheme 52) [166, 281].

3.4 1-(α-Aminoalkyl/aryl)-benzotriazoles

Aminomethylbenzotriazoles and derivatives are useful synthetic auxiliaries in which the methylene carbon is highly electrophilic. They were primarily used as Mannich acceptors for aminomethylation reactions. Their *N*-acyl derivatives were widely utilized as amidomethylation reagents [5], but will not be described in this review. Aminomethylbenzotriazoles were also used as sources for non-stabilized α-aminocarbanions [282] and as Baylis-Hillman electrophiles. One of the simplest aminomethylbenzotriazole derivatives, *N*,*N*-dimethylaminomethylbenzotriazole (**2d**), is the direct benzotriazole analog of Eschenmoser's salt.

3.4.1 Preparation of 1-(α-Aminoalkyl/aryl)benzotriazoles

N,*N*-Dimethylaminobenzotriazole (**2d**) was originally prepared in 1946 by Bachman and Heisey from benzotriazole, dimethylamine, and formaline in ethanol (Scheme 53) [283]. Since then, the procedure has not evolved tremendously and is applicable for the preparation of a wide variety of 1-(α-aminoalkyl)benzotriazole derivatives **156** [284–287]. For instance, aminoalkylbenzotriazole adducts were prepared by a Mannich reaction between an amine, benzotriazole, and formaldehyde (or an aldehyde) in excellent yields (59–99%) (Scheme 53). A library of

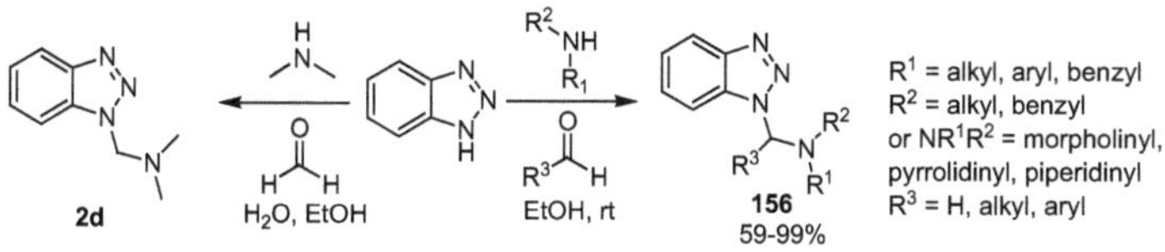

Scheme 53 General procedure for the preparation of a library of aminoalkylbenzotriazole adducts

diverse benzotriazole adducts was easily prepared and reported by Katritzky in 1989 [287]. For aldehydes where $R^3 \neq H$, a Dean-Stark apparatus, or the addition of *p*-TsOH and/or molecular sieve, was necessary to displace the equilibrium towards the formation of the corresponding aminoalkylbenzotriazole adducts [288]. The benzotriazole adducts derived from formaldehyde exist in solution as mixtures of the benzotriazol-1-yl and benzotriazol-2-yl isomers, although they display similar reactivity [287, 289].

3.4.2 Reactivity and Synthetic Utility of 1-(α-Aminoalkyl/aryl) benzotriazoles

Compounds bearing a benzotriazolyl group in α to an amino moiety are precursors of highly reactive iminium species in solution and thus behave as Mannich electrophiles for the aminomethylation of a variety of substrates [284, 285, 287, 290–307]. Katritzky studied the reaction of 1-(aminoalkyl)benzotriazoles **156** with various organometallic reagents [284, 287, 290, 291, 294, 296, 298, 299, 301, 307]. For instance, the reaction of *N*,*N*-dialkyl-*N*-[benzotriazolylalkyl- (or arylalkyl-)]amines **156** with alkynyllithium reagents led to tertiary propargylamines (Scheme 54a) [307]. A decade later, Katritzky also described the reaction of 1-(α-aminoalkyl)benzotriazoles **156** with sodium dialkynyldiethylaluminates towards the preparation of *N*,*N*-dialkylpropargylamines in high yields (Scheme 54a) [296]. Katritzky also devised a general protocol for the preparation of tertiary and secondary amines using the reaction of **156** and Grignard reagents or sodium borohydride (Scheme 54b) [287]. He extended the method towards the preparation of allylamines in high yields (82–94%) from *N*-(α-benzotriazolylalkyl)amines **156** and vinyl Grignard reagents [294]. Cyclic enol phosphonates were also successfully aminomethylated with **156** [291]. Organozinc reagents were successfully employed as well [299] and enabled the preparation of α-fluorinated-γ-amino esters **157** via the Reformatsky reaction with ethyl bromofluoroacetates (Scheme 54c) [298].

A variety of other organometallic species were aminomethylated with *N*-(α-aminoalkyl)benzotriazoles, such as lithiated primary and secondary nitriles [292] and enolates [304]. Zhang reported the allylation and propargylation of *N*-(α-aminoalkyl)benzotriazoles using a Barbier-type addition of allyl and propargyl bromides in the presence of gallium and potassium iodide in THF at 70°C. The corresponding homoallyl and homopropargyl amines were obtained in high yields (65–85%) [301]. **2d** was used for the aminomethylation of cyclopropane

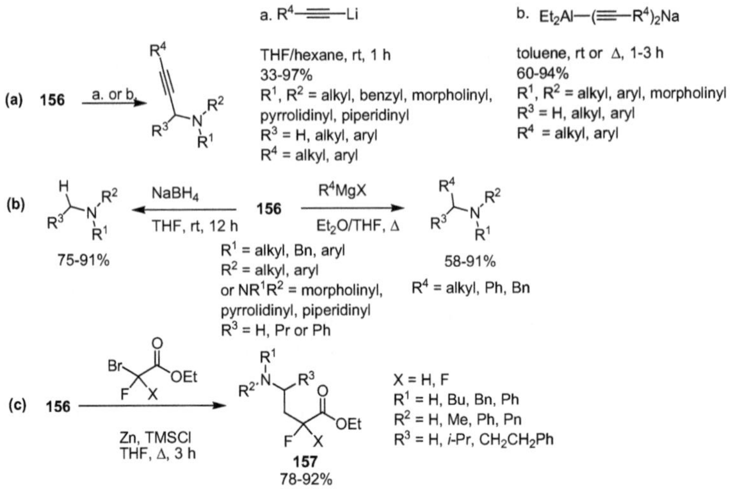

Scheme 54 (**a**) Reaction of *N*,*N*-dialkyl-*N*-[benzotriazolylalkyl- (or arylalkyl-)]amines with alkynyllithium reagents or dialkynyldiethylaluminates; (**b**) Katritzky's general procedure for the preparation of tertiary and secondary amines from *N*,*N*-dialkyl-*N*-[benzotriazolylalkyl- (or arylalkyl-)]amines; (**c**) preparation of α-fluorinated-γ-amino esters

carbenoids in high yields and excellent stereoselectivity [293]. An *N*-(-α-benzotriazolylalkyl) derivative was reacted as a Mannich acceptor with tributylstannyl lithium in a key step towards the preparation of 1,3-dihydro-1,3-azaborine [290]. Recently, Le Gendre disclosed the reaction of α-aminobenzotriazoles with *π*-allyltitanium complexes to give homoallylic amines in good yields (58–92%) [284].

Other nucleophiles than organometallic species were successfully aminomethylated with *N*-(α-aminoalkyl)benzotriazole derivatives [286]. For instance, 1-[(dialkylamino)methyl]benzotriazole derivatives **156** reacted with allyl and propargyltrimethylsilanes in the presence of aluminum chloride to afford 4-(trimethylsilyl)aminoalkanes **160** or 4-(trimethylsilyl)aminoalkenes **161**, respectively [295]. The reaction most likely proceeded via a 1,5-hydride shift in the initial aminomethylated intermediate **158**. In the absence of a methylene group in α of the amine, i.e., when the 1,5-hydride shift is not possible, the expected homoallylamines and α-allenylamines were obtained in low yields (3–33%). For some examples, silylated azetidines and pyrrolidines were isolated instead (Scheme 55).

The final outcome of the reaction of *N*-(α-aminoalkyl)benzotriazole derivatives and enols, enol ethers, enamines, enamides, and other alkenes utterly depends on the substitution pattern on the α-nitrogen [285, 288, 300, 302, 305, 306]. A few unsaturated compounds successfully trapped the iminium cation derived from *N*,*N*-dialkyl-1*H*-benzotriazole-1-methanamines **156a** under acidic conditions, leading to the formation of a stable adduct **162**. Among the successful unsaturated candidates, enamines such as *N*-vinylcarbazole and *N*-vinylamides reacted with *N*,*N*-dialkyl-1*H*-benzotriazole-1-methanamines **156a** to form the corresponding adducts **162**

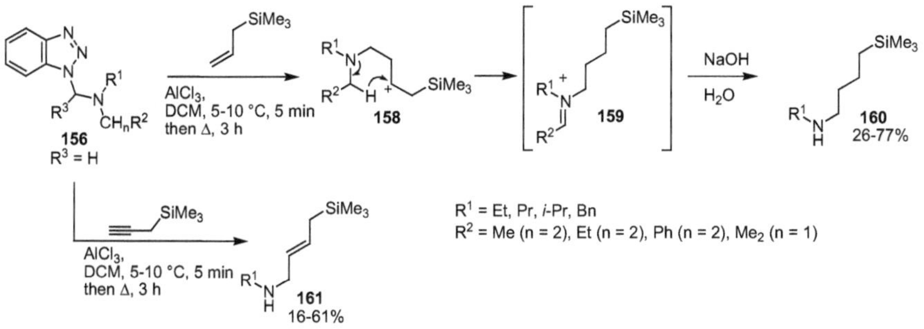

Scheme 55 Preparation of 4-(trimethylsilyl)aminoalkanes or 4-(trimethylsilyl)aminoalkenes

(Scheme 56a) [302]. Further functionalization was successfully achieved by nucleophilic displacement of the benzotriazole moiety with $LiAlH_4$ or with various Grignard reagents, affording substituted unsymmetrical 1,3-diamines **163**. The reaction of *N*-vinylmethylacetamide or 1-vinyl-pyrrolidinone with *N*,*N*-dialkyl-1*H*-benzotriazolyl-1-methanamines **156a** further enhanced the scope of this procedure.

Contrastingly, *N*-alkyl-*N*-aryl-1*H*-benzotriazolyl-1-methanamines **156b** reacted with *N*-vinylamides and *N*-vinylcarbazole in the presence of *p*-TsOH to furnish the corresponding 4-(dialkylamino)tetrahydroquinolines **165**. The reaction most likely proceeded through the formation of a reactive iminium species **164**, which then underwent an intramolecular Friedel-Crafts condensation (Scheme 56b) [288]. Further functionalization of the 4-(dialkylamino)tetrahydroquinolines **165** was successfully achieved as well [288]. Similarly, enol ethers reacted with *N*-alkyl-*N*-aryl-1*H*-benzotriazolyl-1-methanamines **156b** to give substituted 1,2,3,4-tetrahydroquinolines in high yields [305], while their reactions with *N*,*N*-dialkyl-1*H*-benzotriazolyl-1-methanamines **156a** led to the corresponding functionalizable 1-benzotriazolyl-3-aminoalkyl ethers [306]. The condensation of enolizable aldehydes and **156b** furnished substituted 1,2,3,4-tetrahydroquinolines in high yields as well [285]. Non-activated alkenes required the presence of a stronger Lewis acid such as stannic chloride or boron trifluoride etherate to react with **156b**, affording the corresponding 3,4-substituted 1,2,3,4-tetrahydroquinolines in high yields [300]. A similar contrasting reactivity was reported for the iminium salts derived from **156a,b** with dienes in the presence of lithium tetrafluoroborate, leading to 1,2,5,6-tetrahydropyridinium salts or substituted 1,2,3,4-tetrahydroquinolines, respectively [308].

Katritzky reported that *N*-(α-aminoalkyl)benzotriazoles **156** reacted with Li/LiBr or SmI_2 to give the corresponding α-aminocarbanions **166**, which were then condensed with a variety of electrophiles. For instance, their reaction with ketones afforded β-aminoalcohols **167** in moderate to good yields (Scheme 57) [282].

Aminomethylbenzotriazole derivatives **156** were utilized as Baylis-Hillman electrophiles with ethyl acrylate in the presence of $TiCl_4$. The benzotriazole moiety

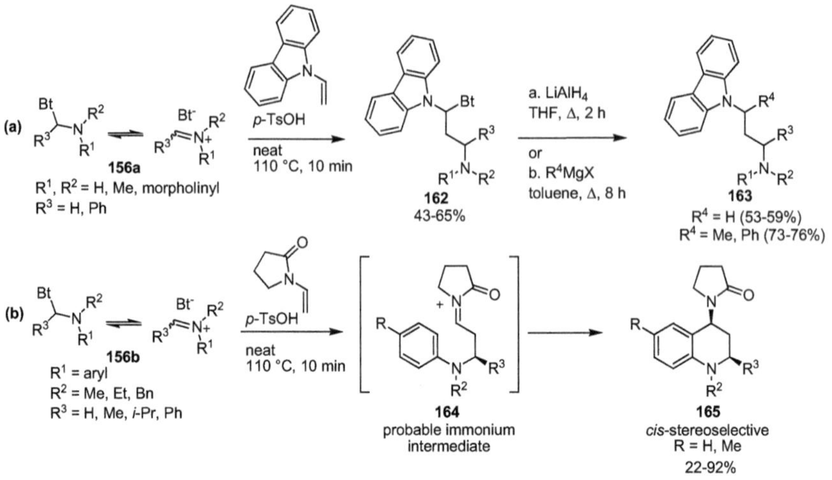

Scheme 56 (**a**) Preparation of unsymmetrical 1,3-diamines from *N*,*N*-dialkyl-1*H*-benzotriazole-1-methanamines; (**b**) preparation of 4-(dialkylamino)tetrahydroquinolines from *N*-alkyl-*N*-aryl-1*H*-benzotriazolyl-1-methanamines

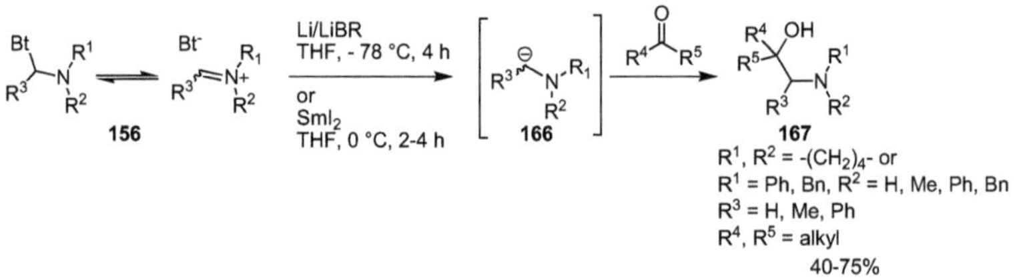

Scheme 57 Preparation of non-stabilized α-aminocarbanions and their reaction with ketones

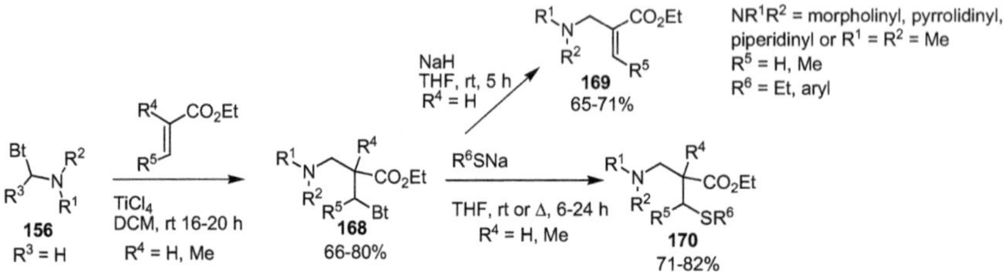

Scheme 58 Baylis-Hillman reaction of aminomethylbenzotriazoles and ethyl acrylates

on the Baylis-Hillman adducts **168** enabled further diversification either by (a) elimination under basic conditions towards olefins **169** or (b) nucleophilic displacement with nucleophiles such as thiolates towards sulfides **170** (Scheme 58) [309].

3.5 2-(1H-Benzotriazol-1-yl)acetonitrile

2-(1*H*-Benzotriazol-1-yl)acetonitrile (**2e**) is an active methylene reagent suitable for Knoevenagel condensations on electrophilic substrates. Besides, 2-(1*H*-benzotriazol-1-yl)acetonitrile is an ambivalent carbonyl synthon and a versatile building block for the preparation of various nitrogen-containing heterocyclic systems.

3.5.1 Preparation of 2-(1*H*-Benzotriazol-1-yl)acetonitrile

The original procedure for the preparation of 2-(1*H*-benzotriazol-1-yl)acetonitrile (**2e**) was reported by Katritzky in 1987 starting from of 1-chloromethylbenzotriazole (**2a**) and sodium cyanide (Scheme 59, left) [191]. A few years later, the procedure was revised [310]: **2e** was obtained by the reaction of sodium benzotriazolate and chloroacetonitrile in refluxing toluene in the presence of 18-C-6 (Scheme 59, right). This procedure led exclusively to the formation of the 1*H*-benzotriazole isomer [311], while a mixture of 1*H*- and 2*H*-isomers was obtained in DMF at 100°C [312].

3.5.2 Reactivity and Synthetic Utility of 2-(1*H*-Benzotriazol-1-yl) acetonitrile

2-(1*H*-benzotriazol-1-yl)acetonitrile (**2e**) was commonly used as an active methylene reagent for Knoevenagel condensations with aldehydes [142, 313–316], ketones [317], and acylbenzotriazoles [142]. **2e** was utilized for the preparation of antiproliferative acrylonitriles [315] and new agents against tuberculosis [314]. More recently, Zhao et al. used a Knoevenagel condensation between **2e** and salicylaldehyde derivatives **171** as a key step for the preparation of novel photoluminescent compounds **172** (Scheme 60) [313].

2-(1*H*-Benzotriazol-1-yl)acetonitrile (**2e**) is also a synthon for the construction of various substituted pyridines scaffolds [310, 318–320]. For instance, Katritzky developed a versatile entry towards polysubstituted 2-aminopyridines. The reaction of **2e** with chalcones **173** in the presence of secondary amines in refluxing ethanol led to a library of 2-(disubstituted amino)pyridines **174** in moderate to good yields (Scheme 61a) [310]. A few years later, this strategy was adapted to solid-phase conditions and enabled the preparation of 2-alkylamino- and 2-dialkylamino-4,6-diarylpyridines **177** [320]. The key step involved the condensation of polymer-supported chalcones, **2e,** and various secondary amines. The reaction sequence most likely involved the condensation of **2e** with secondary amines to form intermediate amidines **175**, which then reacted with the supported chalcones (Scheme 61b). Paluchowska et al. applied Katritzky's solution phase strategy for the preparation of 5-HT_{2A} receptor antagonists [318]. Al-Omran et al. used a similar

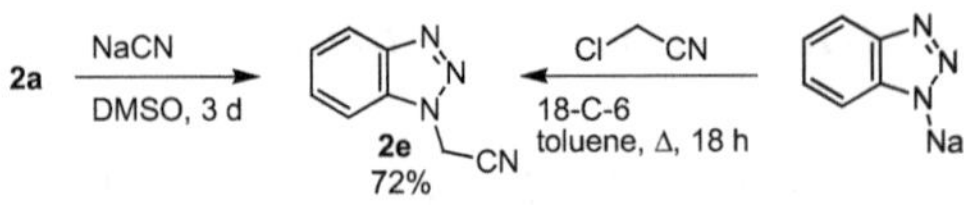

Scheme 59 Preparation of 2-(1*H*-benzotriazol-1-yl)acetonitrile

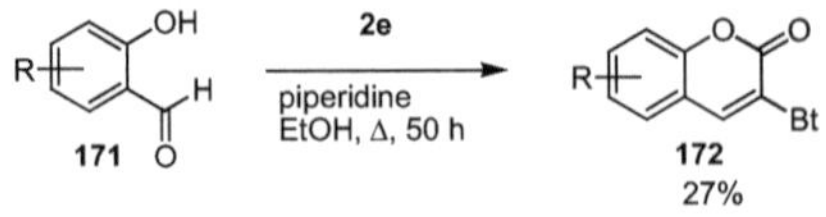

Scheme 60 Knoevenagel condensation of 2-(1*H*-benzotriazol-1-yl)acetonitrile and aromatic aldehydes

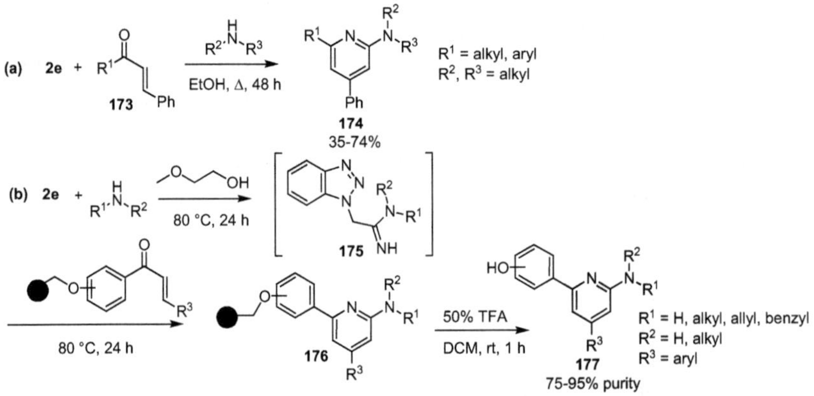

Scheme 61 (**a**) Solution phase strategy towards substituted 2-aminopyridines; (**b**) solid phase strategy towards substituted 2-aminopyridines

strategy to prepare various nitrogen-containing heteroaromatic compounds from **2e**, such as dihydropyridines, aminopyrazoles, and pyrimidine derivatives [321].

Wang studied azole-*N*-acetonitriles as ambivalent carbonyl synthons for the preparation of *N*-containing heteroaryl amides from the corresponding heteroaryl halides and developed a one-pot procedure [322]. The anion derived from **2e** was first reacted with 2-chloroquinoxaline (**178**) to give intermediate cyanohydrins, which were then oxidized to *N*-acylbenzotriazoles. The last step relied on the acylating power of acylbenzotriazoles, which reacted with primary and secondary amines to give the desired heteroaryl amides **179** (Scheme 62).

Marginally, 2-(1*H*-benzotriazol-1-yl)acetonitrile (**2e**) was reacted with sodium azide in the presence of ammonium chloride in refluxing DMF to give 1-((1*H*-tetrazol-5-yl)methyl)-1*H*-benzotriazole (**180**) [323, 324], which was used for the preparation of antimicrobial derivatives (Scheme 63a) [323]. The photochemical behavior of **2e** in the presence of *N*-phenylmaleimide was studied at 254 nm [325]. The corresponding dihydropyrrolo[3,4,*b*]indole **181** was obtained in low yield (Scheme 63b). It was postulated that the photolysis of **2e** generated a

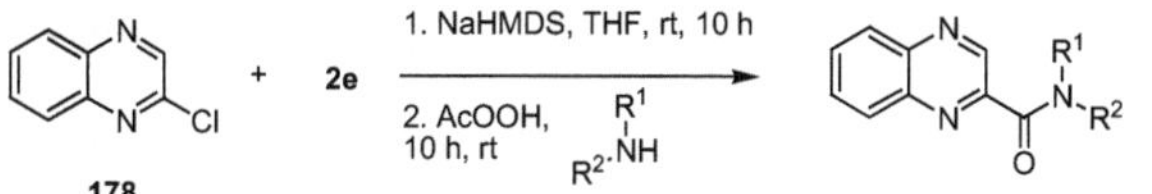

Scheme 62 One-pot preparation of *N*-containing heteroaryl amides from the corresponding heteroaryl halides

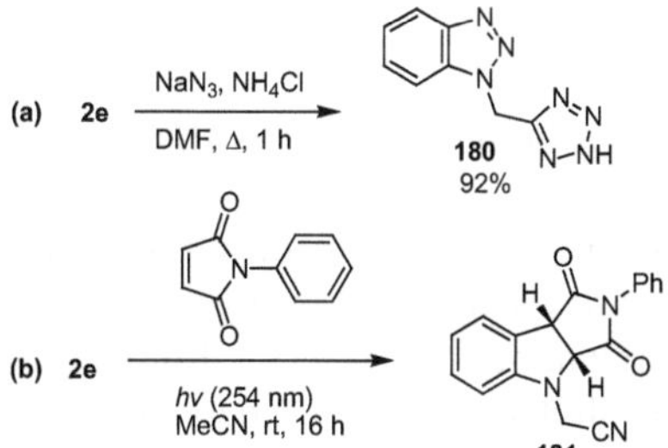

Scheme 63 (**a**) Preparation of 1((1*H*-tetrazol-5-yl)methyl)-1*H*-benzotriazole; (**b**) photochemical behavior of 2-(1*H*-benzotriazol-1-yl)acetonitrile in the presence of *N*-phenylmaleimide

1,3-diradical that underwent intermolecular cyclization in the presence of *N*-phenylmaleimide.

3.6 1-Allylbenzotriazole

1-Allylbenzotriazole (**2f**) is a versatile reagent for inter- and intramolecular allylaminations, for Heck couplings, and for the preparation of various heterocycles.

3.6.1 Preparation of 1-Allylbenzotriazole

One of the earliest reports on 1-allylbenzotriazole (**2f**) by Gasparini and coworkers disclosed a one-pot procedure from 1-trimethylsilylbenzotriazole (**1b**) and allyl bromide [80]. Katritzky reported a direct procedure from benzotriazole and allyl bromide in water in the presence of potassium hydroxide, affording 1- and 2-allylbenzotriazole in 50% and 15%, respectively (after chromatography) [326]. The formation of 1- and 2-allylbenzotriazole (71/29 ratio) was also mentioned in a study of the thermal decarboxylation of 1-alkoxycarbonylbenzotriazoles [327]. More interestingly, the exclusive preparation of 1-allylbenzotriazole (**2f**) was reported by Katritzky via a Mitsunobu reaction [328]. The reaction between allyl alcohol and benzotriazole in the presence of triphenylphosphine and *N*-bromosuccinimide yielded **2f**. This procedure was recently revisited [329]. A green and high yielding procedure using silica as a promotor for the alkylation of benzotriazole was

Scheme 64 Preparation of 1-allylbenzotriazole

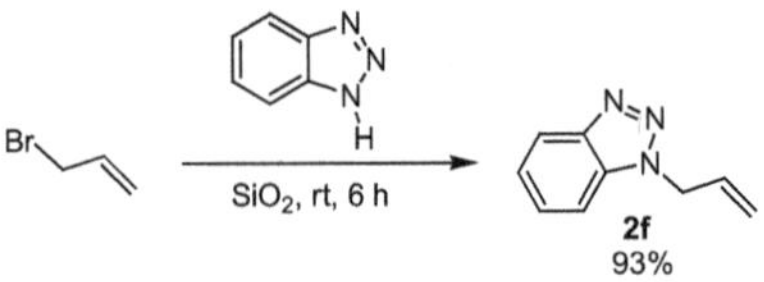

reported by Basu (Scheme 64) [330]. **2f** is easily functionalized to prepare a variety of 1-allylbenzotriazole derivatives [331–334].

3.6.2 Reactivity and Synthetic Utility of 1-Allylbenzotriazole

The main synthetic utility of 1-allylbenzotriazole (**2f**) and its derivatives was revealed when Katritzky devised the Pd-catalyzed coupling of amines and **2f** for the preparation of allylamines [331–333]. Katritzky demonstrated that the benzotriazolyl group in allyl benzotriazole derivatives was sufficiently labile to be substituted by aliphatic and aromatic amines in the presence of a Pd-catalyst. The 2*H*-isomer reacted similarly [331–333]. In the original procedure, the best conditions required methanol as a solvent, potassium hydroxide or carbonate as a base, palladium (II) acetate (3 mol%) and triphenylphosphine (9 mol%) as a catalyst, to avoid the isomerization of 1-allylbenzotriazole (**2f**) to 1-(propenyl) benzotriazole [333]. The method was applied to a variety of 1-allylbenzotriazole derivatives **182** and secondary amines, affording the corresponding allylamines **183** in modest to high yields. Generally speaking, the reaction proceeded regioselectively with substitution at the least substituted carbon (Scheme 65). When primary amines were utilized, a mixture of mono- and di-allylamines was obtained. Katritzky then devised an intramolecular allylamination procedure towards 2-vinylpyrrolidines and 2-vinylpiperidines [332]. Various allylbenzotriazoles **182** were deprotonated and then reacted with dual electrophiles such as 1-bromo-4-chlorobutane to afford the corresponding α-substituted allylbenzotriazoles **184** (Scheme 65). Allylbenzotriazoles **184** were then further reacted with primary amines in the presence of palladium (II) acetate, furnishing 2-vinylpiperidines **185**. The scope of the method was further extended towards the preparation of *N*-allylsulfonamides [331]. Allylbenzotriazoles were also employed for Heck couplings [335].

Katritzky also developed a new *trans*-selective olefination procedure from allyl benzotriazole and its substituted derivatives **182** [334]. Lithiation with *n*-BuLi of allyl benzotriazole derivatives **182** gave anions in α- to the benzotriazolyl moiety, which underwent addition to a variety of aldehydes and ketones. The intermediate *N*-(β-hydroxyalkyl)benzotriazoles **186** were then dehydroxy-benzotriazolylated by a treatment with a low valance titanium (LVT) species to give dienes **187** (Scheme 66). Trienes were obtained upon reaction with α,β-unsaturated aldehydes.

Other heterocyclic compounds than pyrrolidines and piperidines [332] were synthesized from 1-allylbenzotriazole derivatives [336, 337]. For instance, their condensation with isothiocyanates gave 3-functionalized-2-aminothiophenes and

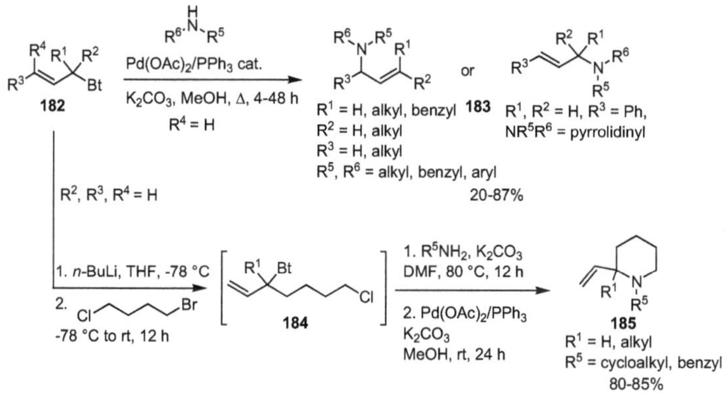

Scheme 65 Katritzky's inter- and intramolecular allylamination

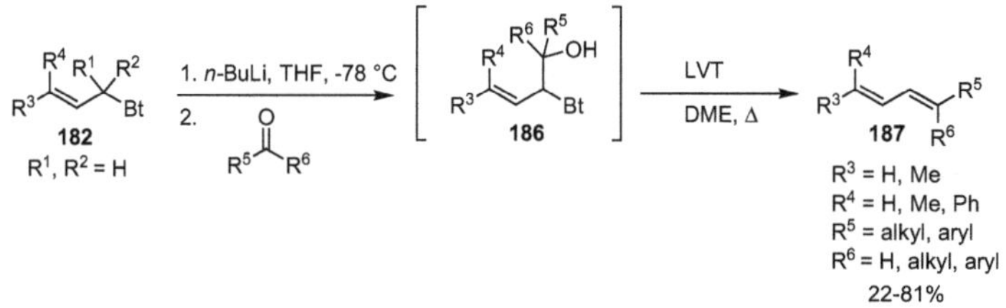

Scheme 66 Olefination procedure from allyl benzotriazole and its substituted derivatives

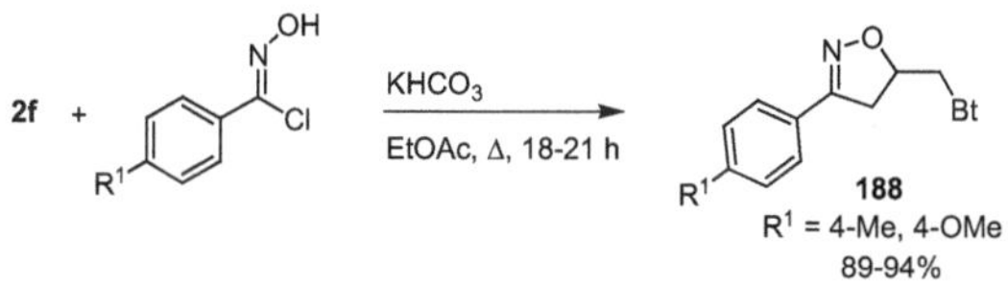

Scheme 67 1,3-Dipolar cycloaddition of allylbenzotriazole and nitrile oxides

1,3-disubstituted-2-methylthiopyrroles in modest to good yields [336]. Katritzky and Button also reported a convenient procedure for the high yielding preparation of polysubstituted isoxazolines **188** from the 1,3-dipolar cycloaddition of 1-allylbenzotriazole (**2f**) and nitrile oxides generated in situ from benzohydroximoyl chlorides (Scheme 67) [337]. Further functionalization of the benzotriazol-1-ylmethyl moiety was effected by nucleophilic displacement of the benzotriazole group or by sequential lithiation/alkylation.

4 Class C Benzotriazole Derivatives 3a–d

*4.1 Bis-(1*H*-benzotriazol-1-yl)-methan(thi)one*

Katritzky studied extensively and demonstrated the wide synthetic utility of bis-(1*H*-benzotriazol-1-yl)-methanone (**3a**) and bis-(1*H*-benzotriazol-1-yl)-methanethione (**3b**). These compounds provided straightforward entries towards substituted ureas, di- and trisubstituted thioureas, respectively. By contrast, similar reactions starting from the corresponding bisimidazolyl reagents often required an extra step for the substitution of the second imidazolyl moiety, i.e., quaternization of the remaining imidazolyl group to enhance its leaving group character. Besides, bis-(1*H*-benzotriazol-1-yl)-methanone (**3a**) and bis-(1*H*-benzotriazol-1-yl)-methanethione (**3b**) are stable and safe surrogates for phosgene and thiophosgene, respectively.

4.1.1 Preparation of Bis-(1*H*-benzotriazol-1-yl)-methan(thi)one

Katritzky reported the first preparation of bis-(1*H*-benzotriazol-1-yl)-methanone (**3a**) according to a straightforward procedure involving 2 eq. of benzotriazole and phosgene. **3a** was obtained in 90% yield and high purity (Scheme 68) [338]. NMR showed that traces of the mono- and bis-benzotriazol-2-yl isomers were present in the crude mixture, but did not affect the overall reactivity with nucleophiles. An excess of benzotriazole to capture HCl lowered the yield [338]. Bis-(1*H*-benzotriazol-1-yl)-methanethione (**3b**) was originally prepared in 1965 from sodium benzotriazolate and thiophosgene [339]. Harpp et al. reported another procedure involving 1-(trimethylsilyl)benzotriazole (**1b**) and thiophosgene, affording **3b** in excellent yields [70]. Katritzky introduced a convenient alternative procedure using an excess of benzotriazole (4 eq.) and thiophosgene (Scheme 68) [340]. The excess of benzotriazole advantageously avoided the preliminary preparation of either 1-(trimethylsilyl)benzotriazole (**1b**) or sodium benzotriazolate.

4.1.2 Reactivity and Synthetic Utility of Bis-(1*H*-benzotriazol-1-yl)-methan(thi)one

Katritzky developed a one-pot versatile strategy for the preparation of both symmetrical and unsymmetrical ureas based on bis-(1*H*-benzotriazol-1-yl)-methanone (**3a**) [338, 341]. **3a** enabled the selective formation of unsymmetrical ureas without the formation of contaminating symmetrical ureas: consecutive treatment of **3a** with two amines (1 eq. each) gave successively the corresponding carbamoylbenzotriazoles **15** and the unsymmetrical ureas **189** (Scheme 69a). Bis-(1*H*-benzotriazol-1-yl)-methanone (**3a**) is compatible with both solution and solid phase syntheses to access ureas in high yields [338, 341]. A more recent example was reported by Nevalainen and coworkers for the preparation of fatty acid amide

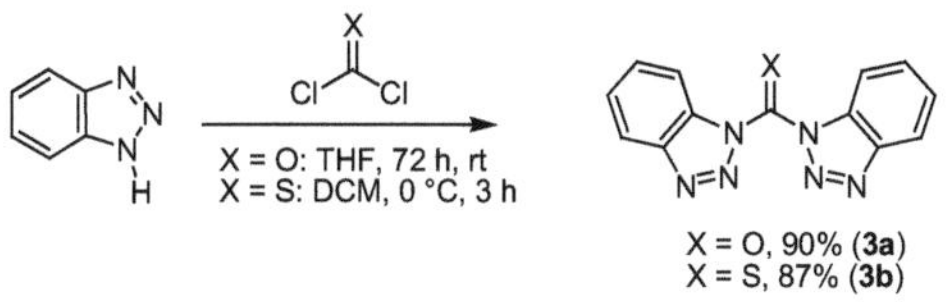

Scheme 68 Preparation of bis-(1*H*-benzotriazol-1-yl)-methanone and bis-(1*H*-benzotriazol-1-yl)-methanethione

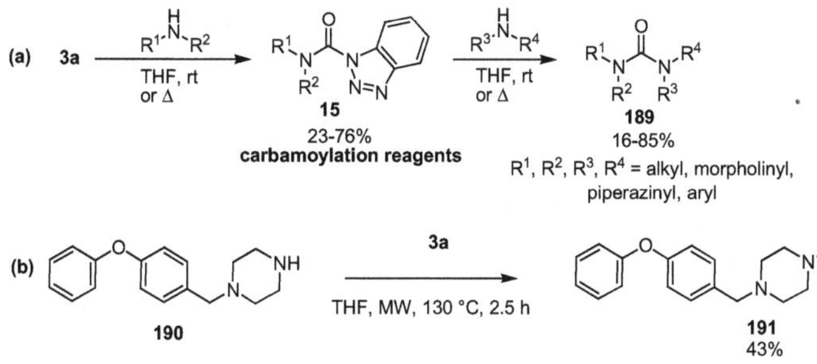

Scheme 69 (**a**) Katritzky's procedure for the preparation of both symmetrical and unsymmetrical ureas; (**b**) preparation of FAAH and MAGL inhibitors

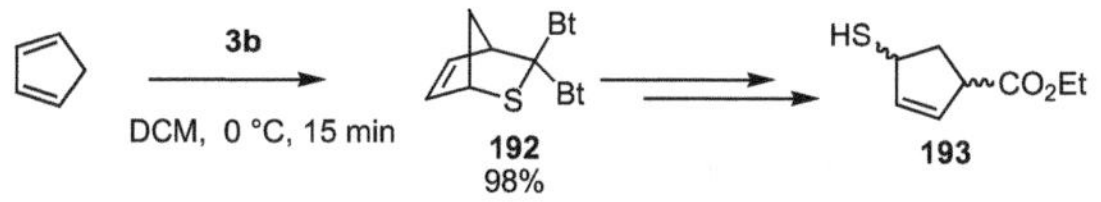

Scheme 70 Cycloaddition of bis-(1*H*-benzotriazol-1-yl)-methanethione with cyclopentadiene

hydrolase (FAAH) and monoacylglycerol lipase (MAGL) inhibitors bearing a urea moiety such as **191** (Scheme 69b) [342].

Harpp was the first to report the synthesis and application of bis-(1*H*-benzotriazol-1-yl)-methanethione (**3b**) as a thiocarbonyl transfer reagent [70, 343]. He also used it as a heterodienophile for Diels-Alder reactions with a variety of dienes, affording the corresponding thiocycloadducts **192** with excellent yields. Cycloadduct **193** was used as a synthon for the preparation of *cis*-3,5-fused cyclopentene derivative **193** (Scheme 70).

Similarly to **3a**, bis-(1*H*-benzotriazol-1-yl)-methanethione (**3b**) was successfully implemented as a versatile reagent for the preparation of symmetrical and unsymmetrical thioureas [31, 67, 69, 344, 345]. Katritzky started his seminal work on the synthetic application of **3b** in the early 2000s [31, 69]. Not only is **3b** a stable and nontoxic equivalent of thiophosgene, but it also has several advantages over 1,1′-thiocarbonyldiimidazole, which is hygroscopic and has a limited stability. **3b** has a long shelf-life and its use on primary amines enabled a selective and almost quantitative reaction (Scheme 71a) [69]. Aryl/alkyl thiocarbamoylbenzotriazoles **194** were successfully obtained from the direct reaction of **3b** and primary or secondary alkyl/arylamines at room temperature in dichloromethane. Reaction with primary aryl amines gave isothiocyanates instead of thiocarbamoylbenzotriazoles

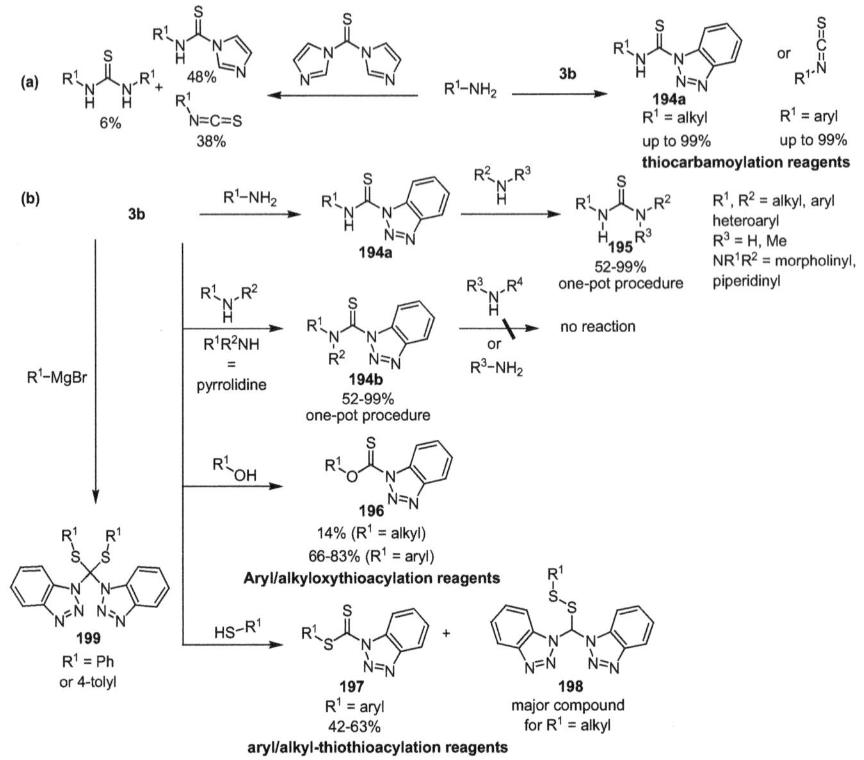

Scheme 71 (**a**) Bis(benzotriazolyl)methanethione vs 1,1′-thiocarbonyldiimidazole; (**b**) Bis-(1*H*-benzotriazol-1-yl)-methanethione led to the development of numerous other benzotriazole-based thioacylation, thiocarbamoylation, aryl/alkyloxythioacylation and aryl/alkyl-thiothioacylation reagents

[31]. These reagents are stable equivalents of isocyanates and can be used for the preparation of unsymmetrical thioureas (Scheme 71b) [31, 69]. A one-pot procedure was successfully developed: **3b** was first reacted with primary amines to give aryl/alkyl thiocarbamoylbenzotriazoles **194**, which were then further reacted with primary or secondary amines. The method was successfully utilized for the preparation of symmetrical and unsymmetrical di- and trisubstituted thioureas **195**, but failed for the preparation of unsymmetrical *N*,*N*-tetrasubstituted thioureas. Bis-(1*H*-benzotriazol-1-yl)-methanethione (**3b**) led to the development of numerous other benzotriazole-based reagents for aryl/alkyloxythioacylation such as **196** and aryl/alkyl-thiothioacylation such as **197**. The direct reaction of **3b** with Grignard reagents unexpectedly gave bis(benzotriazol-1-yl)diarylsulfidemethanes **199** (Scheme 71b), rather than the expected alkyl/aryl thiocarbonylbenzotriazole reagents **10** (see Scheme 2 and Sect. 2.1.1) [31].

Besides their use for the preparation of thioureas, the intermediate thiocarbamoylbenzotriazoles were used as thiocarbamoylation reagents on various substrates, leading to libraries of thioamides, thiocarbamates, and dithiocarbamates

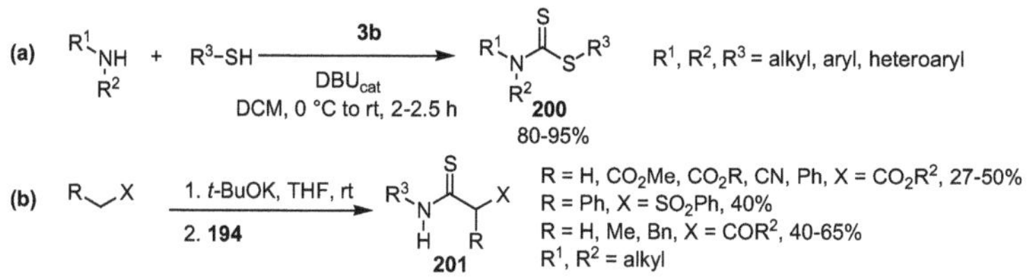

Scheme 72 (**a**) Preparation of homo- and hetero-substituted alkyl, aryl, and heteroaryl dithiocarbamate libraries; (**b**) 1-(alkyl- or arylthiocarbamoyl)benzotriazoles are *C*-thiocarbamoylation reagents for esters, sulfones and ketones

with good to excellent yields [31]. For instance, Tiwari et al. developed a one-pot procedure under mild conditions for the preparation of homo- and hetero-substituted alkyl, aryl, and heteroaryl dithiocarbamate **200** libraries by the reaction of **3b** and mixtures of thiols and amines in the presence of a catalytic amount of DBU or DABCO (Scheme 72a) [346]. Katritzky also reported the high-yielding preparation of libraries of thiosemicarbazides and *N*-hydroxythioureas by the reaction of **3b** with substituted hydrazines and hydroxylamines, respectively [68]. He further extended the use of 1-(alkyl- or arylthiocarbamoyl)benzotriazoles **194** as *C*-thiocarbamoylation reagents for esters (27–50%), sulfones (40%), and ketones (40–65%) (Scheme 72b) [347]. Last but not least, **3b** was also utilized for the preparation of guanidines (see Sect. 4.2) [348].

Bis-(1*H*-benzotriazol-1-yl)-methanethione (**3b**) also enabled the preparation of various heterocyclic compounds [19, 344, 349–351]. For example, Tiwari et al. developed a simple and convenient one-pot procedure for the synthesis of thioquinozolinones **203** by a base-catalyzed reaction of anthranilate esters **202**, **3b**, and various heterocyclic amines under microwave irradiation (Scheme 73a) [344]. Katritzky reported a one-pot procedure for the preparation of 5-(substituted amino)-1,2,3,4-thiatriazoles **205** starting from **3b**, an amine, and sodium azide (Scheme 73b) [351].

Tiwari and coworkers reported the synthesis of 2-*C*/*N*/*S*-substituted benzothiazoles **206–208** from the free-radical ring cleavage/cyclization of the corresponding substituted thiocarbonylbenzotriazoles **10**, **194**, and **197** under mild conditions in the presence of $(TMS)_3SiH$ and AIBN (Scheme 74a) [19, 350]. The strategy was further extended as a modified Barton-McCombie two-step procedure for the regioselective deoxygenation of benzylic alcohols [349]. The first step involved the reaction of compound **209** with **3b** in the presence of triethylamine. The intermediate **210** was then treated with $(TMS)_3SiH$ in the presence of AIBN as a radical initiator under MW or classical thermal conditions in toluene, leading to a formal deoxygenation at the benzylic position. By contrast, the non-benzylic alkoxythioacylbenzotriazole function was converted into the corresponding 2-*O*-substituted benzothiazole via benzotriazole ring cleavage (Scheme 74b).

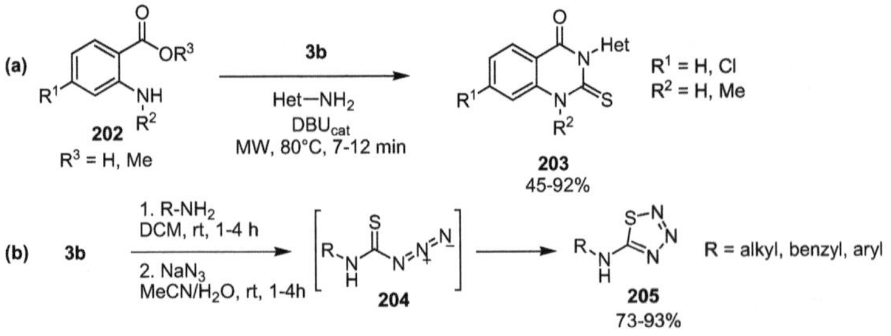

Scheme 73 (**a**) One-pot procedure for the synthesis of thioquinozolinones; (**b**) one-pot procedure for the preparation of 5-(substituted amino)-1,2,3,4-thiatriazoles

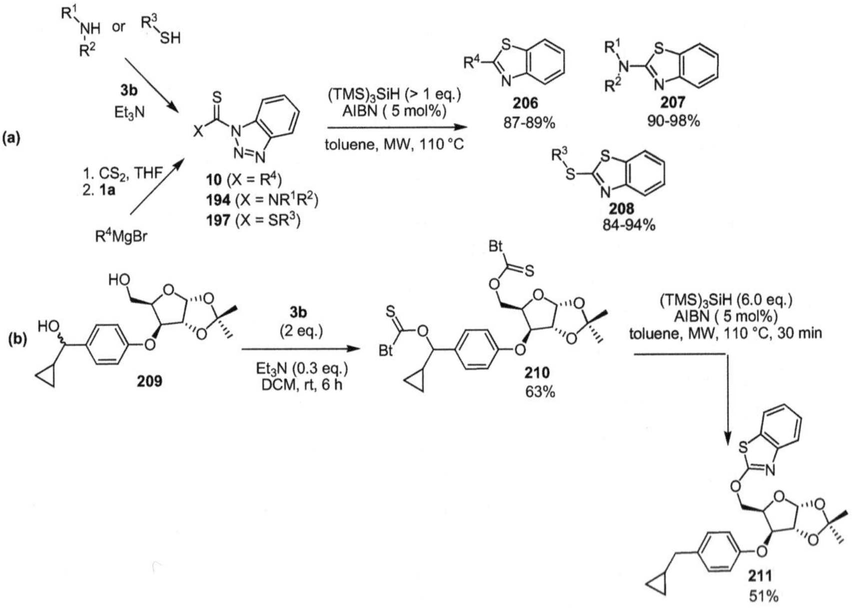

Scheme 74 (**a**) Synthesis of 2-*N*/*S*/*C*-substituted benzothiazoles from the free-radical ring cleavage/cyclization of substituted thiocarbonylbenzotriazoles; (**b**) modified Barton-McCombie deoxygenation for the regioselective deoxygenation of benzylic alcohols

4.2 *Di-(1*H*-benzotriazol-1-yl)methanimine*

Katritzky extensively studied di(1*H*-benzotriazol-1-yl)methanimine (**3c**) and its synthetic application for the preparation of tri- or tetrasubstituted guanidines.

Scheme 75 Preparation of di(1*H*-benzotriazol-1-yl) methanimine

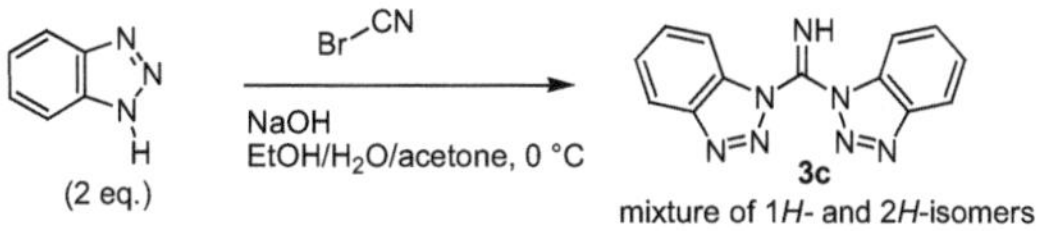

4.2.1 Preparation of Di(1*H*-benzotriazol-1-yl)methanimine

Di(1*H*-benzotriazol-1-yl)methanimine (**3b**) was reported as the main side product contaminating 1-cyanobenzotriazole (**1g**) prepared according to Cava's procedure [42] by reacting 1-chlorobenzotriazole (**1a**) and sodium cyanide in acetonitrile. Di (1*H*-benzotriazol-1-yl)methanimine (**3b**) was prepared according to the original procedure of Ramalingam [352] by the reaction of benzotriazole and cyanogen bromide (Scheme 75). The crude mixture consisted of a mixture of di(1*H*-benzotriazol-1-yl)methanimine and 1*H*-benzotriazol-1-yl(2*H*-benzotriazol-1-y) methanimine [353].

4.2.2 Reactivity and Synthetic Utility of Di(1*H*-benzotriazol-1-yl) methanimine

Di(1*H*-benzotriazol-1-yl)methanimine (**3c**) was originally reported as an electrophilic cyanation agent, but it turned out that the actual cyanation agent was 1-cyanobenzotriazole (**1g**), which was formed in situ by the decomposition of **3c** [352]. Katritzky started to investigate **3c** as a guanylation reagent [353]. The procedure and the guanylation reagents developed by Katritzky since 2000 are particularly appreciable since they enabled the preparation of tri- and tetrasubstituted guanidines under mild and neutral conditions, which are usually difficult to prepare and often require harsh conditions. **3c** reacted with various amines at room temperature in THF through the displacement of a first benzotriazole moiety and formed of *N*-substituted carboximidamides **212** (Scheme 76a). The second benzotriazole moiety was displaced in refluxing THF in the presence of another amine, affording tetrasubstituted guanidines **213**. The intermediate carboximidamides **212** could be isolated without contamination by side products resulting from self-condensation or condensation of a second equivalent of the amine, emphasizing the chemospecificity of the reaction [353]. Di(1*H*-benzotriazol-1-yl)methanimine (**3c**) was also employed for solid phase synthesis [354, 355]: for instance, Katritzky prepared 1*H*-benzotriazole-1-carboximidamide resins **214**, which were then reacted with amines or thiols to give, after cleavage, guanidines **215** or isothioureas **216**, respectively (Scheme 76b) [355]. **3c** reacted with substituted hydrazines to give triazoles [356].

Katritzky's guanidination procedure was considerably extended in the following years with the preparation of polysubstituted acylguanidines and guanylureas [357]. Benzotriazol-1-yl-carboximidamides **212** were obtained as described above. Disubstituted (benzotriazol-1-yl)-carboximidamides **212** reacted smoothly

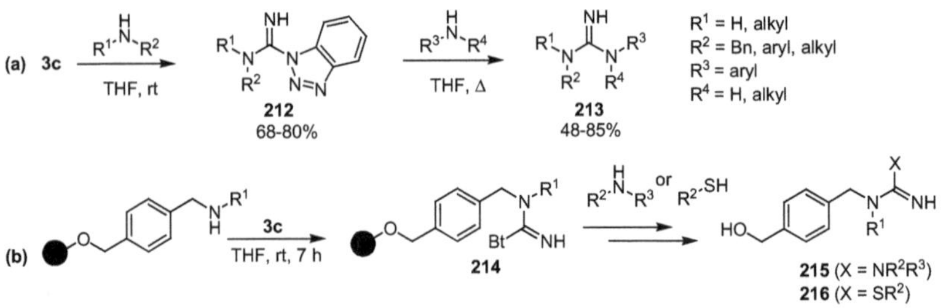

Scheme 76 (**a**) Preparation of tetrasubstituted guanidines from di(1*H*-benzotriazol-1-yl) methanimine and amines; (**b**) preparation of 1*H*-benzotriazole-1-carboximidamide resins and their use for the synthesis of isothioureas and guanidines

with acyl chlorides (62–94%) to give *N*-acyl benzotriazol-1-yl-carboximidamides **217**, while, surprisingly, the same reaction with monosubstituted (benzotriazol-1-yl)-carboximidamides **212** ($R^2 = H$) failed. However, mono- and disubstituted (benzotriazol-1-yl)-carboximidamides **212** reacted with isocyanates to afford *N*-carbamoylguanidines **218** with good to excellent yields (65–89%). Finally, the remaining benzotriazole moiety was displaced with various amines to give the corresponding acylguanidines and guanylureas **219** (Scheme 77) [357]. Substituted triazoles **220** were obtained upon treatment of acylated (benzotriazol-1-yl)-carboximidamides **217** with various hydrazines [358].

Other guanylation reagents were prepared from bis-(1*H*-benzotriazol-1-yl)-methanethione (**3b**) [347, 348]: the reaction of **3b** and triphenylphosphine ylides gave symmetrical guanylation reagents **221** (Fig. 2). Compounds **222** were prepared according to a similar procedure from **212** [347, 348]. Both series of compounds **221**,**222** further complemented Katritzky's palette of guanylation tools.

Unsymmetrical *N*-hydroxyguanidines **223**,**224** were obtained by refluxing **212** or **222** with substituted hydroxylamine hydrochlorides in toluene for 4–12 h in the presence of triethylamine (Scheme 78a) [348]. Similarly, unsymmetrical *N*-aminoguanidines **225**,**226** were prepared by the reaction of **212** or **222** with substituted hydrazines in refluxing toluene for 4–12 h in the presence of triethylamine (Scheme 78b) [348]. Symmetrical dihydroxyguanidines **227** and diaminoguanidines **228** were also synthesized by refluxing **3c** or **221** with hydroxylamine hydrochloride (3 eq.) or substituted hydrazines (3 eq.) in toluene for 30–45 min in the presence of triethylamine (3 eq.) (Scheme 78c) [348, 359, 360]. *C*-aminoimidoylation was successfully attempted on various substrates such as ester enolates, sulfones and ketones [347].

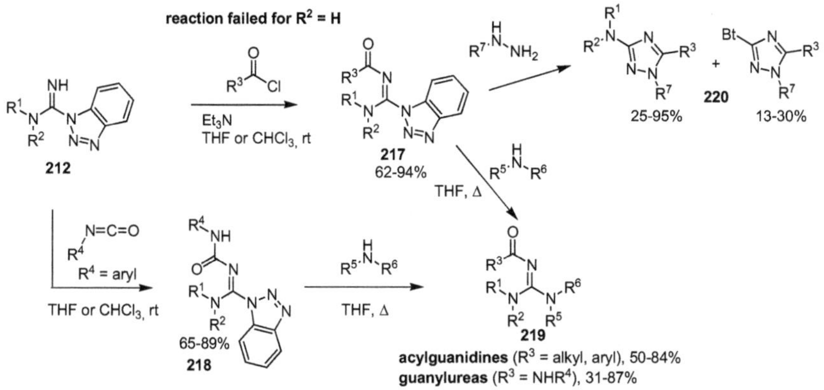

Scheme 77 Preparation of acylguanidines and guanylureas

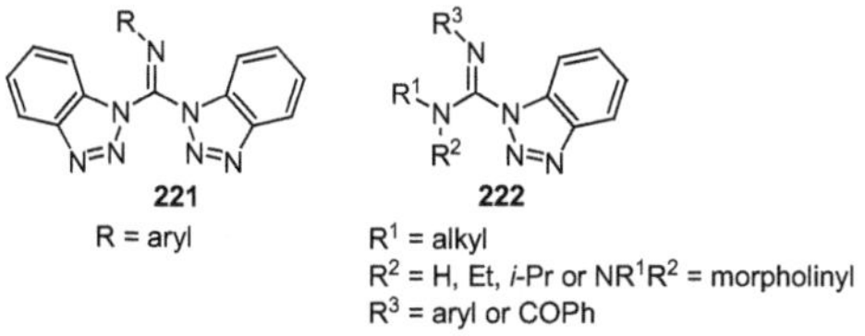

Fig. 2 Other guanylation reagents related to **3c**

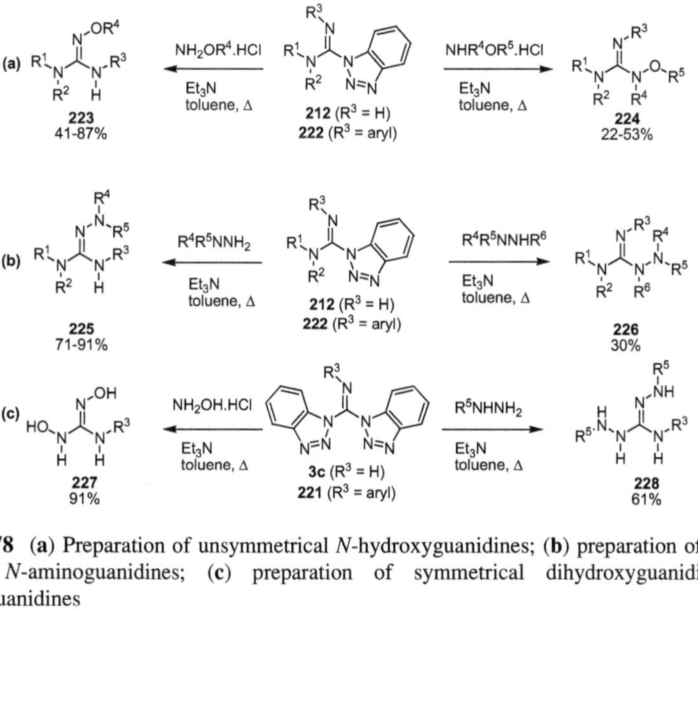

Scheme 78 (**a**) Preparation of unsymmetrical *N*-hydroxyguanidines; (**b**) preparation of unsymmetrical *N*-aminoguanidines; (**c**) preparation of symmetrical dihydroxyguanidine and diaminoguanidines

Scheme 79 Preparation of *N*,*N*-dimethylaminobenzotriazol-1-ylmethyleniminium chloride

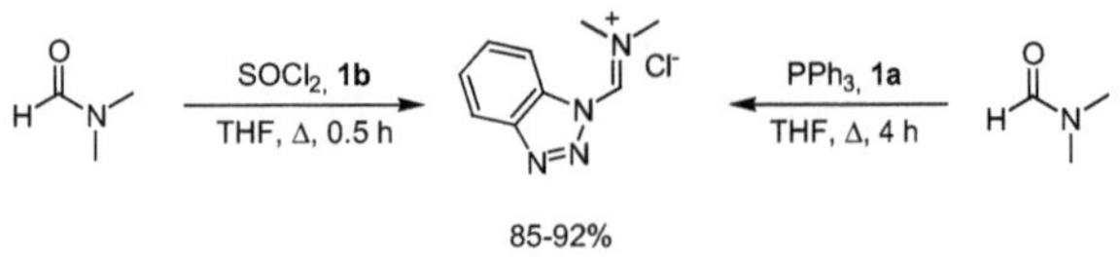

4.3 N,N-*Dimethylaminobenzotriazol-1-ylmethyleniminium Chloride*

N,*N*-Dimethylaminobenzotriazol-1-ylmethyleniminium (**3d**) chloride, *aka* benzotriazole Vilsmeier reagent, is a stable synthetic chloroiminium salt equivalent and can be handled without special precautions – unlike the original Vilsmeier's salt. It will react with a base to release the corresponding *N*-heterocyclic carbene.

4.3.1 Preparation of *N*,*N*-Dimethylaminobenzotriazol-1-ylmethyleniminium Chloride

Two procedures were reported for the preparation of *N*,*N*-dimethylaminobenzotriazol-1-ylmethyleniminium (**3d**) chloride. The first procedure involved refluxing equimolar amounts of 1-trimethylsilylbenzotriazole (**1b**), thionyl chloride, and DMF in THF, affording **3d** chloride in 92% yield (Scheme 79, left) [74, 361]. The second method proceeded with equimolar amounts of triphenylphosphine, 1-chlorobenzotriazole (**1a**), and DMF in THF, giving **3d** in 85% yield (Scheme 79, right) [74].

4.3.2 Reactivity and Synthetic Utility of *N*,*N*-Dimethylaminobenzotriazol-1-ylmethyleniminium Chloride

One of the earliest reports on *N*,*N*-dimethylaminobenzotriazol-1-ylmethyleniminium (**3d**) chloride by Katritzky concerned its use as a precursor of a new class of *N*-heterocyclic carbenes (NHC) [74]. Upon treatment with a base, **3d** was converted into a transient carbenoic species. Evidences of this NHC included the formation of a dimer **229** and a typical carbene reaction in the presence of phenylisocyanate, leading to hydantoin derivatives such as **230** (Scheme 80). Besides, the aminocarbene derived from **3d** reacted with *trans*-dibenzoylethylene to give a mixture of 4-benzoyl-5-dimethylamino-2-phenylfuran **231** (15%) and 3-benzotriazolyl-4-benzoyl-2-phenylfuran **232** (6%) after 24 h in refluxing benzene [74].

A few years later, Katritzky devised a regioselective method for the preparation of quinolines starting from **3d** (Scheme 81a) [361]. The benzotriazole Vilsmeier salt **3d** reacted regioselectively with imines by attacking the sterically less hindered α-position to give enaminoimine hydrochlorides **233**, which were then thermally

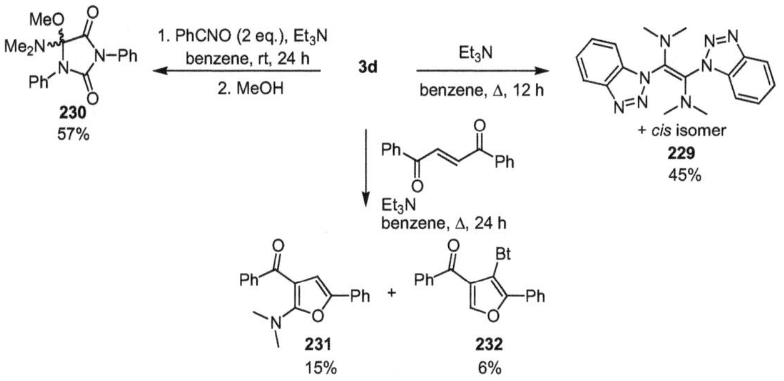

Scheme 80 Typical reactions of the aminocarbene derived from *N*,*N*-dimethylaminobenzotriazol-1-ylmethyleniminium chloride

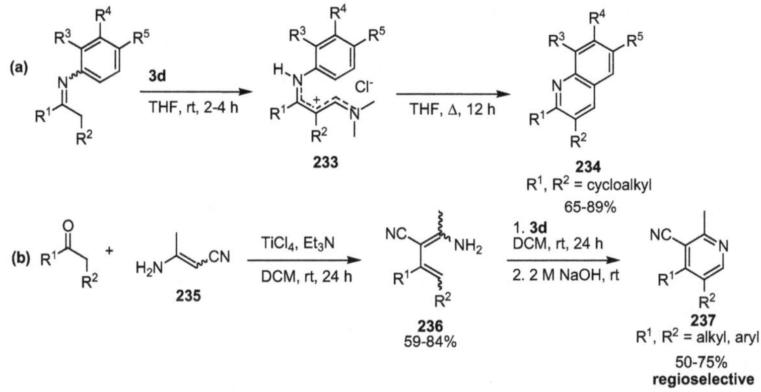

Scheme 81 (**a**) Regioselective method for the preparation of quinolines; (**b**) Two-step procedure for the preparation of nicotinonitriles

transformed into quinolines **234** (65–89%). He then reported a convenient two-step procedure for the preparation of a library of nicotinonitriles **237** (Scheme 81b) [362]. The sequence started with a $TiCl_4$-catalyzed tandem alkylation-elimination involving cyclic or acyclic aryl/alkylketones and β-enaminonitriles **235**. The conjugated β-enaminonitrile intermediates **236** (59–84%) were then further reacted with **3d** to give a variety of diversely substituted nicotinonitriles **237** in good yields (50–75%).

Katritzky designed a simple one-pot entry towards pyrazoles (Scheme 82a) [363]. The procedure started with the reaction of **3d** chloride with various imines. The intermediate enaminoimine hydrochlorides **238** were next reacted in situ with hydrazine, yielding pyrazoles **239** (19–83%). The same year, Katritzky and coworkers reported a convenient procedure for the preparation of dimethylformamidrazones **240** (Scheme 82b) [364]. Typically, stoichiometric mixtures of **3d** and hydrazine derivatives were refluxed in THF to afford the corresponding

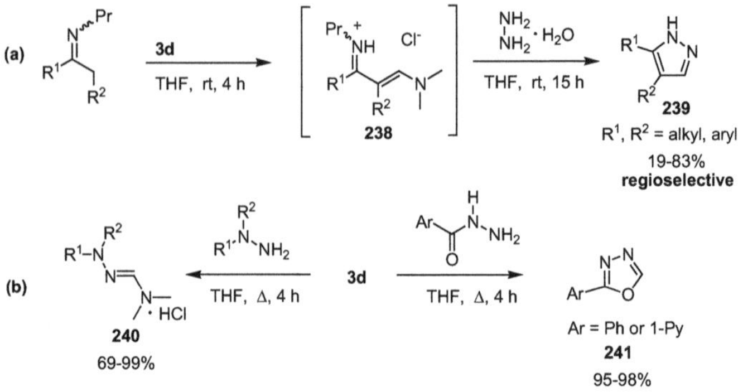

Scheme 82 (**a**) Regioselective entry towards pyrazoles; (**b**) preparation of dimethylformamidrazones and 1,3,4-oxadiazoles

dimethylformamidrazones **240**, with both electron-deficient and electron-enriched aromatic substituents. Interestingly, 1,3,4-oxadiazoles **241** were obtained by the reaction of **3d** and aromatic hydrazides (Scheme 82b, left).

References

1. Panda SS, Hall CD, Scriven E, Katritzky AR (2013) Aldrichimica Acta 46:43
2. Monbaliu JCM, Katritzky AR (2012) Chem Commun 48:11601
3. Katritzky AR, Rachwal S (2011) Chem Rev 111:7063
4. Katritzky AR, Rachwal S (2010) Chem Rev 110:1564
5. Katritzky AR, Manju K, Singh SK, Meher NK (2005) Tetrahedron 61:2555
6. Katritzky AR, Rogovoy BV (2003) Chem Eur J 9:4586
7. Katritzky AR, Lan X, Yang JZ, Denisko OV (1998) Chem Rev 98:409
8. Katritzky AR, Belyakov SA (1998) Aldrichimica Acta 31:35
9. Katritzky AR, Wu H, Xie L, Rachwal S, Rachwal B, Jiang J, Zhang G, Lang H (1995) Synthesis 1315
10. Katritzky AR, Rachwal S, Hitchings GJ (1991) Tetrahedron 47:2683
11. El-Faham A, Albericio F (2011) Chem Rev 111:6557
12. Joullié MM, Lassen KM (2010) ARKIVOC 8:189
13. Valeur E, Bradley M (2009) Chem Soc Rev 38:606
14. Rees CW, Storr RC (1969) J Chem Soc C 1474
15. Monbaliu JCM, Beagle LK, Kovacs J, Zeller M, Stevens CV, Katritzky AR (2012) RSC Adv 2:8941
16. Sasse MJ, Storr RC (1978) J Chem Soc Perkin Trans 1 909
17. Mandal B, Basu B (2014) RSC Adv 4:13854
18. Panmand DS, Tiwari AD, Panda SS, Monbaliu JCM, Beagle LK, Asiri AM, Stevens CV, Steel PJ, Hall CD, Katritzky AR (2014) Tetrahedron Lett 55:5898
19. Kumar D, Mishra BB, Tiwari VK (2014) J Org Chem 79:251
20. Fraix A, Le Gall T, Berchel M, Denis C, Lehn P, Montier T, Jaffrès PA (2013) Org Biomol Chem 11:1650
21. Gnamm C, Jeanguenat A, Dutton AC, Grimm C, Kloer DP, Crossthwaite AJ (2012) Bioorg Med Chem Lett 22:3800

22. Reddie KG, Humphries WH, Bain CP, Payne CK, Kemp ML, Murthy N (2012) Org Lett 14:680
23. Hunter R, Msutu A, Dwyer CL, Emslie ND, Hunt RC, Bezuidenhoudt BCB (2011) Synlett 2011:2335
24. Stellenboom N, Hunter R, Caira MR (2010) Tetrahedron 66:3228
25. Stellenboom N, Hunter R, Caira MR, Szilágyi L (2010) Tetrahedron Lett 51:5309
26. Pietrzak M, Wehling JP, Kong S, Tolstoy PM, Shenderovich IG, López C, Claramunt RM, Elguero J, Denisov GS, Limbach H-H (2010) Chem Eur J 16:1679
27. Teague SJ, Barber S (2010) Tetrahedron Lett 51:4720
28. Katritzky AR, Widyan K, Gyanda K (2008) Synthesis 1201
29. Katritzky AR, Vincek AS, Steel PJ (2008) Heterocycles 76:1401
30. Hunter R, Caira M, Stellenboom N (2006) J Org Chem 71:8268
31. Katritzky AR, Witek RM, Rodriguez-Garcia V, Mohapatra PP, Rogers JW, Cusido J, Abdel-Fattah AAA, Steel PJ (2005) J Org Chem 70:7866
32. Katritzky AR, Abdel-Fattah AAA, Vakulenko AV, Tao H (2005) J Org Chem 70:9191
33. Katritzky AR, Hayden AE, Kirichenko K, Pelphrey P, Ji Y (2004) J Org Chem 69:5108
34. Katritzky AR, Rodriguez-Garcia V, Nair SK (2003) J Org Chem 69:1849
35. Gibson CL, La Rosa S, Suckling CJ (2003) Org Biomol Chem 1:1909
36. Greci L, Castagna R, Carloni P, Stipa P, Rizzoli C, Righi L, Sgarabotto P (2003) Org Biomol Chem 1:3768
37. Ponce MA, Tarzi OI, Erra-Balsells R (2003) J Heterocycl Chem 40:419
38. Katritzky AR, Vakulenko AV, Jain R (2003) Arkivoc 2003:131
39. Katritzky AR, Rogovoy B, Klein C, Insuasty H, Vvedensky V, Insuasty B (2001) J Org Chem 66:2854
40. Katritzky AR, Nichols DA, Voronkov MV (2000) Arkivoc 2000:667
41. Katritzky AR, Voronkov MV, Pastor A, Tatham D (1999) Heterocycles 51:1877
42. Hughes TV, Hammond SD, Cava MP (1998) J Org Chem 63:401
43. Bonesi SM, Erra-Balsells R (1997) J Heterocycl Chem 34:891
44. Katritzky AR, Ignatchenko AV, Lan X, Lang H, Stevens CV (1994) Tetrahedron 50:6005
45. Katritzky AR, Ignatchenko AV, Lang H (1995) Heterocycles 41:131
46. Katritzky AR, Belyakov SA, Lam JN, Dupont Durst H, Karpenko DV (1994) Heterocycles 39:73
47. Toth JE, Ray J, Deeter J (1993) J Org Chem 58:3469
48. Fujihara H, Oi N, Erata T, Furukawa N (1990) Tetrahedron Lett 31:1019
49. Barker SJ, Storr RC (1990) J Chem Soc Perkin Trans 1 3:485
50. Kim JN, Ryu EK (1990) Synth Commun 20:1373
51. De Rosa M, Quesada PA, Dodsworth DJ (1987) J Org Chem 52:173
52. Ganapathy K, Gurumurthy R, Mohan N, Sivagnanam G (1987) Monatsh Chem 118:583
53. Wender PA, Cooper CB (1986) Tetrahedron 42:2985
54. Berti C, Greci L, Andruzzi R, Trazza A (1986) J Chem Soc Perkin Trans 1 4:607
55. Pyne SG (1986) J Org Chem 51:81
56. Akasaka T, Furukawa N, Oae S (1985) Phosphorus Sulfur 21:277
57. Berti C, Greci L, Andruzzi R, Trazza A (1982) J Org Chem 25:4895
58. Keating M, Sasse MJ, Storr RC (1978) J Chem Soc Perkin Trans 1 905
59. Lichman KV (1971) J Chem Soc C 2539
60. Rees CW, Storr RC (1969) J Chem Soc C 1478
61. Niedermann K, Früh N, Senn R, Czarniecki B, Verel R, Togni A (2012) Angew Chem Int Ed 51:6511
62. Katritzky AR, Shobana N, Juliusz P, Afridi AS, Fan WQ (1992) Tetrahedron 48:7817
63. Taboada R, Ordonio GG, Ndakala AJ, Howell AR, Rablen PR (2003) J Org Chem 68:1480
64. Boto A, Hernández D, Hernández R (2010) Eur J Org Chem 2010:6633
65. Lesiv AV, Ioffe SL, Strelenko YA, Tartakovsky VA (2002) Helv Chim Acta 85:3489
66. Birkofer L, Richter P, Ritter A (1960) Chem Ber 93:2804

67. Sureshbabu VV, Chennakrishnareddy G, Hemantha HP (2010) Synlett 2010:715
68. Katritzky AR, Khashab NM, Gromova AV (2006) Arkivoc 2006:226
69. Katritzky AR, Ledoux S, Witek RM, Nair SK (2004) J Org Chem 69:2976
70. Larsen C, Steliou K, Harpp DN (1978) J Org Chem 43:337
71. Boto A, Hernández D, Hernández R, Álvarez E (2007) J Org Chem 72:9523
72. Katritzky AR, Donkor A, Fang Y (2001) J Org Chem 66:4041
73. Katritzky AR, Stevens CV, Zhang GF, Jiang J (1995) Heterocycles 40:231
74. Katritzky AR, Cheng D, Leeming P, Ghiviriga I, Hartshorn CM, Steel PJ (1996) J Heterocycl Chem 33:1935
75. Bourry A, Pitard F, Rigo B, Sanz G, Camus F, Norberg B, Durant F, Couturier D (2002) J Heterocycl Chem 39:109
76. Katritzky AR, Soloducho J, Musgrave RP, Breytenbach JC (1995) Tetrahedron Lett 36:5491
77. Katritzky AR, Hong Q, Yang Z (1995) J Org Chem 60:3405
78. Katritzky AR, Hong Q, Yang Z (1994) J Org Chem 59:7947
79. Gasparini JP, Gassend R, Maire JC (1981) J Organochem Chem 208:309
80. Gasparini JP, Gassend R, Maire JC (1980) J Organomet Chem 188:141
81. Boto A, Hernández D, Hernández R (2010) Eur J Org Chem 2010:3847
82. Denmark SE, Collins WR (2007) Org Lett 9:3801
83. Niedenzu K, Woodrun KR (1989) Inorg Chem 28:4022
84. Campbell CD, Rees CW (1969) J Chem Soc C 752
85. Knight DW, Little PB (2000) J Chem Soc Perkin Trans 1 15:2343
86. Tadross PM, Stoltz BM (2012) Chem Rev 112:3550
87. Wenk HH, Winkler M, Sander W (2003) Angew Chem Int Ed 42:502
88. Pellissier H, Santelli M (2003) Tetrahedron 59:701
89. Sanji T, Mori T, Sakurai H (2006) J Organomet Chem 691:1169
90. Mochida K, Kayamori T, Wakasa M, Hayashi H, Egorov MP (2000) Organometallics 19:3379
91. Birkett MA, Knight DW, Little PB, Mitchell MB (2000) Tetrahedron 56:1013
92. Harsanyi MC, Lay PA, Norris RK, Witting PK (1996) Aust J Chem 49:581
93. Whitney SE, Winters M, Rickborn B (1990) J Org Chem 55:929
94. Rigby JH, Holsworth DD, James K (1989) J Org Chem 54:4019
95. Del Mazza D, Reinecke MG (1988) J Org Chem 53:5799
96. Rickborn B, Whitney SE (1988) J Org Chem 53:5595
97. Cresp TM, Wege D (1986) Tetrahedron 42:6713
98. Ando W, Tsumuraya T (1986) Tetrahedron Asymmetry 27:3251
99. Anthony IJ, Wege D (1984) Aust J Chem 37:1283
100. Houghton PG, Rees CW (1980) J Chem Res Synop 303
101. Fan J, Brown SM, Tu Z, Kharasch ED (2011) Bioconjug Chem 22:752
102. Furusawa M, Hashimoto T, Noma Y, Asakawa Y (2006) Chem Pharm Bull 54:861
103. Callot HJ, Cromer R, Louati A, Gross M (1986) J Chem Soc Chem Commun 767
104. Perona A, Sanz D, Claramunt RM, Elguero J (2006) Molecules 11:453
105. El Kaim L, Meyer C (1996) J Org Chem 61:1556
106. Mester B, Claramunt RM, Elguero J, Atienza J, Barrio AG, Escario JA (1991) Chem Pharm Bull 39:1990
107. Johns AM, Liu Z, Hartwig JF (2007) Angew Chem Int Ed 46:7259
108. Denton JR (2010) Synthesis 775
109. Katritzky AR, Moutou JL, Yang Z (1995) Synlett 99
110. Cid BM, Lopez-Cantarero J, Duce S, Garcia Ruano JL (2009) J Org Chem 74:431
111. Katritzky AR, Hanci S, Meher NK (2009) Arkivoc 2009:115–124
112. Katritzky AR, Khelashvili L, Le KNB, Mohapatra PP, Steel PJ (2007) J Org Chem 72:5805
113. Katritzky AR, Monteux DA, Tymoshenko DO (1999) Org Lett 1:577
114. Katritzky AR, He HY, Suzuki K (2000) J Org Chem 65:8210
115. Micó XA, Ziegler T, Subramanian LR (2004) Angew Chem Int Ed 43:1400

116. Katritzky AR, El Khatib M, Bol'shakov O, Khelashvili L, Steel PJ (2010) J Org Chem 75:6532
117. Sitnikov NS, Kokisheva AS, Fukin GK, Neudörfl JM, Sutorius H, Prokop A, Fokin VV, Schmalz HG, Fedorov AY (2014) Eur J Org Chem 2014:6481
118. Roman BI, Monbaliu JCM, De Coen LM, Verhasselt S, Schuddinck B, Van Hoeylandt E, Stevens CV (2014) Eur J Org Chem 2014:2594
119. Lebedyeva IO, Biswas S, Goncalves K, Sileno SM, Jackson AR, Patel K, Steel PJ, Katritzky AR (2014) Chem Eur J 20:11695
120. Panda SS, Ibrahim MA, Küçükbay H, Meyers MJ, Sverdrup FM, El-Feky SA, Katritzky AR (2013) Chem Biol Drug Des 82:361
121. Baglan M, Atılgan S (2013) Chem Commun 49:5325
122. Frederich JH, Harran PG (2013) J Am Chem Soc 135:3788
123. El Khatib M, Elagawany M, Çalışkan E, Davis EF, Faidallah HM, El-Feky SA, Katritzky AR (2013) Chem Commun 49:2631
124. Bajaj K, Panda SS, Ibrahim MA, El-Feky SA, Katritzky AR (2013) J Pept Sci 19:110
125. Panda SS, Bajaj K, Meyers MJ, Sverdrup FM, Katritzky AR (2012) Org Biomol Chem 10:8985
126. El-Nachef C, Bajaj K, Koblick J, Katritzky AR (2012) Eur J Org Chem 4412–4419
127. Liu T, Nair SJ, Lescarbeau A, Belani J, Peluso S, Conley J, Tillotson B, O'Hearn P, Smith S, Slocum K, West K, Helble J, Douglas M, Bahadoor A, Ali J, McGovern K, Fritz C, Palombella VJ, Wylie A, Castro AC, Tremblay MR (2012) J Med Chem 55:8859
128. Štěpnička P, Solařová H, Císařová I (2011) J Organomet Chem 696:3727
129. Xia Z, Lv X, Wang W, Wang X (2011) Tetrahedron Lett 52:4906
130. Sheikh MC, Takagi S, Sakai M, Mori T, Hayashi N, Fujie T, Ono S, Yoshimura T, Morita H (2011) Org Biomol Chem 9:1244
131. De Wael F, Muccioli GG, Lambert DM, Sergent T, Schneider Y-J, Rees J-F, Marchand-Brynaert J (2010) Eur J Med Chem 45:3564
132. Vitoriano BC, Carvalho LCR, Estevão MS, Sekera MH, Marques MMB (2010) Synlett 753
133. Katritzky AR, Khelashvili L, Kovacs J, Shanab K (2009) Chem Biol Drug Des 73:396
134. Wang X, Li Z, Zhu X, Mao H, Zou X, Kong L, Li X (2008) Tetrahedron 64:6510
135. Katritzky AR, Chen Q-Y, Tala SR (2008) Org Biomol Chem 6:2400
136. Kondo T, Sugimoto I, Nekado T, Ochi K, Ohtani T, Tajima Y, Yamamoto S, Kawabata K, Nakai H, Toda M (2007) Bioorg Med Chem 15:2715
137. Katritzky AR, Abdel-Fattah AAA, Akhmedova RG (2005) Arkivoc 2005:329
138. Katritzky AR, Hoffman S, Suzuki K (2004) Arkivoc 2004:14
139. Katritzky AR, Wang Z, Wang M, Wilkerson CR, Hall CD, Akhmedov NG (2004) J Org Chem 69:6617
140. Katritzky AR, Shestopalov AA, Suzuki K (2004) Synthesis 1806
141. Katritzky AR, Zhang Y, Singh SK, Steel PJ (2003) Arkivoc 2003:47
142. Katritzky AR, Abdel-Fattah AAA, Wang M (2003) J Org Chem 68:4932
143. Katritzky AR, Suzuki K, Singh SK, He HY (2003) J Org Chem 68:5720
144. Katritzky AR, Yang H, Zhang S, Wang M (2002) Arkivoc 2002:39
145. Katritzky AR, Wang M, Zhang S, Akhmedov NG (2002) Arkivoc 2002:134
146. Saravanan V, Muthu Ramalingam B, Mohanakrishnan AK (2014) Eur J Org Chem 2014:1266
147. Micó XÀ, Vagin SI, Subramanian LR, Ziegler T, Hanack M (2005) Eur J Org Chem 4328
148. Katritzky AR, Zhang G, Wu J (1994) Synth Commun 24:205
149. Wyatt P, Hudson A, Charmant J, Orpen AG, Phetmung H (2006) Org Biomol Chem 4:2218
150. Ryan MD, Harpp DN (1997) Tetrahedron Lett 38:8829
151. Ryan MD, Harpp DN (1992) Tetrahedron Lett 33:2129
152. Hermes ME, Marsh FD (1967) J Am Chem Soc 89:4760
153. Katritzky AR, Akue-Gedu R, Vakulenko AV (2007) Arkivoc 2007:512

154. Wang YF, Qiu J, Kong D, Gao Y, Lu F, Karmaker PG, Chen FX (2015) Org Biomol Chem 13:365
155. Adams GL, Carroll PJ, Smith AB (2013) J Am Chem Soc 135:519
156. Adams GL, Carroll PJ, Smith AB (2012) J Am Chem Soc 134:4037
157. Suzuki Y, Shimawaki M, Miyazaki E, Osaka I, Takimiya K (2011) Chem Mater 23:795
158. Hughes TV, Cava MP (1999) J Org Chem 64:313
159. Cushman M, Mohanakrishnan AK, Hollingshead M, Hamel E (2002) J Med Chem 45:4748
160. Canterbury DP, Herrick IR, Um J, Houk KN, Frontier AJ (2009) Tetrahedron 65:3165
161. Drechsler U, Sandman DJ, Foxman BM (2001) J Chem Soc Perkin Trans 2 581
162. Katritzky AR, Kirichenko N, Rogovoy BV (2003) Arkivoc 2003:8
163. Katritzky AR, Kirichenko N, Rogovoy BV, Kister J, Tao H (2004) Synthesis 1799
164. Burckhalter JH, Stephens VC, Hall LAL (1952) J Am Chem Soc 74:3868
165. Katritzky AR, Rachwal S, Rachwal B (1989) J Org Chem 54:6022
166. Dumas AM, Bode JW (2012) Org Lett 14:2138
167. Cho SD, Hwang J, Kim HK, Yim HS, Kim JJ, Lee SG, Yoon YJ (2007) J Heterocycl Chem 44:951
168. Katritzky AR, Palenik GJ, Anders E, Tropsch JG, Vanden Eynde J-J, Zhang Z (1990) Chem Ber 123:1185
169. Joshi AA, Viswanathan CL (2006) Bioorg Med Chem Lett 16:2613
170. Quiclet-Sire B, Zard SZ (2014) Chem Commun 50:5990
171. Verma AK, Jha RR, Chaudhary R, Tiwari RK, Danodia AK (2013) Adv Synth Catal 355:421
172. Potapov AS, Chernova NP, Ogorodnikov VD, Petrenko TV, Khlebnikov AI (2011) J Org Chem 7:1526
173. Cai M, Liang Y, Zhou F, Liu W (2011) J Mater Chem 21:13399
174. Petit S, Duroc Y, Larue V, Giglione C, Léon C, Soulama C, Denis A, Dardel F, Meinnel T, Artaud I (2009) ChemMedChem 4:261
175. Greco MN, Hawkins MJ, Powell ET, Almond HR, de Garavilla L, Hall J, Minor LK, Wang Y, Corcoran TW, Di Cera E, Cantwell AM, Savvides SN, Damiano BP, Maryanoff BE (2007) J Med Chem 50:1727
176. Mobinikhaledi A, Foroughifar N, Goli R (2005) Phosphorus Sulfur Silicon Relat Elem 180:2549
177. Rádl S, Obadalová I (2005) Arkivoc 2005:4
178. Gao G-Y, Li DJ, Keung WM (2003) Bioorg Med Chem 11:4069
179. Katritzky AR, Kirichenko K, Ji Y, Steel PJ, Karelson M (2003) Arkivoc 2003:49
180. Katritzky AR, Button MAC, Denisenko SN (2001) Heterocycles 54:301
181. Katritzky AR, Qiu G, Long Q, He H, Steel PJ (2000) J Org Chem 65:9201
182. Katritzky AR, Fang Y (2000) Heterocycles 53:1783
183. Katritzky AR, Serdyuk L, Xie L (1998) J Chem Soc Perkin Trans 1 1059
184. Lu G, Zhan Z, Zhang Y (1998) Synth Commun 28:3657
185. Zhou J, Bao W, Zhang Y, Wang J (1996) Synth Commun 26:3283
186. Katritzky AR, Wu H, Xie L, Jiang J (1995) J Heterocycl Chem 32:595
187. Katritzky AR, Ghiviriga I, Cundy DJ (1994) Heterocycles 38:1041
188. Katritzky AR, Yang Z, Cundy DJ (1993) Synth Commun 23:3061
189. Katritzky AR, Hughes CV, Rachwal S (1989) J Heterocycl Chem 26:1579
190. Katritzky AR, Drewniak-Deyrup M, Lan X, Brunner F (1989) J Heterocycl Chem 26:829
191. Katritzky AR, Rachwal S, Caster KC, Mahni F, Law KW, Rubio O (1987) J Chem Soc Perkin Trans 1 781
192. Katritzky AR, Gordeev MF, Greenhill JV, Steel PJ (1992) J Chem Soc Perkin Trans 1 1111
193. He W, Wang Q, Wang Q, Zhang B, Sun X, Zhang S (2009) Synlett 2009:1311
194. Wei Y, He W, Liu Y, Liu P, Zhang S (2012) Org Lett 14:704
195. Li M, Ji N, Lan T, He W, Liu R (2014) RSC Adv 4:20346
196. Katritzky AR, Manju K, Steel PJ (2003) J Org Chem 68:407
197. Katritzky AR, Ji Y, Fang Y, Prakash I (2001) J Org Chem 66:5613

198. Pleynet DPM, Dutton JK, Johnson AP (1999) Tetrahedron 55:11903
199. Katritzky AR, Yao J, Bao W, Qi M, Steel PJ (1999) J Org Chem 64:346
200. Katritzky AR, Kirichenko K, Ji Y, Prakash I (2002) Chem Heterocycl Compd 38:156
201. Katritzky AR, Lam JN (1990) Heteroat Chem 1:21
202. Voronkov MG, Trofimova OM, Brodskaya EI, Bolgova YI, Larina LI, Klyba LV, Zelbst EA, Chernov NF (2003) Arkivoc 2003:125
203. Roman BI, De Ryck T, Dierickx L, Vanhoecke BWA, Katritzky AR, Bracke M, Stevens CV (2012) Bioorg Med Chem 20:4812
204. Clark RC, Sang YL, Boger DL (2008) J Am Chem Soc 130:12355
205. Sang YL, Clark RC, Boger DL (2007) J Am Chem Soc 129:9860
206. Zrig S, Andrioletti B, Rose E, Colin J (2005) Tetrahedron Lett 46:1103
207. Katritzky AR, Vvedensky V, Tymoshenko DO (2001) J Chem Soc Perkin Trans 1 2483
208. Katritzky AR, Zhang S, Hussein AHM, Fang Y, Steel PJ (2001) J Org Chem 66:5606
209. Katritzky AR, Tymoshenko DO, Monteux D, Vvedensky V, Nikonov G, Cooper CB, Deshpande M (2000) J Org Chem 65:8059
210. Katritzky AR, Zhang S, Fang Y (2000) Org Lett 2:3789
211. Johnson AP, Dutton JK, Pleynet DPM (1994) Heterocycles 37:1913
212. Jha RR, Singh J, Tiwari RK, Verma AK (2013) Arkivoc 2013:228
213. Tabernor J, Jones LF, Heath SL, Muryn C, Aromi G, Ribas J, Brechin EK, Collison D (2004) Dalton Trans 975
214. Thomas CM, Therrien B, Neels A, Stœckli-Evans H, Süss-Fink G (2002) J Organomet Chem 658:251
215. Thomas CM, Neels A, Stœckli-Evans H, Süss-Fink G (2001) Eur J Inorg Chem 2001:3005
216. Thomas CM, Neels A, Stœckli-Evans H, Süss-Fink G (2001) J Organomet Chem 633:85
217. Davie CP, Danheiser RL (2005) Angew Chem Int Ed 44:5867
218. Katritzky AR, Rachwal S, Rachwal B (1987) J Chem Soc Perkin Trans 1 791
219. Nicolaou KC, Shi L, Lu M, Pattanayak MR, Shah AA, Ioannidou HA, Lamani M (2014) Angew Chem Int Ed 53:10970
220. Inoue A, Kanematsu M, Mori S, Yoshida M, Shishido K (2013) Synlett 24:61
221. Lowe JT, Lee MD, Akella LB, Davoine E, Donckele EJ, Durak L, Duvall JR, Gerard B, Holson EB, Joliton A, Kesavan S, Lemercier BC, Liu H, Marié JC, Mulrooney CA, Muncipinto G, Welzel-O'Shea M, Panko LM, Rowley A, Suh BC, Thomas M, Wagner FF, Wei J, Foley MA, Marcaurelle LA (2012) J Org Chem 77:7187
222. Uno M, Koma Y, Ban HS, Nakamura H (2012) Bioorg Med Chem Lett 22:5169
223. Fresneda MÁ, Alibés R, Font J, Bayón P, Figueredo M (2012) J Org Chem 77:5030
224. Li Z, Liu Z, Cho DW, Zou J, Gong M, Breece RM, Galkin A, Li L, Zhao H, Maestas GD, Tierney DL, Herzberg O, Dunaway-Mariano D, Mariano PS (2011) J Inorg Biochem 105:509
225. Weixler R, Hehn JP, Bach T (2011) J Org Chem 76:5924
226. Petronijevic FR, Wipf P (2011) J Am Chem Soc 133:7704
227. Hehn JP, Herdtweck E, Bach T (2011) Org Lett 13:1892
228. Mahapatra T, Nanda S (2010) Tetrahedron Asymmetry 21:2199
229. Yoshida M, Shoji Y, Shishido K (2010) Tetrahedron 66:5053
230. Mahapatra T, Bhunya R, Nanda S (2009) Tetrahedron Lett 50:5392
231. Degennaro L, Capriati V, Carlucci C, Florio S, Luisi R, Nuzzo I, Cuocci C (2009) Tetrahedron 65:8745
232. Rehbein J, Leick S, Hiersemann M (2009) J Org Chem 74:1531
233. Deguest G, Bischoff L, Fruit C, Marsais F (2007) Org Lett 9:1165
234. De la Reberdière A, Lachaud F, Chuburu F, Cadiou C, Lemercier G (2012) Tetrahedron Lett 53:6115
235. Altman RA, Nilsson BL, Overman LE, De Alaniz JR, Rohde JM, Taupin V (2010) J Org Chem 75:7519
236. Katritzky AR, Singh S, Mohapatra PP, Clemens N, Kirichenko K (2005) Arkivoc 2005:63
237. Katritzky AR, Huang TB, Voronkov MV, Wang M, Kolb H (2000) J Org Chem 65:8819

238. Pearson WH, Stevens EP (1998) J Org Chem 63:9812
239. Katritzky AR, Ignatchenko AV, Lang H (1995) J Org Chem 60:4002
240. Katritzky AR, Yao G, Lan X, Zhao X (1993) J Org Chem 58:2086
241. Katritzky AR, Gordeev MF (1991) J Chem Soc Perkin Trans 1 2199
242. Katritzky AR, Latif M, Urogdi L (1990) J Chem Soc Perkin Trans 1 667
243. Katritzky AR, Rao MSC (1990) Synthesis 663
244. Katritzky AR, Rao MSC (1989) J Chem Soc Perkin Trans 1 2297
245. Aurrecoechea JM, Bustos F, López B, Saornil C, Suero R (2009) Arkivoc 2009:94
246. Aurrecoechea JM, Suero R, De Torres E (2006) J Org Chem 71:8767
247. Dong Y, Sun J, Wang X, Xu X, Cao L, Hu Y (2004) Tetrahedron Asymmetry 15:1667
248. Lee C-H, Yang ID, Lee J-D, Nakamura H, Ko J, Kang SO (2004) Synlett 1799
249. Katritzky AR, Jain R, Akhmedova R, Xu YJ (2003) Arkivoc 2003:4
250. Bustos F, Gorgojo JM, Suero R, Aurrecoechea JM (2002) Tetrahedron 58:6837
251. Abonia R, Albornoz A, Insuasty B, Quiroga J, Meier H, Hormaza A, Nogueras M, Sanchez A, Cobo J, Low JN (2001) Tetrahedron 57:4933
252. Locher C, Peerzada N (2000) Arkivoc 2000:14
253. Walz AJ, Sundberg RJ (2000) J Org Chem 65:8001
254. Locher C, Peerzada N (1999) J Chem Soc Perkin Trans 1 179
255. Katritzky AR, Belyakov SA, Sorochinsky AE, Steel PJ, Schall OF, Gokel GW (1996) J Org Chem 61:7585
256. Katritzky AR, Denisko OV, Belyakov SA, Schall OF, Gokel GW (1996) J Org Chem 61:7578
257. Katritzky AR, Allin SM (1995) Synth Commun 25:2751
258. Sechi M, Rizzi G, Bacchi A, Carcelli M, Rogolino D, Pala N, Sanchez TW, Taheri L, Dayam R, Neamati N (2009) Bioorg Med Chem 17:2925
259. Katritzky AR, Serdyuk L, Xie L, Ghiviriga I (1997) J Org Chem 62:6215
260. Katritzky AR, Toader D, Xie L (1996) J Org Chem 61:7571
261. Katritzky AR, El-Zemity S, Lang H (1995) J Chem Soc Perkin Trans 1 3129
262. Katritzky AR, Xie L, Cundy D (1995) Synth Commun 25:539
263. Katritzky AR, Lan X, Lam JN (1991) Chem Ber 124:1819
264. Katritzky AR, Lan X, Lam JN (1991) Chem Ber 124:1809
265. Katritzky AR, Lan X, Lam JN (1990) Synthesis 341
266. Katritzky AR, Drewniak M (1988) J Am Chem Soc 2339
267. Katritzky AR, He H-Y, Jiang R, Long Q (2001) Tetrahedron Asymmetry 12:2427
268. Katritzky AR, Abonia R, Yang B, Qi M, Insuasty B (1998) Synthesis 1487
269. Juliusz P, Lucyna M (1994) Heterocycles 37:311
270. Quan ZJ, Ren RG, Jia XD, Da YX, Zhang Z, Wang XC (2011) Tetrahedron 67:2462
271. Katritzky AR, Zhao X, Shcherbakova IV (1991) J Chem Soc Perkin Trans 1 3295
272. Katritzky AR, Huang Z, Fang Y, Prakash I (1999) J Org Chem 64:2124
273. Katritzky AR, Zhang Z, Lang H, Xie L (1998) Heterocycles 47:187
274. Katritzky AR, Wang J, Karodia N, Li J (1997) J Org Chem 62:4142
275. Katritzky AR, Wang Z, Lang H, Feng D (1997) J Org Chem 62:4125
276. Katritzky AR, Xie L, Serdyuk L (1996) J Org Chem 61:7564
277. Katritzky AR, Wang Z, Lang H (1996) Organometallics 15:486
278. Katritzky AR, Yang Z, Moutou JL (1995) Tetrahedron Lett 36:841
279. Degl'Innocenti A, Capperucci A, Oniciu DC, Katritzky AR (2000) J Org Chem 65:9206
280. Hammaecher C, Bouillon JP, Portella C (2009) Tetrahedron 65:5527
281. Noda H, Erős G, Bode JW (2014) J Am Chem Soc 136:5611
282. Katritzky AR, Qi M, Feng D, Nichols DA (1997) J Org Chem 62:4121
283. Bachman GB, Heisey LV (1946) J Am Chem Soc 68:2494
284. Toulot S, Bonnin Q, Comte V, Adriaenssens L, Richard P, Le Gendre P (2013) Eur J Org Chem 2013:736
285. Katritzky AR, Rachwal B, Rachwal S (1995) J Org Chem 60:7631
286. Katritzky AR, Fan WQ, Long QH (1993) Synthesis 229

287. Katritzky AR, Yannakopoulou K, Lue P, Rasala D, Urogdi L (1989) J Chem Soc Perkin Trans 1 225
288. Katritzky AR, Rachwal B, Rachwal S (1995) J Org Chem 60:3993
289. Katritzky AR, Yannakopoulou K, Kuzmierkiewicz W, Aurrecoechea JM, Palenik GJ, Koziol AE, Szczesniak M (1987) J Chem Soc Perkin Trans 1 2673
290. Xu S, Zakharov LN, Liu S-Y (2011) J Am Chem Soc 133:20152
291. Piller FM, Bresser T, Fischer MKR, Knochel P (2010) J Org Chem 75:4365
292. Katritzky AR, Abdel-Fattah AAA, Steel PJ (2006) Tetrahedron Lett 47:1465
293. Kopp F, Sklute G, Polborn K, Marek I, Knochel P (2005) Org Lett 7:3789
294. Katritzky AR, Nair SK, Qiu G (2002) Synthesis 199
295. Churlaud C, Pornet J, Oniciu DC, Katritzky AR (1999) Organometallics 18:4270
296. Ahn JH, Joung MJ, Yoon NM, Oniciu DC, Katritzky AR (1999) J Org Chem 64:488
297. Love BE, Nguyen BT (1998) Synlett 1123
298. Katritzky AR, Nichols DA, Qi M (1998) Tetrahedron Lett 39:7063
299. Katritzky AR, Strah S, Belyakov SA (1998) Tetrahedron 54:7167
300. Katritzky AR, Nichols DA, Qi M, Yang B (1997) J Heterocycl Chem 34:1259
301. Bao W, Zhang Y (1997) Synth Commun 27:615
302. Katritzky AR, Rachwal S, Rachwal B (1994) J Org Chem 59:5206
303. Katritzky AR, Belyakov SA, Dupont Durst H, Xu R, Dalal NS (1994) Can J Chem 72:1849
304. Bulman Page PC, Allin SM, Collington EW, Carr RAE (1993) J Org Chem 58:6902
305. Katritzky AR, Rachwal B, Rachwal S (1993) J Org Chem 58:812
306. Katritzky AR, Rachwal S, Rachwal B, Steel PJ (1992) J Org Chem 57:4932
307. Katritzky AR, Gallos JK, Yannakopoulou K (1989) Synthesis 31
308. Katritzky AR, Gordeev MF (1993) J Org Chem 58:4049
309. Katritzky AR, Abdel-Fattah AAA, Celik I (2007) Arkivoc 2007:96
310. Katritzky AR, Belyakov SA, Sorochinsky AE, Henderson SA, Chen J (1997) J Org Chem 62:6210
311. Al-Soud YA, Al-Masoudi NA, Ferwanah AERS (2003) Bioorg Med Chem 11:1701
312. Sanna P, Carta A, Nikookar MER (2000) Eur J Med Chem 35:535
313. Yu T, Zhao M, Li A, Zhao Y, Zhang H, Fan D (2013) Res Chem Intermed 39:2259
314. Das J, Rao CVL, Sastry TVRS, Roshaiah M, Sankar PG, Khadeer A, Kumar MS, Mallik A, Selvakumar N, Iqbal J, Trehan S (2005) Bioorg Med Chem Lett 15:337
315. Carta A, Sanna P, Palomba M, Vargiu L, La Colla M, Loddo R (2002) Eur J Med Chem 37:891
316. Carta A, Sanna P, Bacchi A (2002) Heterocycles 57:1079
317. Dutton JK, Pleynet DPM, Johnson AP (1999) Tetrahedron 55:11927
318. Paluchowska MH, Bojarski AJ, Bugno R, Charakchieva-Minol S, Wesołowska A (2003) Arch Pharm 336:104
319. Henke BR, Drewry DH, Jones SA, Stewart EL, Weaver SL, Wiethe RW (2001) Bioorg Med Chem Lett 11:1939
320. Katritzky AR, Serdyuk L, Chassaing C, Toader D, Wang X, Forood B, Flatt B, Sun C, Vo K (2000) J Comb Chem 2:182
321. Al-Omran F, Abd El-Hay OY, El-Khair AA (2000) J Heterocycl Chem 37:1617
322. Zhang Z, Yin Z, Kadow JF, Meanwell NA, Wang T (2004) Synlett 2323
323. Ali OM, Amr AE-GE, Mostafa EE (2014) Res Chem Intermed 40:1545
324. Katritzky AR, Aslan D, Shcherbakova IV, Chen J, Belyakov S (1996) J Heterocycl Chem 33:1107
325. Al-Jalal NA, Ibrahim MR, Al-Awadi NA, Elnagdi MH, Ibrahim YA (2014) Molecules 19:20695
326. Katritzky AR, Li J, Malhotra N (1992) Liebigs Ann Chem 843
327. Katritzky AR, Zhang GF, Fan WQ, Wu F, Pernak J (1993) J Phys Org Chem 6:567
328. Katritzky AR, Oniciu DC, Ghiviriga I (1997) Synth Commun 27:1613
329. Yan W, Liao T, Tuguldur O, Zhong C, Petersen JL, Shi X (2011) Chem Asian J 6:2720

330. Basu B, Paul S, Nanda AK (2009) Green Chem 11:1115
331. Katritzky AR, Yao J, Denisko OV (2000) J Org Chem 65:8063
332. Katritzky AR, Yao J, Yang B (1999) J Org Chem 64:6066
333. Katritzky AR, Yao J, Qi M (1998) J Org Chem 63:5232
334. Katritzky AR, Cheng D, Henderson SA, Li J (1998) J Org Chem 63:6704
335. Katritzky AR, Ferwanah AERS, Denisenko SN (1999) Heterocycles 50:767
336. Katritzky AR, Wang X, Denisenko A (2001) J Org Chem 66:2850
337. Katritzky AR, Button MAC, Denisenko SN (2000) J Heterocycl Chem 37:1505
338. Katritzky AR, Pleynet DPM, Yang B (1997) J Org Chem 62:4155
339. Orth RE, Soedigdo S (1965) J Heterocycl Chem 2:486
340. Katritzky AR, Tao H, Kirichenko K (2007) Arkivoc 2007:142
341. Nieuwenhuijzen JW, Conti PGM, Ottenheijm HCJ, Linders JTM (1998) Tetrahedron Lett 39:7811
342. Korhonen J, Kuusisto A, van Bruchem J, Patel JZ, Laitinen T, Navia-Paldanius D, Laitinen JT, Savinainen JR, Parkkari T, Nevalainen TJ (2014) Bioorg Med Chem 22:6694
343. Harpp DN, MacDonald JG, Larsen C (1985) Can J Chem 63:951
344. Tiwari VK, Kale RR, Mishra BB, Singh A (2008) Arkivoc 2008:27
345. Tiwari VK, Singh DD, Hussain HA, Mishra BB, Singh A (2008) Monatsh Chem 139:43
346. Tiwari VK, Singh A, Hussain HA, Mishra BB, Tripathi V (2007) Monatsh Chem 138:653
347. Katritzky AR, Khashab NM, Haase DN, Yoshioka M, Ghiviriga I, Steel PJ (2007) J Org Chem 72:6742
348. Katritzky AR, Khashab NM, Bobrov S, Yoshioka M (2006) J Org Chem 71:6753
349. Kumar D, Singh AS, Tiwari VK (2015) RSC Adv 5:31584
350. Kumar D, Mishra A, Mishra BB, Bhattacharya S, Tiwari VK (2013) J Org Chem 78:899
351. Katritzky AR, Meher G, Narindoshvili T (2009) Collect Czechoslov Chem Commun 74:1061
352. Thambidurai S, Jeyasubramanian K, Ramalingam SK (1996) Polyhedron 15:4011
353. Katritzky AR, Rogovoy BV, Chassaing C, Vvedensky V (2000) J Org Chem 65:8080
354. Schell P, Richards MP, Hanson K, Berk SC, Makara GM (2005) J Comb Chem 7:96
355. Katritzky AR, Rogovoy BV, Vvedensky V, Kovalenko K, Forood B (2002) J Comb Chem 4:290
356. Katritzky AR, Vvedensky V, Cai X, Rogovoy BV, Steel PJ (2002) Arkivoc 2002:82
357. Katritzky AR, Rogovoy BV, Cai X, Kirichenko N, Kovalenko KV (2004) J Org Chem 69:309
358. Katritzky AR, Rogovoy BV, Vvedensky V, Kovalenko K, Steel PJ, Markov VI, Forood B (2001) Synthesis 897
359. Katritzky AR, Huang L, Chahar M, Sakhuja R, Hall CD (2012) Chem Rev 112:1633
360. Schade D, Kotthaus J, Klein N, Kotthaus J, Clement B (2011) Org Biomol Chem 9:5249
361. Katritzky AR, Arend M (1998) J Org Chem 63:9989
362. Katritzky AR, Denisenko A, Arend M (1999) J Org Chem 64:6076
363. Katritzky AR, Denisenko A, Denisenko SN (2000) J Heterocycl Chem 37:1309
364. Katritzky AR, Huang TB, Voronkov MV (2000) J Org Chem 65:2246

Top Heterocycl Chem (2016) 43: 67–94
DOI: 10.1007/7081_2015_199

Published online: 28 February 2016

Acylbenzotriazoles: New Allies for Short Linear and Cyclic Peptide Constructs

Danniebelle N. Haase

Abstract Peptides and proteins have tremendous potential as medicinal drugs, but their potential has largely been untapped due to several problems such as their fast degradation by enzymes and limited bioavailability. Major advances in the development of peptide synthesis methods, availability of a wide array of coupling reagents and other augments have lowered the barrier to entry for peptide drugs. However, greater access to peptides and proteins is needed in order to realize the possibilities for diagnosing, treating and monitoring critical diseases. Acylbenzotriazoles are an expedient gateway to cyclic and difficult peptides. With this overview, we now describe the use of acylbenzotriazoles as allies for the construction of peptides and peptide conjugates. The efficacy of *N*-(protected-α aminoacyl) benzotriazoles for the preparation of an extensive range of enantiopure peptides with good yields and purities on solid-phase or in solution, as well as challenges encountered and the way forward is addressed in the chapter.

Keywords Acylbenzotriazole • Amino acids • Isopeptide • Microwave • Peptide synthesis • Peptides • Solid-phase

Contents

D.N. Haase (✉)
The Dow Chemical Company, 400 Arcola Road, Collegeville, PA 19426, USA
e-mail: dhaase@dow.com

1 Introduction

Peptides and proteins are ubiquitous in nature. Peptides and proteins are of interest due to their wide applicability in several areas, including endocrinology, neurology, and hematology. Despite their undeniable utility, peptides and proteins have limited applicability as therapeutics due to their in vivo instability [1]. Since they are composed of naturally occurring amino acids, peptides and proteins suffer from rapid degradation, poor bioavailability, and other problems. Therefore, the synthesis of peptides and proteins is an active area of research as chemists race to develop peptide and proteins with greater in vivo stability and bioavailability and to confirm the structure of proteins and peptides isolated from natural sources [2]. Additionally, there is considerable interest in studying protein–protein interactions and peptide and protein conformations to understand how amino acid sequence controls protein and peptide structure and function in order to understand the role that peptides and proteins play in diseases such as Alzheimer's, Parkinson's, and Creutzfeldt–Jakob [3]. Better understanding of protein and peptide folding, aggregation, or misfolding can facilitate the design of efficacious drugs to monitor and treat a number of diseases. However, useful methods for the preparation of peptides and proteins are necessary.

Notwithstanding more than a century of major breakthroughs in peptide synthesis, in the last four decades, many advances in the efficient preparation of peptides and peptide mimics with greater chemical stability have promoted greater understanding of peptide and protein behavior as well as the development of peptide-based therapeutics [4–6]. This has largely been facilitated by several major developments, including, but not limited to, the availability of improved and efficient coupling reagents, advances in solid-phase peptide synthesis (SPPS), *O*- and *S*-acyl isopeptide synthetic strategies, and the native chemical ligation (NCL) peptide synthesis method [7–10]. This chapter will address the contributions of acylbenzotriazoles, in particular, *N*-(protected-α aminoacyl)benzotriazoles to the field of solid- and solution-phase peptide synthesis of linear peptides, short difficult peptides, *O*- and *S*-isopeptides, and cyclic peptides.

1.1 Benzotriazole-Based Fmoc-, Boc-, and Alloc-Protected Amino Acids for Chemical Transformations

The preparation and use of *N*-acylbenzotriazoles were widely described in the literature. Reacting carboxylic acids with 1*H*-benzotriazole facilitated the smooth conversion to acylbenzotriazoles, versatile synthetic auxiliaries that have been useful for a host of chemical transformations.

Solution- or solid-phase peptide synthesis requires protection of the amino moiety. The most utilized amino protecting groups are 9-fluorenylmethoxycarbonyl (Fmoc) [11, 12], *tert*-butoxycarbonyl (Boc) [12–17], benzyloxycarbonyl (Cbz) [12, 18, 19], and allyloxycarbonyl (Alloc) [11]. Easy removal of the Fmoc protecting group occurs under mild, basic conditions, while the Boc protecting group requires acidolysis [11]. The Fmoc protection strategy has gained traction over Boc protection as deprotection of the Fmoc group can occur under facile conditions, even in the presence of acid-labile protecting groups on the amino acid side chain. As the Alloc group is stable under both acidic and basic conditions, a selective and facile palladium-catalyzed transfer of the allyl group to nucleophiles is required for its removal [11].

Generally, protection of the amino group is achieved by reacting free amino acids with the haloformate or carbonate of the particular protecting group under Schotten–Baumann conditions [20]. Other protecting group strategies employed include the *N*-(9-fluorenylmethoxycarbonyl-oxy)succinimide (Fmoc-OSu) [21–24], 9-fluorenylmethyl pentafluorophenyl carbonate (Fmoc-Pfp) [23], 9-fluorenylmethyl 1-benzotriazolyl carbonate (Fmoc-OBt) [21, 25, 26] and 5-norbonene-2,3-dicarboximido-Fmoc derivatives [27]. More recently, *N*-Fmoc-or-Alloc-protected amino acids were prepared from Fmoc-or-Alloc-cyanopyridyloxime carbonates by way of a two-step procedure [28–30]. For several of these methods, the starting material is unstable, thus limiting the shelf life. Additionally, the reaction yield is often lowered due to the formation of undesired protected di- and tripeptide side products. Azide-based methodologies could minimize the formation of peptide side products; however, their use in protecting the NH_2 of amino acids is limited due to hazards associated with their use [31–34].

Recently, Fmoc-, Boc-, and Alloc-benzotriazoles have gained prominence as selective protecting groups for the synthesis of the corresponding Fmoc-, Boc-, and Alloc-protected amino acids. *N*-(9-fluorenylmethoxycarbonyl)benzotriazole (Fmoc-Bt) and *N*-(allyloxycarbonyl)benzotriazole (Alloc-Bt) are readily available from the Fmoc-Cl and Alloc-Bt with benzotriazole, while *N*-(*tert*-butoxycarbonyl) benzotriazole (Boc-Bt) can be prepared from Boc-anhydride and benzotriazole (Figs. 1 and 2, Table 1) [11, 14, 35, 36].

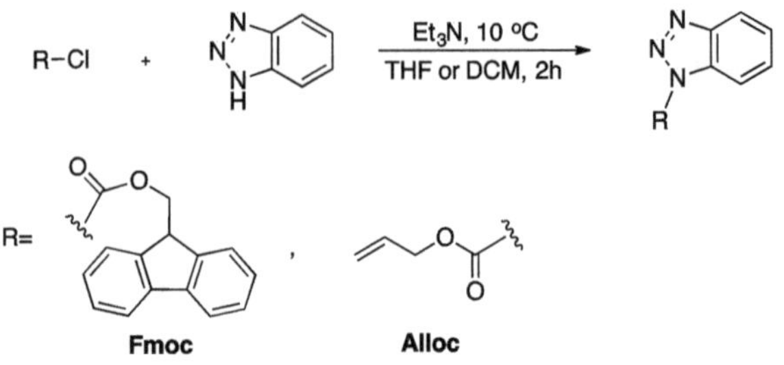

Fig. 1 Preparation of Fmoc- and Alloc-benzotriazoles from 1*H*-benzotriazole [11]

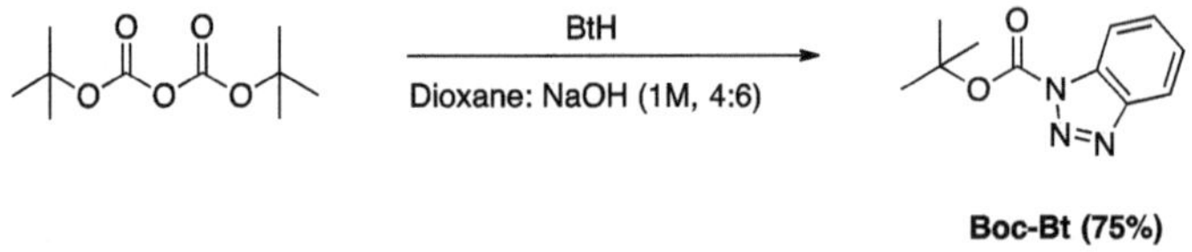

Fig. 2 Preparation of Boc-Bt from Boc-anhydride and 1*H*-benzotriazole

Table 1 Yields of Fmoc-Bt and Alloc-Bt from the corresponding acid chlorides and 1*H*-benzotriazole [11]

Entry	R	Product	mp (°C)	Yield (%)
1	Fmoc	Fmoc-Bt	90–91	88
2	Alloc	Alloc-Bt	105–107[a]	92

[a]Oil in the Lit.

Having successfully prepared the Fmoc-Bt, Alloc-Bt, and Boc-Bt, Katritzky and coworkers used these selective amino protecting reagents in the preparation of 16 Fmoc-, Boc, and Alloc-protected amino acids in yields of 77–94% [11]. The substrate scope is broad and can facilitate the reaction of unprotected serine and tyrosine with the Fmoc-, Alloc-, and Boc-benzotriazole reagents. Additionally, the reaction conditions are mild, with no observable side-product formation. To summarize, Fmoc-, Alloc-, and Boc-benzotriazoles can be used for the selective *N*-acylation of amino acids, which are precursors to *N*-acylbenzotriazoles.

2 Construction of Short Linear Peptides

2.1 Introduction

Short linear peptides are abundant in nature and are quite useful for studying the secondary structure of peptides and proteins. Short linear peptide sequences, frequently referred to as short linear motifs (SLiMs, minimotifs, linear motifs) in structural biology, are typically composed of 3–11 amino acid units (average of 6 amino acid units) [37]. SLiMs facilitate protein–protein interactions and are integrally involved in recognition, targeting, and other functional properties [38]. As short linear peptides display a wide range of biological activity including antimicrobial, antimalarial, and cytotoxic activity, it is imperative that methods facilitating the preparation of a wide array of peptide and peptide conjugates are available.

Several coupling reagents are available for solid-phase and solution peptide synthesis [39]. Acylbenzotriazoles are efficacious synthetic auxiliaries that have been successfully applied to both the solid- and solution-phase preparation of various short peptide sequences. Now presented are the acylbenzotriazole-mediated syntheses of several classes of peptides.

2.2 *Oligopeptides from* N*-Acylbenzotriazoles*

Peptides containing 2–9 amino acid residues were prepared by coupling *N*-(protected-α aminoacyl)benzotriazoles with (1) monofunctionalized amino acids (Ala, Gly, Leu, Phe, Val), (2) unprotected functionalized amino acids with hydroxy (Ser, Tyr) and sulfanyl (Cys, Met) functionalities, (3) amino acids with dicarboxylic acid moieties (Asp, Glu), and (4) amino- (Lys) and guanidino (Arg)-containing amino acids.

The peptides were prepared using several methods, including solution- and solid-phase peptide synthesis strategies [40]. Solid- and solution-phase methods display several advantages, and both have a unique niche in the field of peptide synthesis. The solution-phase preparation of peptides is known for its versatility, especially in the preparation of peptides containing unnatural amino acids and peptide conjugates. Convergent synthesis can easily be performed when solution-phase methods are employed, and side products are quickly detected. However, this method is time-consuming and rigorous, as the growing peptide must be purified after each step in the sequence [41]. Today, solid-phase peptide synthesis (SPPS) strategies are more popular than solution-phase methods because the peptide can be rapidly prepared and more easily separated from reagents. As such, in SPPS, coupling reactions can often be driven to completion by using a large excess of reagents. In SPPS, side products are not readily detected and/or separated from the product during the process, and this ultimately leads to lower yields than solution-phase methods [41].

2.2.1 Syntheses of Dipeptides

Solution-phase preparation of dipeptides was achieved by reacting free L- and D-amino acids with *N*-(protected-α aminoacyl)benzotriazoles in aqueous acetonitrile (MeCN: H_2O, 2:1) for 1 h at 20°C in the presence of triethylamine. Separation of the dipeptides from the BtH byproduct by washing the mixture with 4N HCl afforded the enantiopure products in yields of 78–98% without further purification [42–48].

Diastereomeric mixtures of select dipeptides were prepared with DL-amino acids to confirm the enantiopurity of the synthesized peptides via HPLC and NMR spectroscopy. All LL- and LD-dipeptides showed no detectable racemization (<1%).

2.2.2 Preparation of Tri- and Tetrapeptides via Solution- and Solid-phase Assembly

Tripeptides can be prepared via the stepwise synthetic route. Typically, the solution-phase, *N*-acylbenzotriazole-mediated, stepwise assembly of tripeptides begins with the coupling of a free amino acid and a *N*-(protected α-aminoacyl) benzotriazoles (see Fig. 3). The resulting *N*-protected dipeptide is easily converted to the corresponding *N*-(protected dipeptidoyl)benzotriazoles, after which, the *N*-(protected dipeptidoyl)benzotriazoles are reacted with a free amino acid to afford tripeptides in good yields and with minimal racemization. Examples of tripeptides prepared by the solution-phase, stepwise method are illustrated in Fig. 4 and Table 1.

Alternatively, Katritzky and coworkers demonstrated the utility of *N*-(protected α-aminoacyl)benzotriazoles in the microwave-assisted, solid-phase preparation of tripeptides (Fig. 5). Peptide amides were synthesized in fair yields with excellent purity by anchoring the growing peptide to the Rink amide resin. Following the stepwise or fragment coupling of *N*-acylbenzotriazoles, global deprotection and cleavage of the peptide from the resin using trifluoroacetic acid (TFA) gave the tripeptides. An example of tripeptides made via the stepwise and segment condensation routes is presented in Table 2 [49, 50].

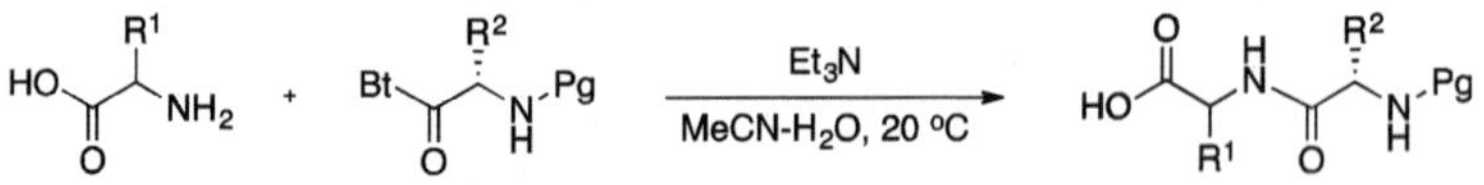

Fig. 3 General solution-phase synthesis of dipeptides from *N*-(protected-α aminoacyl) benzotriazoles

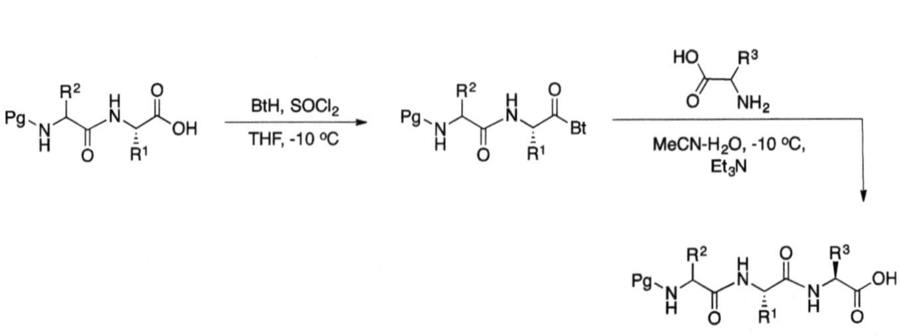

Fig. 4 Synthesis of tripeptides starting from *N*-(protected α-aminoacyl)benzotriazoles

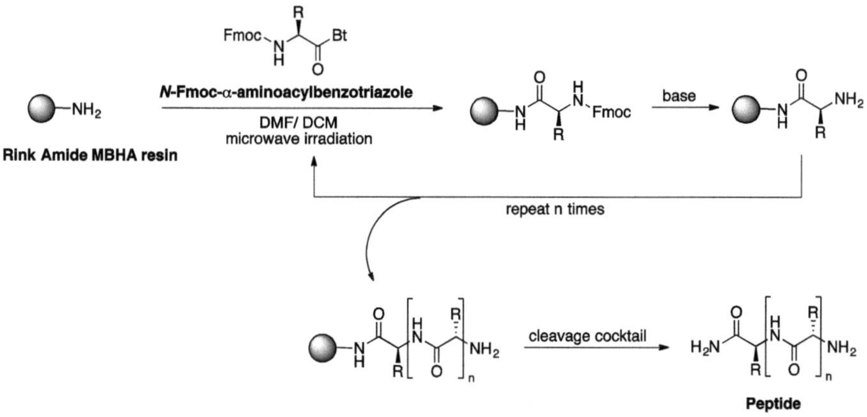

Fig. 5 General scheme for the preparation of peptides via solid-phase peptide synthesis

Table 2 Tripeptides synthesized by solid-phase peptide synthesis [49, 50]

Sequence (N-C terminus)	t_R (min)	Crude product % purity	Crude product % yield	Product after preparative HPLC % purity	Product after preparative HPLC % yield
H-L-Ala-L-Phe-L-Ala-NH_2[a]	5.17	90	77[b]	>99	45
H-L-Leu-L-Ala-L-Leu-NH_2[c]	8.49	n.r.	n.r	99	68

[a]Tripeptide from stepwise solid-phase peptide synthesis
[b]Yield quoted is calculated based on the amount of product obtained multiplied by the % purity obtained from HPLC
[c]Tripeptide from segment condensation on solid phase
n.r. not reported

Table 3 Examples of tetrapeptides from the solution- and solid-phase peptide syntheses

Sequence (N-C terminus)	Method	% purity	% yield[a]	References
H-L-Pro-L-Trp-L-Met-L-Trp-NH_2	Solid phase, stepwise	>95	39	[49]
H-L-Trp-L-Ala-L-Met-L-Ala-NH_2[b]	Solid phase, fragment	>99	26	[50]
Z-L-Ala-L-Phe-Gly-L-Leu-OH	Solution phase, fragment	>99	85	[42]
Z-L-Phe-L-Ala-Gly-L-Leu-OH	Solution phase, fragment	>99	86	[42]

[a]Isolated yields
[b]Crude yields and purity are 73% and 82%, respectively

Tetrapeptides can be prepared by (1) the stepwise coupling of amino acid residues in either solution or solid phase as previously described (Table 3) or (2) the solution- or solid-phase coupling of fragments (Table 3). The fragments included free dipeptides and *N*-(protected dipeptidoyl)benzotriazoles. As expected, solution-phase peptide affords higher yields than solid-phase peptide synthesis.

Table 4 Solid-phase peptide synthesis of oligopeptides containing 4–7 amino acid residues from *N*-(protected α-aminoacyl)benzotriazoles and *N*-(protected dipeptidoyl)benzotriazoles

Sequence (N-C terminus)	Solid-phase method	% purity[a]	% yield[b]	t_R (min)	HRMS $[M+1]^+$	References
H-L-Leu-L-Met-Gly-L-Phe-L-Ala-NH_2[c]	Stepwise	>99	24	11.67	537.2626	[49]
H-L-Pro-L-Leu-L-Met-Gly-L-Phe-L-Ala-NH_2[c]	Stepwise	>99	31	12.42	634.3678	[49]
H-L-Pro-L-Met-L-Phe-Gly-L-Leu-L-Ala-L-Phe-NH_2[c]	Stepwise	>99	27	15.97	781.3959	[49]
H-L-Trp-L-Phe-L-Met-L-Leu-L-Ala-NH_2	Segment condensation	99	21[d]	15.45	666.3432	[50]
H-L-Leu-L-Ala-L-Met-L-Phe-L-Phe-L-Met-NH_2	Segment condensation	72	20	16.43	758.3707	[50]
H-L-Leu-L-Ala-L-Met-L-Ala-L-Phe-L-Ala-L-Gly-NH_2	Segment condensation	99	30	11.29	679.3589	[50]

[a]Purity after HPLC purification
[b]Isolated yield after HPLC purification
[c]See reference for crude yields and purities
[d]Isolated yield of the major peak and an impurity not derived from racemization

With solid-phase peptide synthesis, stepwise coupling offers higher yields than segment condensation.

2.2.3 Penta-, Hexa-, and Heptapeptides via Solid-Phase Peptide Synthesis

Stepwise and fragment coupling procedures were used to prepare penta-, hexa-, and heptapeptides. Following the established *N*-acylbenzotriazole protocol, the enantiopure peptide were obtained in good yields. Several examples are shown in Table 4.

2.3 *Sterically Hindered Peptides from* N-*(Protected α-Aminoacyl)Benzotriazoles*

The coupling reactions of α,α-disubstituted amino acids are of interest due to their unique conformation [51] and *N*-methyl peptides are useful in the drug discovery process as they can improve the pharmacokinetic properties of biologically active peptides [52]. On this premise, Katritzky and coworkers synthesized di- and tripeptides featuring a sterically hindered amino acid residue such as *N*-methyl

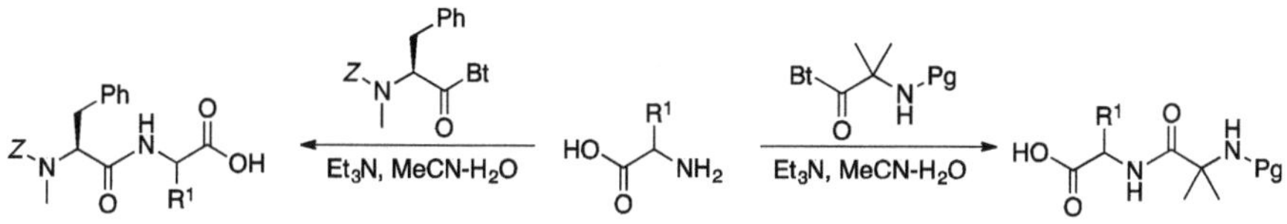

Fig. 6 Preparation of dipeptides containing a C-activated sterically hindered amino acid residue

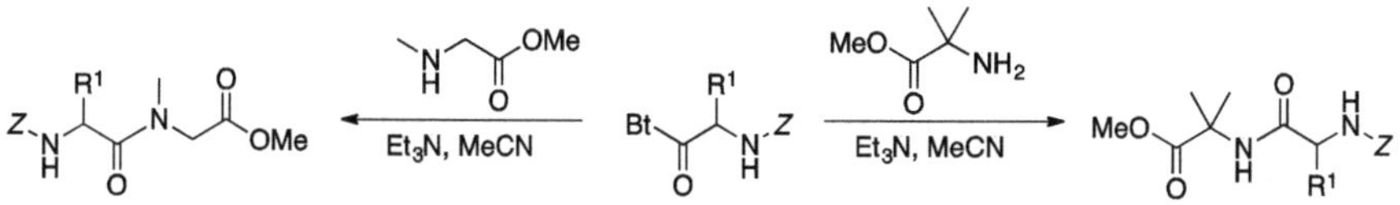

Fig. 7 Preparation of dipeptides by linking the amino group of a sterically hindered amino acid with *N*-(Z-α-aminoacyl)benzotriazoles

amino acids *N*(Me)-Phe and *N*(Me)-Gly, as well as the α,α-disubstituted amino acid, α-aminoisobutyric acid (Aib) (Figs. 6 and 7).

The coupling reaction of C-activated sterically hindered and free amino acids proceeded smoothly under the typical conditions for benzotriazole-mediated amide link formation (see Sect. 2.2.1, Figs. 3 and 5). However, the coupling of the amino group of sterically hindered amino acids requires different conditions to that of proteinogenic amino acids. Using the general conditions for benzotriazole-mediated amide bond formation and modified conditions (different reaction times, temperatures, solvent systems, additives and bases), the coupling reactions at the N-terminus of sterically hindered free amino acids with *N*-(Z-α-aminoacyl) benzotriazoles resulted largely in the hydrolysis of the *N*-(Z-α-aminoacyl) benzotriazoles [48]. Thus, water insoluble hindered amino acid esters were coupled with the *N*-(Z-α-aminoacyl)benzotriazoles in the presence of anhydrous THF or MeCN at 20°C for 24–36 h to yield the corresponding dipeptide (Fig. 7). Employing microwave irradiation was found to reduce the reaction time to approximately 1 h. In all cases, enantiopure dipeptides were obtained in fair to excellent yields of 67–93% [48].

2.4 *Difficult Peptides from the Microwave-Assisted Coupling of* N*-(Protected α-Aminoacyl)Benzotriazoles on the Solid Phase*

Solution- and solid-phase syntheses of biologically active peptides and proteins are of particular importance due to their potential use as medicines. Katritzky and coworkers have demonstrated the effectiveness of the *N*-(protected α-aminoacyl) benzotriazole auxiliaries in the fast assembly of peptides in solution- and solid-phase syntheses by preparing over 215 distinct peptides using diverse methods [42–48, 50, 53–55].

When compared to solution-phase syntheses, solid-phase syntheses of peptides offer several advantages. Formation of the amide linkage via solid-phase methodologies affords significant enhancements in reaction yields and purities, and this is often associated with a concomitant decrease in reaction times [40, 56, 57]. In solid-phase peptide synthesis (SPPS), reactions are driven to completion by utilizing a large excess of reagents, and the completion of each coupling reaction can be conveniently and rapidly monitored by the qualitative Kaiser (ninhydrin) test. Additionally, SPPS eliminates the majority of product losses associated with conventional solution-phase methodologies (filtration, recrystallization, etc.), and it offers fast and easy separation of the product from excess reagents [40, 56, 57]. With better coupling reagents, improved yields, and shorter coupling times, SPPS has become the dominant method for the rapid assembly of peptides over the last 35 years [39].

Microwave irradiation has revolutionized the preparation of a wide variety of organic compounds [58]. The use of microwave irradiation in organic synthesis is primarily known for accelerating reaction times. As such, microwave-assisted synthesis of organic compounds has been applied in SPPS to minimize coupling times (6–20 min) and to improve peptide product yields [59–62].

One challenge in the syntheses of peptides is the preparation of difficult sequences. The primary reason for the occurrence of difficult peptide sequences are incomplete aminoacylation and/or deprotection during the assembly of the peptide. In particular, challenges in amide link formation arise due to steric effects when the two amino acids being linked contain β-branched side chains, for example, isoleucine, valine, and threonine. Secondly, difficult sequences arise when intra- and interchain hydrogen bonding promote the formation of secondary structures. These synthesized peptides may be partially racemized and/or contaminated with deletion sequences, aspartimides, and related side products [56, 63]. Several strategies to alleviate the problems associated with the preparation of difficult sequences have been described in the literature [56, 57].

Having demonstrated wide applicability of the benzotriazole-based auxiliaries in peptide synthesis, Katritzky and coworkers further expanded the scope of this methodology in preparing several difficult peptides. The difficult peptide sequences were assembled on the Rink amide resin via the microwave-accelerated coupling of *N*-(Fmoc-α-aminoacyl)benzotriazoles (Fig. 5). Examples of the difficult peptides, both model and biologically relevant peptides, are illustrated in Table 5 [57].

Table 5 Difficult peptides from the microwave-assisted, benzotriazole-mediated SPPS

		Crude		After preparative HPLC		
Sequence (N-C terminus)	t_R (min)	% purity	% yield[a]	% purity	% yield	HRMS $[M+1]^+$
H-Val-Val-Ser-Val-Val-NH_2	7.57	60	63	>99	7	501.3385
H-Val-Val-Val-Ser-Val-Val-NH_2	9.79	60	34	nd[b]	nd[b]	600.4058
H-Val-Ile-Val-Ile-Gly-OH[c]	10.82	70	40	>99	24	500.3643
H-Thr-Val-Thr-Val-Thr-Val-NH_2	8.70	50	35	89	16	618.3864
H-Val-Lys-Asp-Gly-Tyr-Ile-NH_2	8.31	30	68	>99	24	693.3916
H-Val-Lys-Asp-Val-Tyr-Ile-NH_2	9.10	37	64	>99	29	735.4419
H-Tyr-Gly-Gly-Phe-Leu-NH_2	11.54	80	86	>99	37	555.2948
H-Leu-Met-Val-Gly-Gly-Val-Val-Ile-Ala-NH_2	13.82	28	42	89	22	857.5277

[a]Calculated from the amount of product obtained multiplied by the purity from HPLC
[b]*nd* not determined
[c]Peptide amide hydrolyzed to the corresponding carboxylic acid on purification

3 Cyclic Peptides

Peptides, peptide analogues, and proteins play significant roles in numerous biochemical processes and continue to be attractive targets for novel therapeutics and biomaterials [64, 65]. Recent advances in the preparation of linear peptides, peptide analogues, and proteins have to some extent circumvented several barriers to their utility [39, 56, 66, 67]. Therefore, there has been tremendous growth in the number of previously unattainable peptides and related compounds available in the medicinal chemistry toolbox. However, even with major synthetic breakthroughs, the portion of peptide-based drugs on the market has lagged behind. The primary reason for this underutilization of peptide-based drugs is their lack of in vivo stability [68].

On the contrary, their cyclic counterparts have attracted considerable interest given their tunable lipophilicity and resistance to enzymatic degradation. Cyclic peptides display a wide array of biological activity, including but not limited to immunosuppressive, anti-tumor, and antibacterial properties. Additionally, cyclic peptides that contain bioactive hormone sequences are suitable models for studying conformation–activity relationships [69].

Several synthetic methods are available for the preparation of cyclic peptides. Commonly used methods are (1) the intramolecular head-to-tail condensation between the amino and carboxyl groups at the N- and C-termini of the matching linear peptide, (2) intermolecular coupling of two symmetrical peptides and (3) methods devoid of peptide coupling [64, 65].

In exploiting several areas of expertise, Katritzky and coworkers have synthesized a plethora of cyclic oligopeptides. Now presented are the benzotriazole-mediated preparations of cyclic peptides.

3.1 Benzotriazole-Mediated Preparation of 2,5-Diketopiperazines

2,5-Diketopiperazines (2,5-DKPs) represent the simplest class of naturally occurring, bioactive cyclic peptides and are obtained from the condensation of two α-amino acids [69–71]. These cyclodipeptides, also known as 2,5-dioxopiperazines or anhydride dipeptides, are biosynthesized by several organisms, for example, bacteria, fungi, plants, and mammals. The 2,5-DPK motifs can be found alone or as a part of larger, more complex structures. Additionally, they are frequently made from the degradation of polypeptides in processed goods and are observed as undesired side products resulting from the synthesis of peptide construct [69–71].

2,5-DPKs have garnered widespread attention as their particular heterocyclic subunit makes them a promising source for new drug candidates. Diversity can be introduced at all six positions of these small, conformationally restricted heterocycles with stereocontrol at up to four positions [69–71]. They are stable to proteolysis and have the ability to bind to multiple types of receptors. As a result of these characteristics, they possess cytotoxic, neuroprotective, antifungal, antimalarial, antibacterial, antiviral, plant retardation, and other properties [72–85]. They have been used as scaffolds for the construction of combinatorial libraries, and recent breakthroughs in SPPS have made them even more attractive drug targets.

2,5-DKPs are readily available from the head-to-tail condensation of α-amino acids. However, this reaction is challenging as the cyclic dimer can be formed exclusively or in conjunction with the desired cyclic monomer [64]. Gottesfeld and coworkers found that carrying out the reaction under dilute conditions could reduce the formation of the cyclic dimer [86]. A difficulty that poses more of a challenge is being able to bring the N- and C-termini in close proximity by prompting a β-turn in the peptide. However, amino acid residues such as Pro and Gly, D-amino acids, *N*-methyl amino acids, and other unnatural amino acids or non-proteinogenic are known to form *cis* peptide linkages and can promote β-turns in peptide sequences [71, 87–90].

Head-to-tail condensation of dipeptides typically requires long reaction times and harsh conditions [91, 92]. Generally, the 2,5-DKPs are formed in low yields with low enantiopurity and are often contaminated with other side products [93–99]. Historically, traceless Staudinger ligation has been a useful tool for the preparation of medium and large cyclic peptides until Katritzky and coworkers pioneered the application of the traceless Staudinger ligation for the synthesis of small cyclopeptides, for example, 2,5-DKPs [69].

Table 6 Optimization of the Staudinger cyclization for the preparation of cyclo(glycyl-L-leucyl)

	Conditions					
Entry	X_3	Solvent	Equiv. of H_2O	rt or MW	Time (min)	Yield (%)
1	Bu	CH_2Cl_2	–	rt	720	55
2[a,b]	Bu	CH_2Cl_2	–	MW	30	66
3[b]	Bu	CH_2Cl_2	5	MW	5	78
4[b]	Bu	THF	5	MW	5	70
5[b]	Bu	MeCN	5	MW	5	65
6[b]	Bu	Toluene	5	MW	5	63
7[b]	Ph	CH_2Cl_2	5	MW	5	58

[a]Reaction was quenched with water
[b]Reaction was carried out at 50°C and 50 W

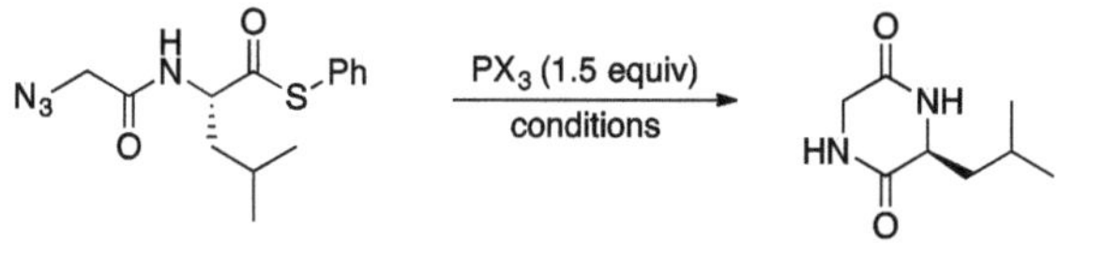

Fig. 8 Head-to-tail cyclization of azido glycyl-L-leucyl thiophenyl ester

Katritzky and coworkers have utilized the traceless Staudinger ligation for the preparation of 2,5-DKPs via solid- and solution-phase methods. This atom-economical yet inexpensive approach features mild reaction conditions for the fast and expedient tandem deprotection and cyclization steps [69].

By modifying established procedures, the starting material, azido glycyl-L-leucyl thiophenyl ester, was prepared in 65% yield over three steps from chloroacetyl chloride and L-Leu. Further reaction of the azido-protected dipeptide thioester with phosphines under microwave acceleration afforded the cyclic peptide (Table 6) [69]. The pure cyclic peptide was separated from the reaction mixture by filtration and required no further purification. Examining the effect of reaction time, solvent, and phosphine on the product yield and purity assisted in the optimization of the Staudinger-mediated cyclization. In addition, the reaction was carried out in the presence or absence of water and microwave irradiation (Table 6). The highest yields and shortest reaction times were obtained when tributylphosphine was reacted with the azido thioester for 5 min. in the presence of dichloromethane, five equivalents of water and under microwave irradiation (Fig. 8) [69].

Two additional examples of 2,5-DKPs were prepared using the optimized reaction conditions described above (Fig. 9, Table 7). Following the successful Staudinger-mediated synthesis of small cyclic peptides in solution, Katritzky and coworkers demonstrated the scope of the method by applying it to the solid-phase preparation of several cyclic peptides (Fig. 9, Table 7) [69]. Using standard DCC coupling procedures, *N*-(Boc-protected aminoacyl)benzotriazoles were added to the 3-thiopropanoic acid linker of a modified aminomethyl (AM) resin. Boc

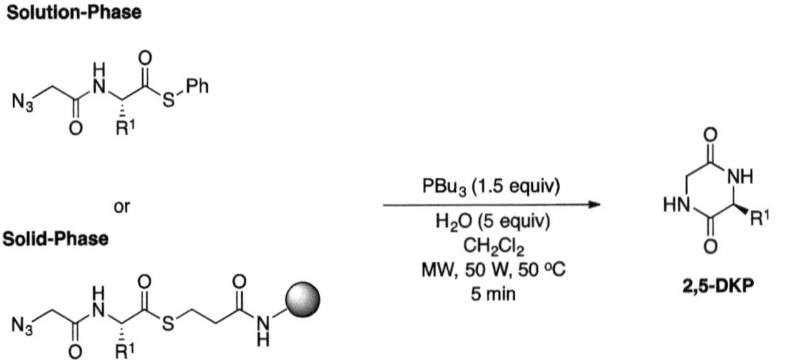

Fig. 9 Solution- and solid-phase preparation of 2,5-diketopiperazines (2,5-DKPs)

Table 7 Examples of 2,5-diketopiperazines prepared by solution- and solid-phase Staudinger cyclization methods

Entry	R^1	Method	Yield (%)
1	*i*-Bu	Solution	78
2[a]	Bn	Solution	74
3	Me	Solid	78
4	*i*-Bu	Solid	79
5	Bn	Solid	72
6	Me	Solid	81
7[a]	Bn	Solid	82

[a]Linker-AM resin was recovered and reused

deprotection and successive coupling with azido glycyl benzotriazole afforded the solid supported linear sequence. The cleavage and subsequent cyclization to the corresponding 2,5-DKP was achieved by subjecting the solid supported material to the optimized solution-phase conditions (Table 6, entry 3) [69]. Filtration, treatment of the residue with hot MeOH, and then recrystallization afforded the enantiopure 2,5-DKP. When compared with the solution-phase Staudinger ligation, the solid-phase method gave similar or slightly better yields. Furthermore, the modified AM resin could be recovered and reused in subsequent cyclization with no significant loss in activity [69].

3.2 *Hetero-2,5-Diketopiperazines from Proline-Containing Cbz-N-Protected Dipeptidoyl Benzotriazoles*

The majority of naturally occurring 2,5-diketopiperazines (2,5-DKPs) are obtained from L-α-amino acids and therefore possess a *cis* configuration (Fig. 10) [70, 71]. In addition to the previously described Staudinger ligation method for the preparation of cyclic peptides, Katritzky and coworkers have developed a *cis*-selective tandem deprotection/cyclization sequence for the synthesis of *cis*-2,5-diketopiperazines [71].

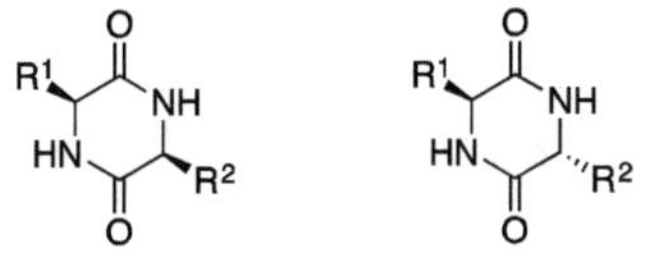

Fig. 10 *cis*- and *trans*-2,5-DKP (diketopiperazine)

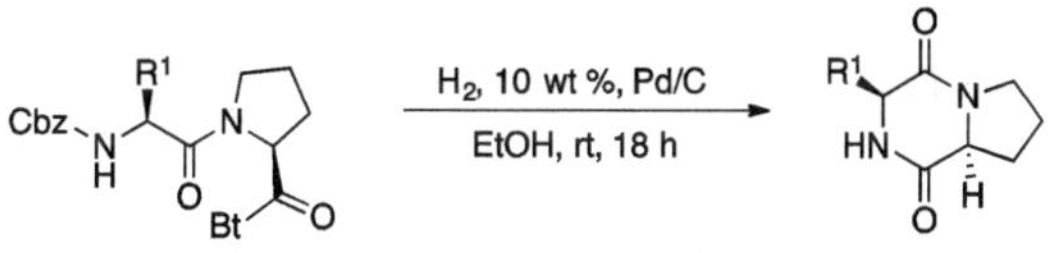

Fig. 11 Preparation of *cis*-DKPs by tandem deprotection/cyclization

Starting with Cbz-dipeptidoyl benzotriazoles, the one-pot, two-step, condensation–deprotection sequence proceeded smoothly to afford the *cis*-DKPs. L-Proline-containing dipeptidoyl benzotriazoles were converted to *cis*-DKPs in the presence of Pd/C (10 wt%) and hydrogen gas in 65–69% yields (Fig. 11) [71]. The stereochemistry of the *cis*-DKPs was confirmed via optical rotation and NMR spectroscopy.

Over the last couple of years, *trans*-diketopiperazines (*trans*-DKPs) have been gaining increasing attention due to their potential as anti-HIV, anticancer, and antibiotic agents [100–102]. The works of several researchers indicate that *trans*-DKPs can serve as foldamer building blocks [103]. Peptide drugs that have *trans*-DKPs incorporated into the backbone may be more potent due to the slower enzymatic degradation of DKPs [104]. Also, peptide drugs with a *trans*-DKPs backbone may be more selective for specific substrates [97].

Several strategies have been used in the synthesis of *trans*-DKPs. The most straightforward method requires the use of unnatural amino acid, D-proline [101, 105–107]. Alternatively, access to *trans*-DKPs can be obtained by epimerization of the corresponding *cis*-DKPs, although epimerization is frequently untended and occurs upon cyclization of the *cis*-DKP precursor or upon further functionalization of the *cis*-DKP [108–111]. Unintentional epimerization is inefficient and leads to *cis*/*trans*-DKP mixtures. Ultimately, the *trans*-DKPs are obtained in low yields.

Katritzky and coworkers introduced a more flexible and inexpensive method for the preparation of *trans*-DKPs. The sequence began with the synthesis of dipeptides from *N*-(Cbz-α-aminoacyl)benzotriazoles and L-proline via established procedures. The dipeptides were then smoothly converted to the corresponding benzotriazole derivatives [71]. In the presence of 1 equiv. of Et_3N, the Cbz-L-aa-L-Pro-Bt derivatives were transformed to the protected *trans*-DKPs in 69–75% yields. As a proof of concept, a protected *trans*-DKP was deprotected with hydrogen in the presence of Pd/C to afford the trans-DKPs (Fig. 12) [71].

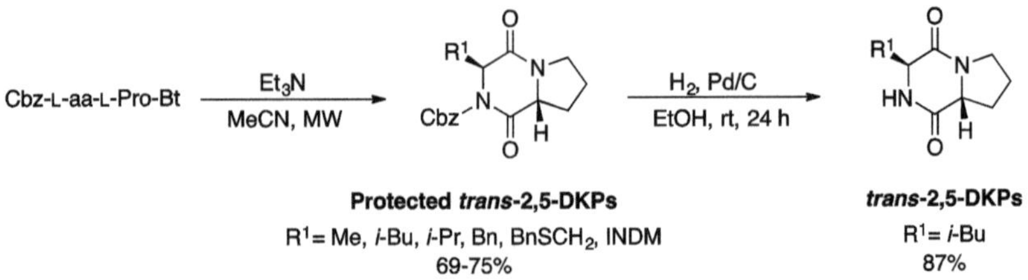

Fig. 12 Tandem cyclization/epimerization of dipeptidoyl benzotriazoles to afford *trans*-DKPs

Computational studies provided insight into the possible mechanisms for the transformations. Two possible mechanisms were proposed: The first was a unimolecular (unassisted) mechanism, in which the triazole leaving group functions as a base. The second scenario involves a bimolecular assisted mechanism where the amine base (Me_3N used as a model for Et_3N) abstracted a proton and was used to induce cyclization. Regardless of the pathway, as expected, the most stable conformation of the dipeptidoyl benzotriazole was twisted into a ready-to-cyclize conformation [71]. However, the bimolecular pathway had a lower activation barrier than the unassisted unimolecular pathway. Generally, the formation of the *trans*-DKP is thermodynamically favored over the *cis*-DKP. The formation of the *trans*-DKP from the dipeptidoyl benzotriazole is thought to occur via enolization and an ensuing protonation on the *Si*-face of the DKPs. In the case of the *cis*-DKP, it was found that the Cbz protecting group played a significant role in lowering the energy difference between the *cis*-DKP and the corresponding enol [71].

Having explored the preparation of DKPs, Katritzky and coworkers expanded the Staudinger-mediated ring closure to the synthesis of difficult medium-sized cyclic peptides. Several, novel 7- and 8-membered cyclic dipeptides and 10-membered cyclic tripeptides were prepared from azido peptide thioesters in yields of 48–75% [112]. Additionally, computational studies provided insight into the reactivity as a function of ring size [112].

4 *O*- and *S*-Acyl Isopeptide Strategies for the Preparation of Difficult Peptides

Since Emil Fisher's [113] pioneering work in the field of peptide synthesis at the turn of the twentieth century, tremendous progress has been made in the syntheses of peptides [4–6, 39]. Despite the plethora of peptide coupling reagents, utilization of microwave acceleration, and other advances in solution- and solid-phase peptide synthesis, the preparation of complex peptides with high purity and yields still poses a challenge. Typically, convergent peptide synthesis methods feature a reduction in the number of synthetic steps and as such are popular due to increased yields and easy purification. Such exploitation of chemoselectivity during peptide

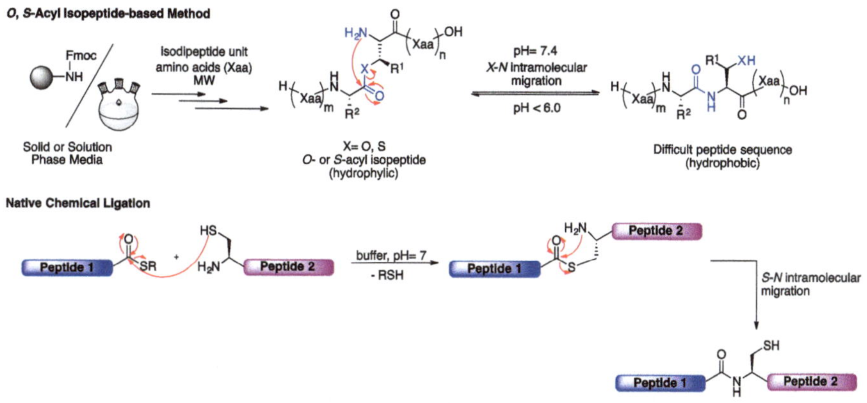

Fig. 13 Representations of *O*- and *S*-acyl isopeptide method and native chemical ligation (NCL)

synthesis affords the rapid assembly of enantiopure peptides under facile conditions that reduces the need for protecting groups. Chemoselective peptide synthesis strategies have focused on the use of naturally occurring amino acids cysteine (Cys, C), serine (Ser, S), and threonine (Thr, T). Cysteine possesses the nucleophilic β-sulfhydryl side chain, while both serine and threonine feature a hydroxyl side group. Over the last 30 years, extensive research effort on chemoselective transformations involving these three amino acid residues has led to the development of two primary methods for easier peptide chain extensions. The methods are the *O*- and *S*-acyl isopeptide-based strategy and native chemical ligation (NCL) [114–120]. When the *O*- and *S*-acyl isopeptide-based strategy for peptide synthesis are used, peptides are prepared using fully or partially protected amino acids and conventional peptide synthesis; however, native chemical ligation involves the condensation of unprotected peptide fragments (Fig. 13). Ultimately, the *O*- and *S*-acyl isopeptide-based strategies and native chemical ligation (NCL) feature an *X*-to-*N*-acyl transfer (X=*O*, *S*) leading to the formation of peptides that possess all native amide linkages [121–126]. This section will cover the *O*- and *S*-acyl isopeptide-based strategies. For further details regarding native chemical ligation, see [127].

4.1 Isopeptides and Isodipeptides in Peptide Synthesis

Isopeptides are frequently referred to as click peptides, switch peptides, or depsipeptides and are characterized by an isopeptide bond [128]. An isopeptide bond is formed when a peptide chain extension occurs at a lateral amino acid side chain (e.g. Ser, Thr, Cys) as opposed to the α-amide backbone of the growing peptide [129]. While Ohfune and coworkers established the basis for *O*- and *S*-acyl isopeptide strategies [130] in the 1990s, it wasn't until the close of the twentieth century that *O*- and *S*-acyl isopeptides emerged as a solution for the syntheses of difficult peptides [131–135]. When compared to native peptides, isopeptides tend to have greater solubility and are therefore less prone to aggregation. The introduction of the (thio)ester linkage into the growing peptide interrupts the usual hydrogen-bonding network and allows additional degrees of freedom [131–135]. As a result, isopeptide linkages have been broadly utilized in the preparation of difficult peptides. Additionally, isopeptide rearrangements have found great utility in polymer science and have provided insight into protein folding and aggregation and the relationship between peptide conformation and catalytic activity.

In the early part of the twenty-first century, Carpino, Kiso, and Mutter elegantly illustrated the application of *O*-isopeptide strategies [131–135]. Generally, the peptide is assembled using standard Fmoc-based SPPS. On completion of the sequence, the isopeptide is cleaved from the resin and stored as the TFA salt as no *O*-to-*N*-acyl shift is anticipated with protonated Ser, Thr, or Cys residues. Subsequent adjustment of the pH to 6.2–7.4 at ambient temperature promotes the rapid, quantitative *O*-to-*N*-acyl transfer (Fig. 13) [116, 136, 137]. At a pH of <6.0, the reverse reaction, the *N*-to-*O*-acyl transfer, will occur [138]. In addition to the pH-triggered *O*-to-*N*-acyl transfer, photolabile protecting groups, for example, 6-nitroveratryloxycarbonyl (Nvoc) introduced by Patchornik, Amit, and Woodward, have been shown to promote the clean *O*-to-*N*-acyl rearrangement in the presence of UV irradiation [139–142].

Generally, *O*-to-*N*-acyl transfers occur more rapidly with Thr as opposed to Ser amino acid residues [143, 144]. More information regarding the kinetics of the process was uncovered by Stella and coworkers [145, 146]. Although advantageous, the classical *O*-to-*N*-acyl strategy was plagued by several obstacles. From Carpino's work, it was apparent that the early inclusion of the isopeptide resulted in an easier synthesis of the difficult peptide; however, racemization could occur in up 23% of the peptide [135]. Along with epimerization, other side reactions included the formation of deletion sequences and 2,5-diketopiperazines (DKPs) (Fig. 14). However, Kiso and coworkers' subsequent improvement to the *O*-to-*N*-acyl strategy via incorporation of an isodipeptide eliminated or substantially reduced side reactions [139, 144]. Of note, using the *O*-acyl isodipeptide method can result in deletion of the Ser or Thr residue in the presence of a base. This reaction appears to be solvent dependent, and it can be avoided via changes in the solvents and/or activation method.

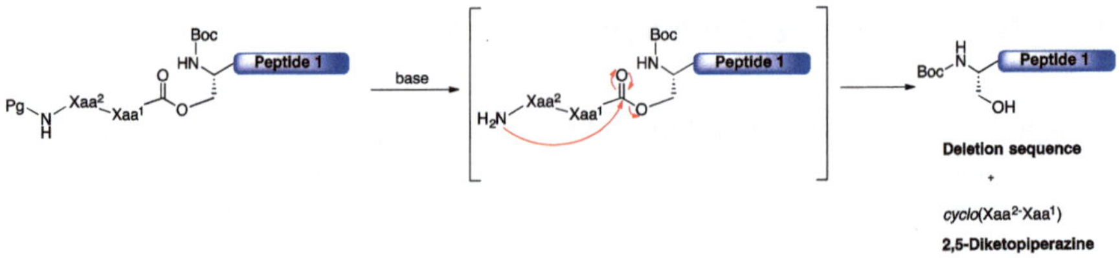

Fig. 14 2,5-Diketopiperazine and deletion sequence formation under classical *O*-acyl isopeptide conditions

Kiso and coworkers further extended the *O*-acyl isopeptide strategies to *S*-acyl isopeptides [147]. Unlike their oxygen counterparts, *S*-acyl isopeptides display a lower tendency for aggregation; therefore, they are easier to prepare and purify. Building on their improvements to the *O*-acyl isopeptide methods, they added to the *S*-acyl isopeptide method by using the allyl-based protective Alloc group for the synthesis of *S*-acyl isopeptides [148].

4.2 Application of N*-(Protected α-Aminoacyl)Benzotriazoles in the Synthesis of Peptides via the* O*- and* S*-Acyl Isopeptide Strategy*

Further contributions by Katritzky and coworkers expanded the *O*-to-*N*-isodipeptide methodology. In general, *O*-to-*N*-acyl rearrangements occur via a 5-membered cyclic transition state. However, Katritzky and coworkers prepared peptides of high yields and purities, facilitated by rearrangements involving 8- and 11-membered transition states [129]. These transformations occurred at 50°C in DMF in the presence of microwave irradiation (50 W). For example, the native peptide yields obtained from the rearrangement of the *O*-acyl isopeptide via the 8-membered transition state were quite notable and ranged from 57 to 78%, indicating that the nature of the amino acid residue acting as the nucleophile and not the nature of the amino acid esterifying the *O*-acyl isopeptide had a tremendous effect on the yield. Thus, the preorganization of the initial *O*-acyl isopeptide significantly affected the success of these long-range rearrangements [129]. In contrast to the *O*-to-*N*-acyl rearrangements, *S*-to-*N*-acyl transfers were facilitated by 17–20-membered macrocyclic transition states [149]. In fact, the formation of 8-, 11-, and 14 membered cyclic transition states posed a challenge. The yields and rates of the *S*-to-*N*-acyl rearrangements were strongly correlated to the size of the macrocyclic transition state [150–153]. As with the *O*-acyl isopeptide, the preorganization of the starting oligo *S*-acyl isopeptide had a large influence on the long-range acyl transfers [154]. Generally, analysis of the experimental data can be problematic as the ligation product may be contaminated by a range of side products [150, 152, 153, 155]. Thus, to facilitate the interpretation of results, a computational study of the long-range rearrangements in *S*-acyl isopeptides was designed [149]. Enthalpic effects were found to be the driving force for reactions involving 5–20-membered transition states. Again, the nature of the amino acid substituent was negligible. Finally, it was determined that the introduction of turn inducers may lower the activation barrier for long-range *S*-to-*N*-acyl rearrangements by promoting hydrogen-bonding interactions [149].

5 Benzotriazole Activation and Coupling of Amino Acids on a Preparative Scale

The widespread use of benzotriazole-mediated methods in the synthesis of peptides and their analogues hinges on the ability to scale up these processes. Frequently, microwave irradiation is required to accelerate the coupling steps, so the large-scale preparation of peptides and peptide conjugates from *N*-(protected-α aminoacyl) benzotriazoles can be difficult. Until recently there was no literature precedence for the use of *N*-(protected-α aminoacyl)benzotriazoles in conjunction with continuous flow methods. Now, as a proof of concept, Stevens and coworkers have reported the large-scale preparation of a library of (α-aminoacyl)amino-substituted heterocycles via a two-step continuous flow process [156]. The first step involves the benzotriazole activation of *N*-(Cbz-α-aminoacyl)benzotriazoles under continuous flow conditions. No racemization of the products was observed, and yields of 71–99% were obtained under optimized reaction conditions (solvent, time, temperature, reagents) [156]. The second step, the reaction of the *N*-(Cbz-α-aminoacyl) benzotriazole derivatives and heterocyclic amines in the presence of DMSO afforded (α-aminoacyl)amino-substituted heterocycles in yields of 40–99% [156].

Additionally, Stevens and coworkers explored telescoping the coupling step with the benzotriazole activation; however, the inconvenient work-up associated with the tandem process resulted in product losses and difficulties with product purification. In addition to increased yields and easier purification, the two-step, continuous flow method facilitates free solvent selection for each step, thereby enabling the maximum possible product yields [156].

6 Conclusion

N-(Protected-α aminoacyl)benzotriazoles are useful reagents for the preparation of a wide range of peptides via solution- and solid-phase peptide synthesis methods. These reagents have made important contributions to the field of peptide synthesis by facilitating the rapid assembly of peptides and peptide conjugates under mild conditions, with good to excellent yields and purities. Importantly, the prepared peptide and peptide conjugates are enantiopure. Microwave-assisted, benzotriazole-mediated synthesis of peptides has provided greater and more rapid access to cyclic and difficult peptides. Additionally, the *O*- and *S*-acyl isopeptide strategies have been an instrumental gateway to large difficult peptides, which are usually prone to aggregation. In order for the benzotriazole-based peptide synthesis platform to be widely applicable, the reagents and methods must be amenable to scale up. Recently, Stevens and coworkers demonstrated the scalability of such processes by preparing a library of (α-aminoacyl)amino-substituted heterocycles.

Despite significant improvements in accessing important peptides and peptide conjugates, our knowledge of protein–protein interactions and our ability to elucidate protein structure, there is still more to do in order to reduce the barrier to drug development for many of the world's critical diseases. Exploiting acylbenzotriazoles as allies for the construction of peptides and peptide conjugates and the development of such methods is a step in the right direction.

References

1. Bruno BJ, Miller GD, Lim CS (2013) Basics and recent advances in peptide and protein drug delivery. Ther Deliv 4(11):1443–1467
2. Dunn BM (ed) (2015) Peptide chemistry and drug design. Wiley, New Jersey
3. Fischer G, Rossmann M, Hyvönen M (2015) Alternative modulation of protein–protein interactions by small molecules. Curr Opin Biotechnol 35:78–85
4. Ramakers BEI, van Hest JCM, Löwik DWPM (2014) Molecular tools for the construction of peptide-based materials. Chem Soc Rev 43(8):2743–2756
5. Palomo JM (2014) Solid-phase peptide synthesis: an overview focused on the preparation of biologically relevant peptides. RSC Adv 4(62):32658–32672
6. Tang W, Becker ML (2014) "Click" reactions: a versatile toolbox for the synthesis of peptide-conjugates. Chem Soc Rev 43(20):7013–7039
7. Chen M, Heimer P, Imhof D (2015) Synthetic strategies for polypeptides and proteins by chemical ligation. Amino Acids 47(7):1283–1299
8. Vanier GS (2013) Microwave-assisted solid-phase peptide synthesis based on the Fmoc protecting group strategy (CEM). Methods Mol Biol 1047:235–249
9. Chow HY, Li X (2015) Development of thiol-independent peptide ligations for protein chemical synthesis. Tetrahedron Lett 56:3715–3720
10. Malins LR, Payne RJ (2015) Synthetic amino acids for applications in peptide ligation–desulfurization chemistry. Aust J Chem 68(4):521–537
11. Ibrahim TS, Tala SR, El-Feky SA, Abdel-Samii ZK, Katritzky AR (2011) Benzotriazole reagents for the syntheses of Fmoc-, Boc- and Alloc-protected amino acids. Synlett 14:2013–2016
12. Ibrahim TS, Tala SR, El-Feky SA, Abdel-Samii ZK, Katritzky AR (2012) Cysteinoyl- and cysteine-containing dipeptidoylbenzotriazoles with free sulfhydryl groups: easy access to N-terminal and internal cysteine peptides. Chem Biol Drug Des 80(2):194–202
13. Butula I, Lj Č, Proštenik MV, Vela V, Zorko F (1977) Reaktion mit 1-chlorocarbonylbenzotriazol; II[1]. Synthese von Carbamidsäure-estern und verwandten. Synthesis 1977(10):704–706
14. Paio A, Zaramella A, Ferritto R, Conti N, Marchioro C, Seneci P (1999) Solid-supported benzotriazoles: synthetic auxiliaries and traceless linkers for the combinatorial synthesis of amine libraries. J Comb Chem 1(4):317–325
15. Dener JM, Rice KD, Newcomb WS, Wang VR, Young WB, Gangloff AR, Kuo EY-L, Cregar L, Putnam D, Wong M (2001) Dibasic inhibitors of human mast cell tryptase. Part 3: identification of a series of potent and selective inhibitors containing the benzamidine functionality. Bioorg Med Chem Lett 11:1629–1633
16. Kumareswaran R, Hassner A (2001) Asymmetric syntheses of *N*-acetyl-(*R*)-coniine and *N*-Boc-(2*R*,6*R*)-solenopsin A via ring-closing metathesis. Tetrahedron Asymmetry 12:2269–2276
17. Wang J, Liang Y-L, Qu J (2006) Boiling water-catalyzed neutral and selective *N*-Boc deprotection. Chem Commun 5144–5146
18. Wünsch E, Graf W, Keller O, Keller W, Wersin G (1986) On the synthesis of benzyloxycarbonyl amino acids. Synthesis 1986(11):958–960

19. Akamatsu M, Fujimoto Y, Kataoka M, Suda Y, Kusumoto S, Fukase K (2006) Synthesis of lipid A monosaccharide analogues containing acidic amino acid: exploring the structural basis for the endotoxic and antagonistic activities. Bioorg Med Chem 14:6759–6777
20. Smith MB, March J (2007) 20 Advanced organic chemistry: reactions mechanism and structure, 6th edn. Wiley, Hoboken
21. Paquet A (1982) Introduction of 9-fluorenylmethyloxycarbonyl, trichloroethoxycarbonyl, and benzyloxycarbonyl amine protecting groups into O-unprotected hydroxyamino acids using succinimidyl carbonates. Can J Chem 60(8):976–980
22. Ten Kortenaar PBW, Van Dijk BG, Peeters JM, Raaben BJ, Adams PJHM, Tesser GI (1986) Rapid and efficient method for the preparation of Fmoc-amino acids starting from 9-fluorenylmethanol. Int J Pept Protein Res 27(4):398–400
23. Lapatsanis L, Milias G, Froussios K, Kolovos M (1983) Synthesis of *N*-2,2,2-(Trichloroethoxycarbonyl)-L-amino acids and *N*-(9-fluorenylmethoxycarbonyl)-L-amino acids involving succinimidoxy anion as a leaving group in amino acid protection. Synthesis 1983 (8):671–673
24. Chinchilla R, Dodsworth DJ, Najera C, Soriano JM (2001) A new polymer-supported reagent for the Fmoc-protection of amino acids. Tetrahedron Lett 42(43):7579–7581
25. Bogdan FM, Chow CS (1998) The synthesis of allyl- and allyloxycarbonyl-protected RNA phosphoramidites. Useful reagents for solid-phase synthesis of RNAs with base-labile modifications. Tetrahedron Lett 39(14):1897–1900
26. Hayakawa Y, Wakabayashi S, Kato H, Noyori R (1990) The allylic protection method in solid-phase oligonucleotide synthesis. An efficient preparation of solid-anchored DNA oligomers. J Am Chem Soc 112(5):1691–1696
27. Henklein P, Heyne H-U, Halatsch W-R, Niedrich H (1987) 5-Norbornene-2,3-dicarboximido carbonochloridate. A new stable reagent for the introduction of amino-protecting groups. Synthesis 1987(2):166–167
28. Hlebowicz E, Andersen AJ, Andersson L, Moss BA (2005) Identification of Fmoc-β-Ala-OH and Fmoc-β-Ala-amino acid-OH as new impurities in Fmoc-protected amino acid derivatives. Synthesis 65(1):90–97
29. Zalipsky S (1998) Alkyl succinimidyl carbonates undergo Lossen rearrangement in basic buffers. Chem Commun 1:69–70
30. Isidro-Llobet A, Just-Baringo X, Ewenson A, Alvarez M, Albericio F (2007) Fmoc-2-mercaptobenzothiazole, for the introduction of the Fmoc moiety free of side-reactions. Biopolymers 88(5):733–737
31. Carpino LA, Han GY (1972) 9-Fluorenylmethoxycarbonyl amino-protecting group. J Org Chem 37(22):3404–3409
32. Grzonka Z, Lammek B (1974) A simple method of preparation of *t*-butoxycarbonyl amino acids. Synthesis 1974(9):661–662
33. Tessier M, Albericio F, Pedroso E, Grandas A, Eritja R, Giralt E, Granier C, Van Rietschoten J (1983) Amino-acids condensations in the preparation of *N*α-9-fluorenylrnethyloxycarbonylamino-acids with 9-fluorenylmethylchloroformate. Int J Pept Protein Res 22(1):125–128
34. Cruz LJ, Beteta NG, Ewenson A, Albericio F (2004) "One-Pot" preparation of *N*-carbamate protected amino acids via the azide. Org Process Res Dev 8(6):920–924
35. Katritzky AR, Fali CN, Li J, Ager DJ, Prakash I (1997) Synthesis of 1-(*T*-butoxycarbonyl) benzotriazole and 1-(*p*-methoxybenzyloxycarbonyl)benzotriazole and their use in the protection of amino acids. Synth Commun 27(9):1623–1630
36. Katritzky AR, Zhang G-F, Fan W-Q, Wu J, Pernak J (1993) Studies on the thermal decarboxylation of 1-alkoxycarbonylbenzotriazoles. J Phys Org Chem 6(10):567–573
37. Yang C-W (2012) A comparative study of short linear motif compositions of the influenza A virus ribonucleoproteins. PLoS One 7(6), e38637
38. Searle MS, Williams DH, Packman LC (1995) A short linear peptide derived from the N-terminal sequence of ubiquitin folds into a water-stable non-native β-hairpin. Nat Struct Biol 2:999–1006

39. El-Faham A, Albericio F (2011) Peptide coupling reagents, more than a letter soup. Chem Rev 111:6557–6602
40. Katritzky AR, Angrish P, Todadze E (2009) Chiral acylation with N-(protected α-aminoacyl) benzotriazoles for advantageous syntheses of peptides and peptide conjugates. Synlett 2009 (15):2392–2411
41. Chandrudu S, Simerska P, Toth I (2013) Chemical methods for peptide and protein production. Molecules 18(4):4373–4388
42. Katritzky AR, Suzuki K, Singh SK (2004) Highly diastereoselective peptide chain extensions of unprotected amino acids with N-(Z-α-aminoacyl)benzotriazoles. Synthesis 2004 (16):2645–2652
43. Katritzky AR, Angrish P, Suzuki K (2006) The efficient preparation of di- and tripeptides by coupling N-(Cbz- or Fmoc-α-aminoacyl)benzotriazoles with unprotected amino acids. Synthesis 2006(3):411–424
44. Katritzky AR, Meher G, Angrish P (2006) Peptides by extension at the N- or C-termini of lysine. Chem Biol Drug Des 68(6):326–333
45. Katritzky AR, Todadze E, Cusido J, Angrish P, Shestopalov AA (2006) Selective peptide chain extension at the N-terminus of aspartic and glutamic acids utilizing 1-(N-protected-α-aminoacyl)benzotriazoles. Chem Biol Drug Des 68(1):37–41
46. Katritzky AR, Todadze E, Shestopalov AA, Cusido J, Angrish P (2006) Selective peptide chain extension at the C-terminus of aspartic and glutamic acids utilizing N-protected (α-aminoacyl)benzotriazoles. Chem Biol Drug Des 68(1):42–47
47. Katritzky AR, Meher G, Narindoshvili T (2008) Efficient synthesis of peptides by extension at the N- and C-termini of arginine. J Org Chem 73(18):7153–7158
48. Katritzky AR, Todadze E, Angrish P, Draghici B (2007) Efficient peptide coupling involving sterically hindered amino acids. J Org Chem 72(15):5794–5801
49. Katritzky AR, Khashab NM, Yoshioka M, Haase DN, Wilson KR, Johnson JV, Chung A, Haskell-Luevano C (2007) Microwave-assisted solid-phase peptide synthesis utilizing N-Fmoc-protected (α-aminoacyl)benzotriazoles. Chem Biol Drug Des 70(5):465–468
50. Katritzky AR, Yoshioka M, Narindoshvili T, Chung A, Khashab NM (2008) N-Fmoc-protected(α-dipeptidoyl)benzotriazoles for efficient solid-phase peptide synthesis by segment condensation. Chem Biol Drug Des 2(3):182–188
51. Isao A, Rolf A, Stefan K, Julius R Jr (1995) Synthesis and coupling reactions of α, α-dialkylated amino acids with nucleobase side chains. Proc Natl Acad Sci 92 (26):12013–12016
52. Chatterjee J, Laufer B, Kessler H (2012) Synthesis of *N*-methylated cyclic peptides. Nat Protoc 7(3):432–444
53. Katritzky AR, Wang M, Yang H, Zhang S, Akhmedov NG (2002) 1-(α-Boc-aminoacyl) benzotriazoles: stable chiral α-aminoacylation reagents. Arkivoc (viii):134–142
54. Katritzky AR, Vincek AS, Suzuki K (2005) Microwave-assisted synthesis of peptidyl phosphorus ylides. Arkivoc (v):116–126
55. Katritzky AR, Angrish P (2006) Efficient microwave assisted access to chiral O-(α-protected-aminoacyl)steroids. Steroids 71(8):660–669
56. Chan WC, White PD (2000) Basic principles. In White PD (ed) Fmoc solid phase peptide synthesis, a practical approach. University Press, New York
57. Katritzky AR, Haase DN, Johnson JV, Chung A (2009) Benzotriazole-assisted solid-phase assembly of Leu-enkephalin, amyloid β segment 34–42, and other "Difficult" peptide sequences. J Org Chem 74(5):2028–2032
58. Monroc S, Feliu L, Planas M, Bardaji E (2006) Microwave-assisted cyclization of peptides on SynPhase™ lanterns. Synlett 2006(9):1311–1314
59. Erdélyia M, Gogoll A (2002) Rapid microwave-assisted solid phase peptide synthesis. Synthesis 2002(11):1592–1596
60. Gorske BC, Jewell SA, Guerard EJ, Blackwell HE (2005) Expedient synthesis and design strategies for new peptoid construction. Org Lett 7(8):1521–1524

61. Yu HM, Chen ST, Wang KT (1992) Enhanced coupling efficiency in solid-phase peptide synthesis by microwave irradiation. J Org Chem 57(18):4781–4784
62. Matsushita T, Hinou H, Fumoto M, Kurogochi M, Fujitani N, Shimizu H, Nishimura S-I (2006) J Org Chem 71(8):3051–3063
63. Mergler M, Dick F, Sax B, Stähelin C, Vorherr TJ (2003) The aspartimide problem in Fmoc-based 56 Part II. J Peptide Sci 9(8):518–526
64. Humphrey JM, Chamberlin AR (1997) Chemical synthesis of natural product peptides: coupling methods for the incorporation of noncoded amino acids into peptides. Chem Rev 97(6):2243–2266
65. Han S-Y, Kim Y-A (2004) Recent development of peptide coupling reagents in organic synthesis. Tetrahedron 60(11):2447–2467
66. Pourvali A, Cochrane JR, Hutton CA (2014) A new method for peptide synthesis in the N→C direction: amide assembly through silver-promoted reaction of thioamides. Chem Commun 50(100):15963–15966
67. Bray BL (2003) Large-scale manufacture of peptide therapeutics by chemical synthesis. Nat Rev Drug Discov 2:587–593
68. Avan I, Hall CD, Katritzky AR (2014) Peptidomimetics *via* modifications of amino acids and peptide bonds. Chem Soc Rev 43(10):3575–3594
69. Beagle LK, Hansen FK, Monbaliu J-CM, DesRosiers MP, Phillips AM, Stevens CV, Katritzky AR (2012) Efficient synthesis of 2,5-diketopiperazines by staudinger-mediated cyclization. Synlett 23(16):2337–2340
70. Borthwick AD (2012) 2,5-Diketopiperazines: synthesis, reactions, medicinal chemistry, and bioactive natural products. Chem Rev 112(7):3641–3716
71. Monbaliu J-CM, Hansen MK, Beagle LK, Panzner MK, Steel PJ, Todadze E, Stevens CV, Katritzky AR (2012) A new benzotriazole-mediated stereoflexible gateway to hetero-2,5-diketopiperazines. Chem Eur J 18(9):2632–2638
72. Balboni G, Guerrini R, Salvadori S, Tomatis R, Bryant SD, Bianchi C, Attila M, Lazarus LH (1997) Opioid diketopiperazines: synthesis and activity of a prototypic class of opioid antagonists. Biol Chem 378(1):19–29
73. Bryant SD, Balboni G, Guerrini R, Salvadori S, Tomatis R, Bianchi C, Attila M, Lazarus LH (1997) Opioid diketopiperazines: refinement of the δ opioid antagonist pharmacophore. Biol Chem 378(2):107–114
74. Crescenzi O, Fraternali F, Picone D, Tancredi T, Balboni G, Guerrini R, Lazarus LH, Salvadori S, Temussi PA (1997) Design and solution structure of a partially rigid opioid antagonist lacking the basic center — models of antagonism. Eur J Biochem 247(1):66–73
75. Hayashi Y, Orikasa S, Tanaka K, Kanoh K, Kiso Y (2000) Total synthesis of anti-microtubule diketopiperazine derivatives: phenylahistin and aurantiamine. J Org Chem 65(24):8402–8405
76. Edmondson S, Danishefsky SJ, Sepp-Lorenzino L, Rosen N (1999) Total synthesis of spirotryprostatin A, leading to the discovery of some biologically promising analogues. J Am Chem Soc 121(10):2147–2155
77. Kanoh K, Kohno S, Katada J, Takahashi J, Uno I, Hayashi Y (1999) Synthesis and biological activities of phenylahistin derivatives. Bioorg Med Chem 7(7):1451–1457
78. Graz CJM, Grant GD, Brauns SC, Hunt A, Jamie H, Milne PJ (2000) Cyclic dipeptides in the induction of maturation for cancer therapy. J Pharm Pharmacol 52(1):75–82
79. Boger DL, Fink BE, Hedrick MP (2000) A new class of highly cytotoxic diketopiperazines. Bioorg Med Chem Lett 10(10):1019–1020
80. Prakash KRC, Tang Y, Kozikowski AP, Flippen-Anderson JL, Knoblach SM, Faden AI (2002) Bioorg Med Chem 10(9):3043–3048
81. Sugie Y, Hirai H, Inagaki T, Ishiguro M, Kim YJ, Kojima Y, Sakakibara T, Sakemi S, Sugiura A, Suzuki Y, Brennan L, Duignan J, Huang LH, Sutcliffe J, Kojima N (2001) A new antibiotic CJ-17,665 from *Aspergillus ochraceus*. J Antibiotics 54(11):911–916

82. Stevens BW, Joska TM, Anderson AC (2006) Progress toward re-engineering non-ribosomal peptide synthetase proteins: a potential new source of pharmacological agents. Drug Dev Res 66(1):9–18
83. Houston DR, Synstad B, Eijsink VGH, Stark MJR, Eggleston IM, van Aalten DMF (2004) Structure-based exploration of cyclic dipeptide chitinase inhibitors. J Med Chem 47 (23):5713–5720
84. Nilanonta C, Isaka M, Kittakoop P, Saenboonrueng J, Rukachaisirikul V, Kongsaeree P, Thebtaranonth Y (2003) New Diketopiperazines from the Entomopathogenic Fungus Verticillium hemipterigenum BCC 1449. J Antibiotics 56(7):647–651
85. Sinha S, Srivastava R, De Clercq E, Singh RK (2004) Synthesis and antiviral properties of arabino and ribonucleosides of 1,3-dideazaadenine, 4-nitro-1,3-dideazapurine and diketopiperazine. Nucleosides Nucleotides Nucleic Acids 23(12):1815–1824
86. Hu F, Chou CJ, Gottesfeld JM (2009) Design and synthesis of novel hybrid benzamide–peptide histone deacetylase inhibitors. Bioorg Med Chem Lett 19(14):3928–3931
87. Fairweather KA, Sayyadi N, Luck IJ, Clegg JK, Jolliffe KA (2010) Synthesis of all-L cyclic tetrapeptides using pseudoprolines as removable turn inducers. Org Lett 12(14):3136–3139
88. Singh EK, Ravula S, Pan C-M, Pan P-S, Vasko RC, Lapera SA, Weerasinghe SVW, Pflum MKH, McAlpine SR (2008) Synthesis and biological evaluation of histone deacetylase inhibitors that are based on FR235222: a cyclic tetrapeptide scaffold. Bioorg Med Chem Lett 18(8):2549–2554
89. Singh EK, Nazarova LA, Lapera SA, Alexander LD, McAlpine SR (2010) Histone deacetylase inhibitors: synthesis of cyclic tetrapeptides and their triazole analogs. Tetrahedron Lett 51(33):4357–4360
90. Nolasco L, Gonzalez MP, Caggiano L, Jackson RFW (2009) Application of negishi cross-coupling to the synthesis of the cyclic tripeptides OF4949-III and K-13. J Org Chem 74 (21):8280–8289
91. Pérez-Picaso L, Escalante J, Olivo HF, Yolanda Rios M (2009) Efficient microwave assisted syntheses of 2,5-diketopiperazines in aqueous media. Molecules 14(8):2836–2849
92. Cini E, Botta CB, Rodriguez M, Taddei M (2009) Microwaves enhance cyclisation of tetrapeptides. Tetrahedron Lett 50(51):7159–7161
93. Kopple KD, Ghazarian HG (1968) Convenient synthesis of 2,5-piperazinediones. J Org Chem 33(2):862–864
94. Ueda T, Saito M, Kato T, Izumiya N (1983) Facile synthesis of cyclic dipeptides and detection of racemization. Bull Chem Soc Jpn 56(2):568–572
95. Lin Q, Blackwell HE (2006) Rapid synthesis of diketopiperazine macroarrays via Ugi four-component reactions on planar solid supports. Chem Commun 27:2884–2886
96. Tullberg M, Grøtli M, Luthman K (2006) Efficient synthesis of 2,5-diketopiperazines using microwave assisted heating. Tetrahedron 62(31):7484–7491
97. Houston DR, Eggleston I, Synstad B, Eijsink VGH, van Aalten DMF (2002) The cyclic dipeptide CI-4 [cyclo-(l-Arg-d-Pro)] inhibits family 18 chitinases by structural mimicry of a reaction intermediate. Biochem J 368(1):23–27
98. Montero A, Beierle JM, Olsen CA, Ghadiri MR (2009) Design, synthesis, biological evaluation, and structural characterization of potent histone deacetylase inhibitors based on cyclic α/β-tetrapeptide architectures. J Am Chem Soc 131(8):3033–3041
99. Saxon E, Armstrong JI, Bertozzi CR (2000) A "Traceless" staudinger ligation for the chemoselective synthesis of amide bonds. Org Lett 2(14):2141–2143
100. Jain HD, Zhang C, Zhou S, Zhou H, Ma J, Liu X, Liao X, Deveau AM, Dieckhaus CM, Johnson MA, Smith KS, Macdonald TL, Kakeya H, Osada H, Cook JM (2008) Synthesis and structure–activity relationship studies on tryprostatin A, an inhibitor of breast cancer resistance protein. Bioorg Med Chem 16(8):4626–4651
101. Pérez-Balado C, de Lera AR (2010) Concise total synthesis and structural revision of (+)-pestalazine B. Org Biomol Chem 8(22):5179–5186

102. Kwon OS, Park SH, Yun B-S, Pyun YR, Kim C-J (2001) Cyclo (D-Pro-L-Val), a specific. BETA.-glucosidase inhibitor produced by *Aspergillus* sp. F70609. J Antibiot 54(2):179–181
103. Gupta S, Macala M, Schafmeister CE (2006) Synthesis of structurally diverse bis-peptide oligomers. J Org Chem 71(23):8691–8695
104. Zhao S, Smith KS, Deveau AM, Dieckhaus CM, Johnson MA, Macdonald TL, Cook JM (2002) Biological activity of the tryprostatins and their diastereomers on human carcinoma cell lines. J Med Chem 45(8):1559–1562
105. Fischer PM (2003) Diketopiperazines in peptide and combinatorial chemistry. J Pept Sci 9 (1):9–35. doi:10.1002/psc.446
106. Lim HJ, Gallucci JC, RajanBabu TV (2010) Annulated diketopiperazines from dipeptides or Schöllkopf reagents via tandem cyclization−intramolecular N-arylation. Org Lett 12 (9):2162–2165
107. Moir EM, Yoshiizumi K, Cairns J, Cowley P, Ferguson M, Jeremiah F, Kiyoi T, Morphy R, Tierney J, Wishart G, York M, Baker J, Cottney JE, Houghton AK, McPhail P, Osprey A, Walker G, Adam JM (2010) Design, synthesis, and structure–activity relationship study of bicyclic piperazine analogs of indole-3-carboxamides as novel cannabinoid CB1 receptor agonists. Bioorg Med Chem Lett 20(24):7327–7330
108. Fdhila F, Vázquez V, Sánchez JL, Riguera R (2003) dd-Diketopiperazines: antibiotics active against vibrio anguillarum isolated from marine bacteria associated with cultures of *Pecten maximus*. J Nat Prod 66(10):1299–1301. doi:10.1021/np030233e
109. Eguchi C, Kakuta A (1974) Cyclic dipeptides. I Thermodynamics of the cis-trans isomerization of the side chains in cyclic dipeptides. J Am Chem Soc 96(12):3985–3989
110. Pichowicz M, Simpkins NS, Blake AJ, Wilson C (2008) Studies towards complex bridged alkaloids: regio- and stereocontrolled enolate chemistry of 2,5-diketopiperazines. Tetrahedron 64(17):3713–3735
111. Sanz-Cervera JF, Stocking EM, Usui T, Osada H, Williams RM (2000) Synthesis and evaluation of microtubule assembly inhibition and cytotoxicity of prenylated derivatives of cyclo-l-Trp-l-Pro. Bioorg Med Chem 8(10):2407–2415
112. Ha K, Monbaliu J-CM, Williams BC, Pillai GG, Ocampo CE, Zeller M, Stevens CV, Katritzky AR (2012) A convenient synthesis of difficult medium-sized cyclic peptides by Staudinger mediated ring-closure. Org Biomol Chem 10:8055–8058
113. Fischer E, Fourneau E (1901) Ueber einige Derivate des Glykocolls. Ber Dtsch Chem Ges 34 (2):2868–2879
114. Payne RJ, Wong C-H (2010) Advances in chemical ligation strategies for the synthesis of glycopeptides and glycoproteins. Chem Commun 46:21–43
115. Rohde H, Seitz O (2010) Ligation-desulfurization: a powerful combination in the synthesis of peptides and glycopeptides. Biopolymers 94(4):551–559
116. Coin I (2010) The depsipeptide method for solid-phase synthesis of difficult peptides. J Pept Sci 16(5):223–230
117. Kent SBH (2009) Total chemical synthesis of proteins. Chem Soc Rev 38:338–351
118. Hackenberger CPR, Schwarzer D (2008) Chemoselective ligation and modification strategies for peptides and proteins. Angew Chem Int Ed 47(52):10030–10074
119. Dirksen A, Dawson PE (2008) Expanding the scope of chemoselective peptide ligations in chemical biology. Curr Opin Chem Biol 12(6):760–766
120. Coltart DM (2000) Peptide segment coupling by prior ligation and proximity-induced intramolecular acyl transfer. Tetrahedron 56(22):3449–3491
121. Deakyne CA, Ludden AK, Roux MV, Notario R, Demchenko AV, Chickos JS, Liebman JF (2010) Energetics of the lighter chalcogen analogues of carboxylic acid esters. J Phys Chem B 114(49):16253–16262
122. Castro EA (2009) Kinetics and mechanism of the aminolysis of thioesters and thiocarbonates in solution. Pure Appl Chem 81(4):685–696
123. Erben MF, Boese R, Della Védova CO, Oberhammer H, Willner H (2006) Toward an intimate understanding of the structural properties and conformational preference of

oxoesters and thioesters: gas and crystal structure and conformational analysis of dimethyl monothiocarbonate, CH3OC(O)SCH3. J Org Chem 71(2):616–622
124. Yang W, Drueckhammer DG (2001) Understanding the relative acyl-transfer reactivity of oxoesters and thioesters: computational analysis of transition state delocalization effects. J Am Chem Soc 123(44):11004–11009
125. Yang W, Drueckhammer DG (2000) Computational studies of the aminolysis of oxoesters and thioesters in aqueous solution. Org Lett 2(26):4133–4136
126. Castro EA (1999) Kinetics and mechanisms of reactions of thiol, thiono, and dithio analogues of carboxylic esters with nucleophiles. Chem Rev 99(12):3505–3524
127. Albers T, Watkins DL, Gameiro AF, Povstyanoy V, Povstyanoy MV, Lebedyeva IO (2015) Benzotriazole-based strategies toward peptidomimetics, conjugates, and other peptide derivatives. Top Heterocycl Chem. doi:10.1007/7081_2015_182
128. Monbaliu J-CM, Katritzky AR (2012) Recent trends in Cys- and Ser/Thr-based synthetic strategies for the elaboration of peptide constructs. Chem Commun 48:11601–11622
129. El Khatib M, Elagawany M, Jabeen F, Todadze E, Bol'shakov O, Oliferenko A, Khelashvili L, El-Feky SA, Asiri A, Katritzky AR (2012) Traceless chemical ligations from *O*-acyl serine sites. Org Biomol Chem 10(25):4836–4838
130. Horikawa M, Shigeri Y, Yumoto N, Yoshikawa S, Nakajima T, Ohfune Y (1998) Syntheses of potent Leu-enkephalin analogs possessing β-hydroxy-α, α-disubstituted-α-amino acid and their characterization to opioid receptors. Bioorg Med Chem Lett 8(15):2027–2032
131. Mutter M, Chandravarkar A, Boyat C, Lopez J, Dos Santos S, Mandal B, Mimna R, Murat K, Patiny L, Saucède L, Tuchscherer G (2004) Switch peptides in statu nascendi: induction of conformational transitions relevant to degenerative diseases. Angew Chem Int Ed 43 (32):4172–4178
132. Sohma Y, Sasaki M, Hayashi Y, Kimura T, Kiso Y (2004) Novel and efficient synthesis of difficult sequence-containing peptides through *O–N* intramolecular acyl migration reaction of *O*-acyl isopeptides. Chem Commun 2004:12–125
133. Sohma Y, Sasaki M, Hayashi Y, Kimura T, Kiso Y (2004) Design and synthesis of a novel water-soluble Aβ1-42 isopeptide: an efficient strategy for the preparation of Alzheimer's disease-related peptide, Aβ1-42, via *O–N* intramolecular acyl migration reaction. Tetrahedron 45(31):5965–5968
134. Sohma Y, Hayashi Y, Skwarczynski M, Hamada Y, Sasaki M, Kimura T, Kiso Y (2004) O-N intramolecular acyl migration reaction in the development of prodrugs and the synthesis of difficult sequence-containing bioactive peptides. Biopolymers 76(4):344–356
135. Carpino LA, Krause E, Sferdean CD, Schümann M, Fabian H, Biernert M, Beyermann M (2004) Synthesis of "difficult" peptide sequences: application of a depsipeptide technique to the Jung-Redemann 10- and 26-mers and the amyloid peptide A(1-42). Tetrahedron Lett 45 (40):7519–7523
136. Bozso Z, Penke B, Simon D, Laczkó I, Juhász G, Szegedi V, Kasza A, Soós K, Hetényi A, Wéber E, Tóháti H, Csete M, Zarándi M, Fülöp L (2010) Controlled in situ preparation of Aβ (1–42) oligomers from the isopeptide "iso-Aβ(1–42)", physicochemical and biological characterization. Peptides 31(2):248–256
137. Sohma Y, Yoshiya T, Taniguchi A, Kimura T, Hayashi Y, Kiso Y (2007) Development of *O*-acyl isopeptide method. Biopolymers 88(2):253–262
138. Carpino LA, Krause E, Sferdean CD, Schümann M, Biernert M, Beyermann M (2005) Dramatically enhanced N→O acyl migration during the trifluoroacetic acid-based deprotection step in solid phase peptide synthesis. Tetrahedron Lett 46(8):1361–1364
139. Vila-Perelló M, Hori Y, Ribó M, Muir TW (2008) Activation of protein splicing by protease- or light-triggered O to N acyl migration. Angew Chem Int Ed 47(40):7764–7767
140. Taniguchi A, Sohma Y, Kimura M, Okada T, Ikeda K, Hayashi Y, Kimura T, Hirota S, Matsuzaki K, Kiso Y (2006) "Click Peptide" based on the "O-Acyl Isopeptide Method": control of Aβ1−42 production from a photo-triggered Aβ1−42 analogue. J Am Chem Soc 128(3):696–697

141. Sohma Y, Kiso Y (2006) "Click Peptides"—chemical biology-oriented synthesis of Alzheimer's disease-related amyloid β peptide (Aβ) analogues based on the "*O*-Acyl Isopeptide Method". ChemBioChem 7(10):1549–1557
142. Shigenaga A, Hirakawa H, Yamamoto J, Ogura K, Denda M, Yamaguchi K, Tsuji D, Itoh K, Otaka A (2011) Design and synthesis of caged ceramide: UV-responsive ceramide releasing system based on UV-induced amide bond cleavage followed by O-N acyl transfer. Tetrahedron 67(22):3984–3990
143. Sohma Y, Taniguchi A, Skwarczynski M, Yoshiya T, Fukao F, Kimura T, Hayashi Y, Kiso Y (2006) *O*-Acyl isopeptide method' for the efficient synthesis of difficult sequence-containing peptides: use of '*O*-acyl isodipeptide unit'. Tetrahedron Lett 47(18):3013–3017
144. Dos Santos S, Chandravarkar A, Mandal B, Mimna R, Murat K, Saucède L, Tella P, Tuchscherer G, Mutter M (2005) Switch-peptides: controlling self-assembly of amyloid β-derived peptides in vitro by consecutive triggering of acyl migrations. J Am Chem Soc 127(34):11888–11889
145. Oliyai R, Stella VJ (1995) Structural factors affecting the kinetics of O, N-acyl transfer in potential O-peptide prodrugs. Bioorg Med Chem Lett 5(22):2735–2740
146. Oliyai R, Siahaan TJ, Stella VJ (1995) The importance of structural factors on the rate and the extent of N, O-acyl migration in cyclic and linear peptides. Pharm Res 12(3):323–328
147. Yoshiya T, Ito N, Kimura T, Kiso Y (2008) Isopeptide method: development of *S*-acyl isopeptide method for the synthesis of difficult sequence-containing peptides. J Pept Sci 14 (11):1203–1208
148. Yoshiya T, Hasegawa Y, Kawamura W, Kawashima H, Sohma Y, Kimura T, Kiso Y (2010) *S*-acyl isopeptide method: Use of allyl-type protective group for improved preparation of thioester-containing *S*-acyl isopeptides by Fmoc-based SPPS. Biopolymers 96(2):228–239
149. Monbaliu J-CM, Dive G, Stevens CV, Katritzky AR (2013) Governing parameters of long-range intramolecular S-to-N acyl transfers within (S)-acyl isopeptides. J Chem Theory Comput 9(2):927–934
150. Ha K, Chahar M, Monbaliu J-CM, Todadze E, Hansen FK, Oliferenko AA, Ocampo CE, Leino D, Lillicotch A, Stevens CV, Katritzky AR (2012) Long-range intramolecular S → N acyl migration: a study of the formation of native peptide analogues via 13-, 15-, and 16-membered cyclic transition states. J Org Chem 77(6):2637–2648
151. Hansen FK, Ha K, Todadze E, Lillicotch A, Frey A, Katritzky AR (2011) Microwave-assisted chemical ligation of *S*-acyl peptides containing non-terminal cysteine residues. Org Biomol Chem 9(20):7162–7167
152. Katritzky AR, Tala SR, Abo-Dya NE, Ibrahim TS, El-Feky SA, Gyanda K, Pandya KM (2011) Chemical ligation of S-acylated cysteine peptides to form native peptides via 5-, 11-, and 14-membered cyclic transition states. J Org Chem 76(1):85–96
153. Katritzky AR, Abo-Dya NE, Tala SR, Abdel-Samii ZK (2010) The chemical ligation of selectively *S*-acylated cysteine peptides to form native peptides via 5-, 11- and 14-membered cyclic transition states. Org Biomol Chem 8:2316–2319
154. Oliferenko AA, Katritzky (2011) Alternating chemical ligation reactivity of S-acyl peptides explained with theory and computations. Org Biomol Chem 9:4756–4759
155. Haase C, Seitz O (2009) Internal cysteine accelerates thioester-based peptide ligation. Eur J Org Chem 2009(13):2096–2101
156. Seghers S, Van Waes FEA, Cukalovic A, Monbaliu J-CM, De Visscher J, Thybaut JW, Heugebaert TSA, Stevens CV (2015) Efficient continuous flow benzotriazole activation and coupling of amino acids. J Flow Chem 5(4):220–227

Top Heterocycl Chem (2016) 43: 95–142
DOI: 10.1007/7081_2015_182

Published online: 30 October 2015

Benzotriazole-Based Strategies Toward Peptidomimetics, Conjugates, and Other Peptide Derivatives

Thomas Albers, Davita L. Watkins, Armanda F. Gameiro, V'yacheslav Povstyanoy, Mykhaylo V. Povstyanoy, and Iryna O. Lebedyeva

Abstract Benzotriazole-mediated routes to peptidomimetics and peptide conjugates have been discussed in detail. The Katritzky group developed a benzotriazole methodology toward the activation of carbonyl groups of amino acids and modified analogs, which allowed the synthesis of cyclic peptides, azapeptides, azidopeptides, aminoxypeptides, oxyazapeptides, depsipeptides, and isopeptides. High-yielding reactions of *N*-, *O*-, *S*-, and *C*-acylated nucleophiles with activated aminoacyl or peptidoyl benzotriazole derivatives have also been reported. Benzotriazole methodology enabled the efficient incorporation of bioactive moieties into peptides, peptidomimetics, amino acids, and other carbonyl-containing compounds. Predominant number of reported products retained chiral purity. Some of the products displayed promising biological activities such as anticancer and antibacterial activity along with the improved stability under physiological conditions.

Keywords Acylation · Benzotriazole · Conjugate · Coupling · Cyclic peptides · Peptidomimetic

T. Albers, A.F. Gameiro, and I.O. Lebedyeva (✉)
Department of Chemistry and Physics, Georgia Regents University, 1120 15th Street SCI W3005, Augusta, GA 30912, USA
e-mail: ilebedyeva@gru.edu

D.L. Watkins
Department of Chemistry and Biochemistry, University of Mississippi, Mississippi, MS 38677, USA

V. Povstyanoy and M.V. Povstyanoy
Department of Organic and Biochemical Synthesis, Kherson National Technical University, Berislavs'ke Highway 24, Kherson 73008, Ukraine

Contents

1 Introduction

Peptidomimetics and peptide conjugates play an increasingly important role as therapeutics in areas including diabetes, oncology, metabolic, cardiovascular, and infectious diseases [1]. Based on the fact that the FDA approved six peptides in 2012 (lucinactant, peginesatide, pasireotide, carfilzomib, linaclotide, and teduglutide), peptide and peptidomimetic therapeutic compositions have spurred additional efforts in drug discovery [2]. The demand for modified peptides with improved stability profiles and pharmacokinetic properties is the driving force of the drug discovery research [3]. Modulation of protease activity with synthetic peptidomimetic inhibitors has proven to be clinically useful for treating human immunodeficiency virus (HIV) infection and hypertension [4]. Medicinal applications of peptidomimetics in cancer; obesity; cardiovascular, inflammatory, and neurodegenerative diseases; and various infectious and parasitic diseases are currently being developed [4]. Exploration of natural peptide inhibitors and synthetic peptidomimetic analogs provided promising compounds that performed successfully in animal studies [5, 6]. Several protease inhibitors are undergoing further evaluation in human clinical trials to treat the abovementioned diseases [4, 6, 7].

Many structural modifications of natural peptides guided by rational drug design and molecular modeling have been provided to develop stable, bioavailable, and effective mimetics of natural peptide sequences [3, 8]. Peptidomimetics are compounds which mimic a natural peptide or protein in 3D space and which retain the ability of the peptide to interact with the biological target and produce the same biological effect. However, clinical applications of bioactive natural peptides have been limited by their susceptibility to rapid hydrolysis by peptidases [9].

The corresponding peptidomimetics are discussed as strategies toward extending the half-life and bioavailability of the peptide and retain the bioactivity profile of the natural product [3].

D-amino acids have been successfully incorporated in the synthesis of peptidomimetics [10]. A stabilized D-amino acid-containing alpha helical peptide that mimics proapoptotic BH3-only proteins, such as BID and BAD, and suppresses the growth of human leukemia in a mouse xenograft model has been reported by Walensky et al. [11]. A dimeric small molecule that mimics the proapoptotic protein Smac mimics the N-terminal linear motif Ala-Val-Pro-Ile and sensitizes the non-small-cell lung cancer cells [12–14]. Peptidomimetic AG7088 is a potent, irreversible inhibitor of human rhinovirus (HRV) 3C protease [15]. Omapatrilat, a peptidomimetic underdevelopment, has revealed significantly greater antihypertensive efficacy than existing agents [16].

In order to proceed to clinical studies, naturally occurring bioactive peptides are often put through structure–activity relationship studies toward the improvement of their stability, solubility, and bioactivity profiles [17, 18]. The development of the peptidomimetic ADC toxin monomethyl auristatin E (MMAE) from the natural peptide dolastatin 10 [19], which gained FDA approval for the treatment of Hodgkin's lymphoma in 2011 [20], represents a successful example of peptidomimetic-based drug discovery and structure–activity relationship (SAR) strategies.

Due to their high bioavailability, peptides are often coupled with the imaging moieties or nanoplatforms using bioconjugation [21, 22] or radiolabeling techniques [23]. Another common way of drug delivery is conjugation of the bioactive molecules to cell-penetrating peptides (CPPs) [24, 25]. Conjugation of various drugs to hydrophilic peptides such as TAT (a cell-penetrating peptide, GRKKRRQRRRPQ) is a widely adopted strategy to improve the drug's solubility, cellular uptake, and potency against the molecular target of the drug [25]. CPPs such as HIV-1 TAT, penetratin, and oligoarginine are useful tools for the intracellular delivery of therapeutic macromolecules [24–28]. CPPs enhance the absorption of biodrugs through the intestinal epithelial membrane and, therefore, overcome the low permeability of therapeutic peptides and proteins through the intestinal membranes [29].

Peptide conjugates, covalently bound to fluorescent tags, have been employed in nontoxic imaging and nuclear targeting in living cells as fluorescent markers [30]. Highly sensitive protease assays using fluorescence quenching of peptide probes based on photoinduced electron transfer have been reported to assay proteolytic enzymes with detection limits in the picomolar range [31]. McGuire et al. reported identification of tumor targeting peptide–dye conjugates effective against non-small-cell lung cancer [32]. Recently reported PEG-coated peptide-conjugated quantum dots for noninvasive near-infrared fluorescence imaging have been developed to target tumor vasculature markers in small animal models [33].

The obvious advantages of peptidomimetics and peptide conjugates over bioactive peptides spurred the development of various synthetic approaches toward novel peptide-resembling chemical entities with improved properties based on naturally occurring precursors [3, 34]. In this chapter, benzotriazole-mediated synthetic routes toward peptidomimetics and peptide conjugates will be discussed in detail.

Katritzky and group have contributed to the development of novel synthetic approaches toward the synthesis of linear [35] and cyclic peptides [36]; peptidomimetics [37]; peptide conjugates with *N*, *O*, *S*, and *C* nucleophiles [38]; and bioactive moieties [39]. New methodology, where benzotriazole was used as carbonyl group activator to conjugate the CO-active moieties to various nucleophiles, was developed and studied in detail [39, 40].

2 Benzotriazole-Mediated Synthesis of Peptidomimetics

2.1 Unnatural Amino Acids in the Benzotriazole-Mediated Synthesis of Peptidomimetics

2.1.1 Introduction of *N*-Methyl Amino Acids into Peptidomimetics

L-Amino acids, D,L-Ala, and Gly **3** were coupled with Cbz-L-*N*(Me)-Phe-Bt **2** in partially aqueous solution (MeCN/H_2O 7:3) in the presence of TEA for 1 h at room temperature. After the extractive workup, the Cbz-protected dipeptides **4** were isolated in 72–93% yields (Scheme 1) [41]. Dipeptides with sarcosine methyl ester at the C-terminus were synthesized with 51–66% yields by acylation of sarcosine methyl ester with L- and DL-amino acids [41].

Synthesis of dipeptides, containing hindered amino acids (AA) such as *N*-Me-AA, using 1*H*-benzotriazole (BtH) methodology allowed fast acylation of various amino acids at room temperature, use of water during the acylation, convenient acid/base water workup, and isolation of dipeptides **8** in yields up to 93% (Scheme 2) [41].

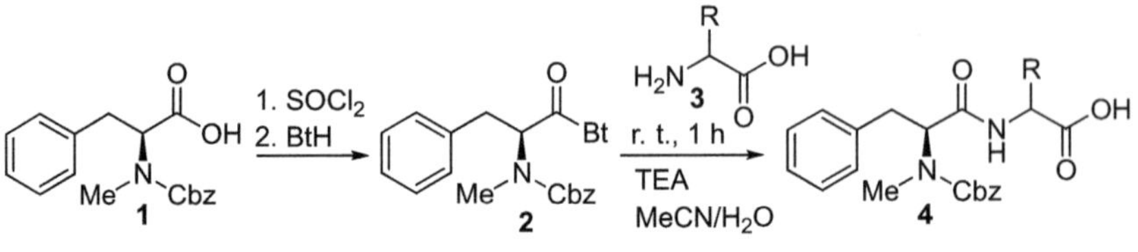

Scheme 1 Incorporation of N-Me-amino acids into peptidomimetics

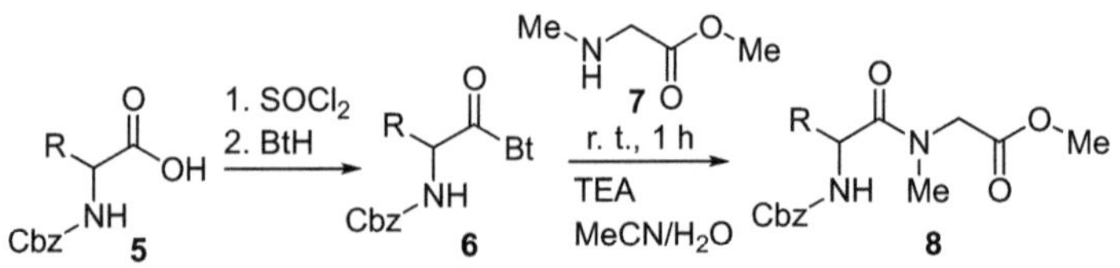

Scheme 2 RCOBt-mediated acylation of N-Me-Gly-OMe **7**

2.1.2 Introduction of Gabapentin into Peptidomimetics

Gabapentin **9** (1-(aminomethyl)cyclohexane acetic acid, marketed by Pfizer as Neurontin) is the first-line treatment for various types of neuropathic pain. Gabapentin is prone to intramolecular cyclization to form the five-membered cyclic lactam 2-aza-spiro[4,5]decan-3-one **15**, which shows LD_{50} of 300 mg/kg in rats, 25 times that of gabapentin [42]. In order to prevent intramolecular cyclization of gabapentin, its *N*-acylated prodrug gabapentin enacarbil entered the pharmaceutical market in 2013. In a similar manner, *N*-acylated gabapentin precursors were synthesized using a series of benzotriazole acylations of gabapentin. Synthesis of gabapentin-containing di-, tri-, and tetrapeptides **12** was performed via acylation of gabapentin with COBt-activated N-protected amino acids and di- and tripeptides **10** in the yields of 77–86% (Scheme 3). The Boc or Cbz groups were then removed to give final gabapentin-containing peptides in quantitative yields [43].

Carbonyl activation of N-protected gabapentin with benzotriazole under different conditions invariably led to the isolation of the cyclic by-product **18**. Boc-, Cbz-, and Phth-protecting groups were used for the N protection of gabapentin (Scheme 4) [43].

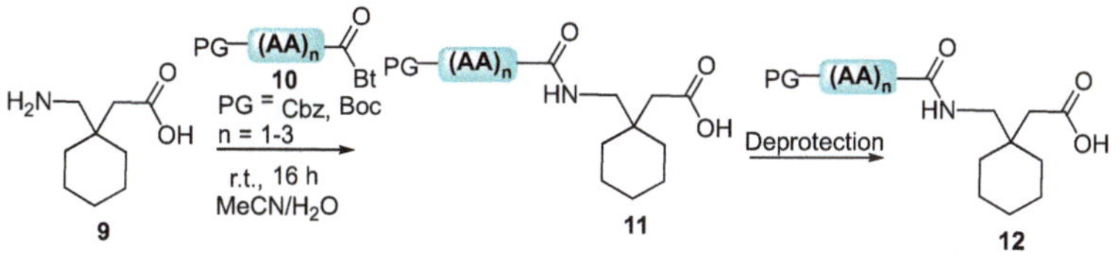

Scheme 3 Acylation of gabapentin at the N-terminus with RCOBt-amino acids and peptides

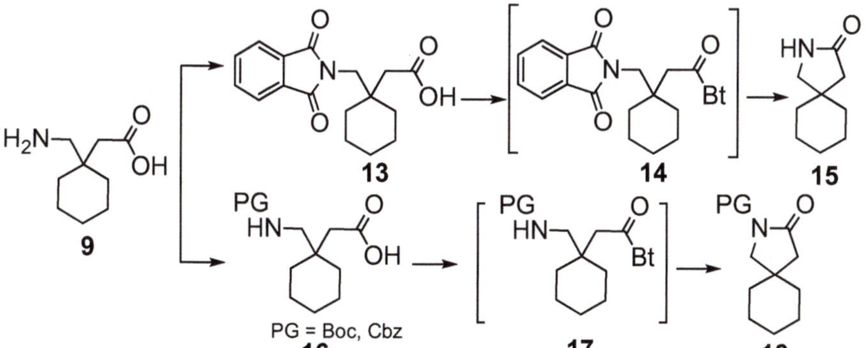

Scheme 4 RCOBt activation of gabapentin

2.1.3 Introduction of Taurine into Peptidomimetics

Taurine or 2-aminoethanesulfonic acid **19** plays a role in the modulation of the intracellular free calcium concentration. Although it is one of the few amino acids not incorporated into proteins, taurine is one of the most abundant amino acids in the brain, retina, muscle tissue, and organs throughout the body [44]. Acylation of taurine at N-terminus resulted in the synthesis of a library of 17 short peptides **20** with Boc-, Cbz-, and Fmoc-protecting group at the N-terminus side of the peptide (Scheme 5). The high hydrophilicity of taurine challenges its synthetic incorporation into peptides. Incorporation of benzotriazole into the sulfone moiety of taurine resulted in sulfonylation of amino acid esters and di- and tripeptide esters **22** in 70–86% yields [45]. Sulfonylation of *N*- and *O*-nucleophiles with *N*-Cbz-protected taurine–SO_2–Bt resulted in the synthesis of target products **23** in 62–78% yields [45]. Thus, moisture-sensitive taurine peptidomimetics **20** with various amino acids and di- and tripeptides have been synthesized and characterized. Taurine-containing peptidomimetics and sulfonopeptides mimic natural peptides and, therefore, represent attractive scaffolds for drug delivery as well as prodrug and tool applications.

2.1.4 Introduction of $C^{\alpha,\alpha}$-Dialkylglycines into Peptidomimetics

$C^{\alpha,\alpha}$-dialkylglycines, such as 2-methylalanine (Aib), diethylglycine (Deg), and isovaline (Iva), are found in peptaibols, a family of naturally occurring antibiotic peptides isolated from *Trichoderma* soil fungi effective against Gram-negative and Gram-positive bacteria [46]. Benzotriazole-mediated carbonyl activation of Aib **24** and consecutive acylation of various L-amino acids and D,L-methionine resulted in a library of dipeptides containing Aib at N-terminus **25** in 67–92% yields (Scheme 6) [41].

Acylation of Aib methyl ester **26** with PG-NH-CHR-COBt L- and D,L-intermediates **6** resulted in a library of dipeptides containing Aib at C-terminus **27** in 41–78% yields (Scheme 7) [41].

Scheme 5 Bt-mediated coupling of taurine with amino acids, peptides, and *N*-, *O*-nucleophiles

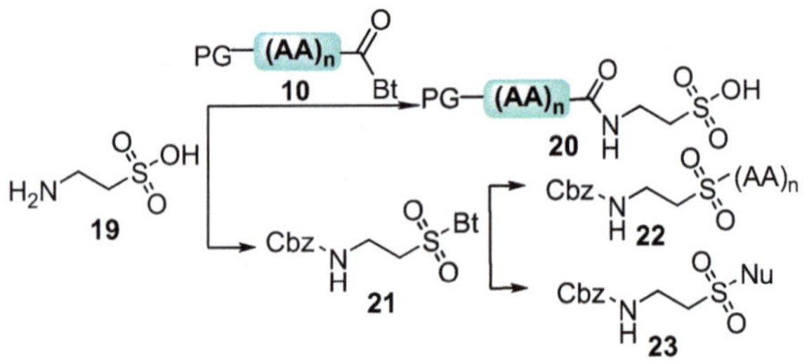

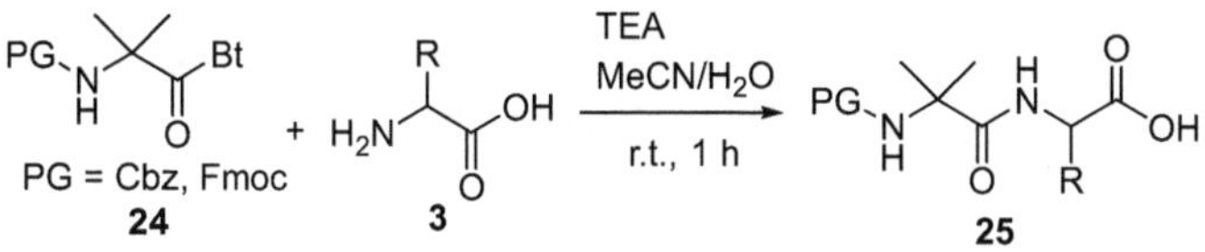

Scheme 6 Coupling of activated Aib with amino acids

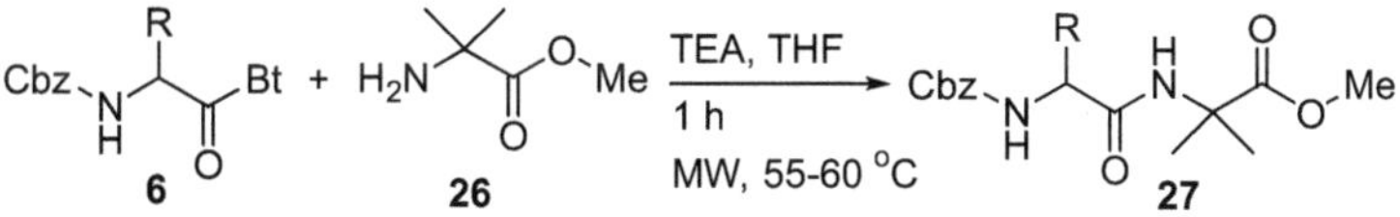

Scheme 7 Coupling of N-(Cbz-α-aminoacyl)benzotriazoles **6** with Aib-methyl ester **26**

2.2 *Incorporation of Modified Amino Acids into Peptidomimetics*

Benzotriazole-mediated coupling via COBt intermediates with amino acids and peptides has proven to be an efficient method for the construction of peptides (**A**, Fig. 1). This method has been expanded to incorporate modified amino acids into peptidomimetics. In order to satisfy the growing demand for modified peptides [3], depsipeptides (**B**, Fig. 1), aminoxypeptides (**C**, Fig. 1), oxyazapeptides (**D**, Fig. 1), hydrazinopeptides (**E**, R^2=H, Fig. 1), and azapeptides (**E**, $R^2 \neq$H, Fig. 1) have been synthesized and characterized. In all cases, the coupling between the RCOBt and the amino acid, peptide, modified amino acid, or peptidomimetic proceeded with the unambiguous formation of one reaction product in good to excellent yields depending on the nature of the product.

2.2.1 Benzotriazole-Mediated Synthesis of Depsipeptides

A depsipeptide is a peptide in which one or more of its amide (C(O)NHR-) groups are replaced by the corresponding ester, -C(O)OR. Naturally occurring bioactive depsipeptides, mainly derived from marine organisms, have shown anticancer [47] and antiviral activity [48].

Carbonyl active N-protected (depsidipeptidoyl)benzotriazoles **29** proved to be effective acylation agents in the reactions with amino acids (*N*-acylation) **3** to give depsidipeptides **30** (47–87%) under mild conditions with the reactions accelerated by microwave irradiation (Scheme 8) [49].

O-Acylations of α-hydroxycarboxylic acids **33** and unprotected depsipeptides **31** with *N*-Cbz (α-aminoacyl)benzotriazoles **6** gave corresponding depsipeptides **34** and **32** in 56–71% yields. All intermediates **31** and reaction products **32** and **34** proved to be optically pure compounds (Scheme 9) [49, 50].

N-Cbz-protected aminoxydipeptide **34** undergoes carbonyl activation under regular conditions (thionyl chloride/1*H*-benzotriazole in THF). RCOBt

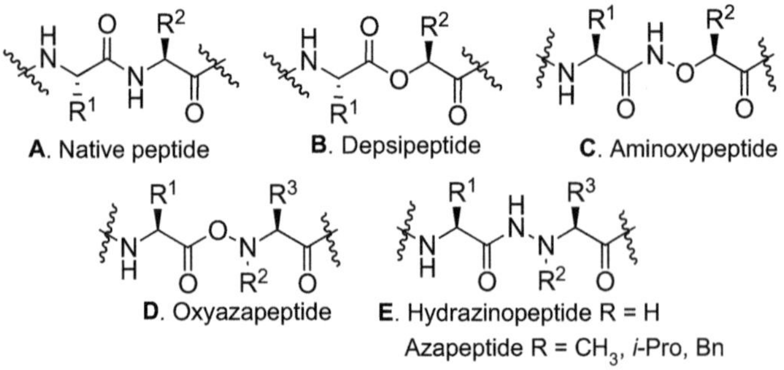

Fig. 1 General structure of a two-unit native peptide (**a**) and peptidomimetics (**b–e**)

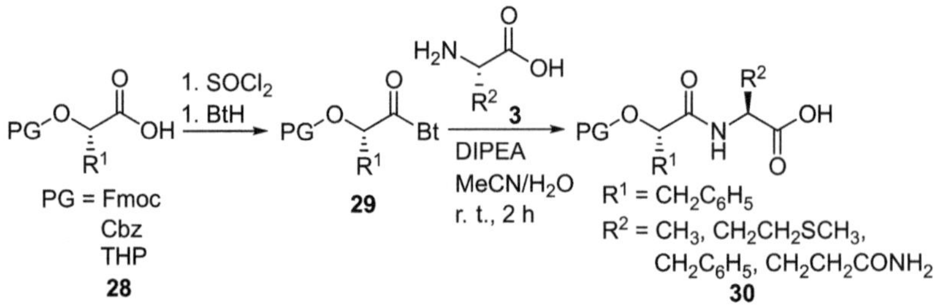

Scheme 8 Synthesis of N-terminus depsidipeptides **30**

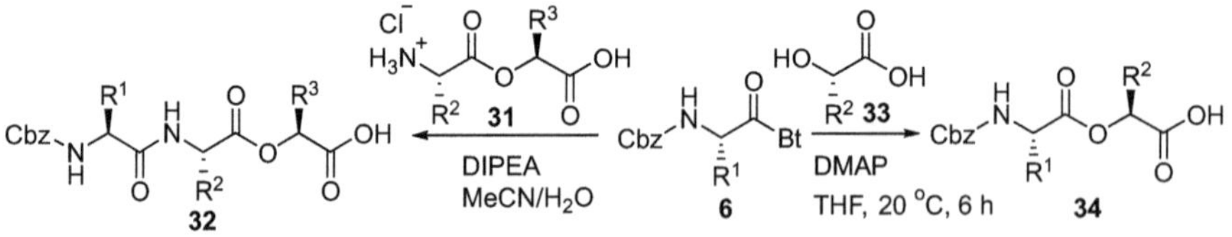

Scheme 9 Synthesis of C-terminus depsidipeptides **34** and depsitripeptides **32**

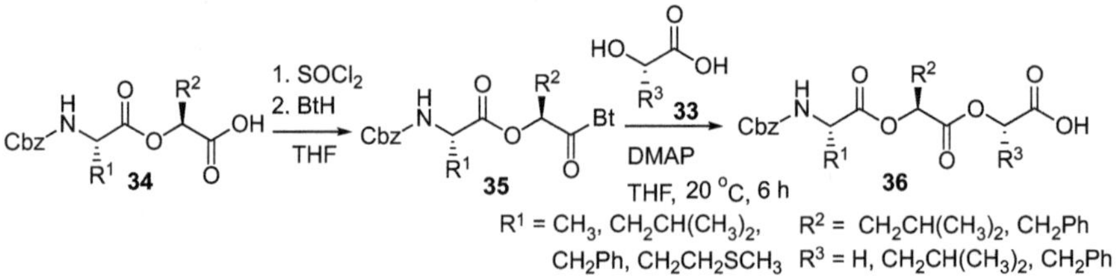

Scheme 10 Synthesis of aminoxydipeptides **35** and depsitripeptides **36**

intermediate **35** was coupled with the α-hydroxy acids (*O*-acylation) to depsitripeptides **36** in the yields of 55–78% (Scheme 10). Chiral purity of the products **36** has been confirmed with chiral HPLC [49].

2.2.2 Benzotriazole-Mediated Synthesis of Aminoxypeptides

Aminoxypeptides of general structure $[-(CO)CHR(ONH)-]_n$ can feature strong intramolecular hydrogen bonds between adjacent amino acid residues. Bioisosteric α-aminoxy acids structurally resemble the folding behavior of β-amino acids [51].

Synthesis of *N*-Cbz aminoxy-COBt **39** intermediates was conducted in three steps. The N-protecting group of phthalimide **37** was exchanged against the Cbz group to form **38**. Activation of the carbonyl group was carried out under regular reaction conditions at 0°C using thionyl chloride followed by addition of benzotriazole to give *N*-Cbz-protected aminoxy benzotriazolides **39** (Scheme 11).

Reactions of *N*-Cbz-protected aminoxy benzotriazolides **39** with non-protected aminoxy acids **40** (Scheme 12), amino acids **3**, and dipeptides **43** (Scheme 13) formed a library of optically pure aminoxy hybrid dipeptides **41** (Scheme 12), **44** (Scheme 13), and tripeptides **45** (Scheme 13) in 50–81% yields [52].

The tetrapeptide goralatide (AcSDKP) enhances the myelopoietic response to granulocyte–macrophage colony-stimulating factor [6], selectively protects murine hematopoietic progenitors and stem cells from hyperthermic damage [6], and prevents doxorubicin-induced toxicity in mice [7]. The half-life of goralatide in plasma is only 4.5 min due to intramolecular hydrolysis. Synthesis of the aminoxy analog of bioactive natural tetrapeptide goralatide employing a convenient benzotriazole methodology for linking two dipeptides to form the final aminoxy tetrapeptide has been reported recently [53]. Aminoxy tetrapeptide AcSD-NH-*O*-KP **51** has been synthesized in fourteen steps. RCO-Bt-activated *N*-acylated O-protected dipeptide **48** was coupled with the aminoxy-Lys-$^{\gamma}N$-Cbz-Pro-*t*-Bu ester **49** (Scheme 14). After the two steps of deprotection, the final L-Lys aminoxytetrapeptide **51** was isolated and its stability was studied. Comparative analysis between the aminoxy tetrapeptide and native goralatide in various solvents showed no decomposition of the analog tetrapeptide during deprotection or a long (7 days) exposure to polar

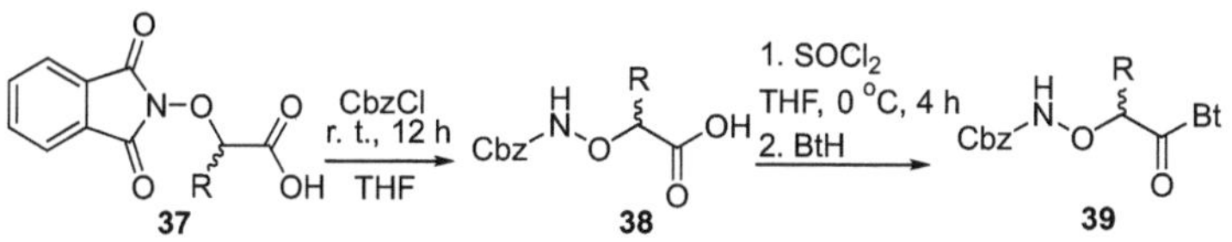

Scheme 11 Synthesis of PG-NH-(α-aminoxyacyl)benzotriazoles **39**

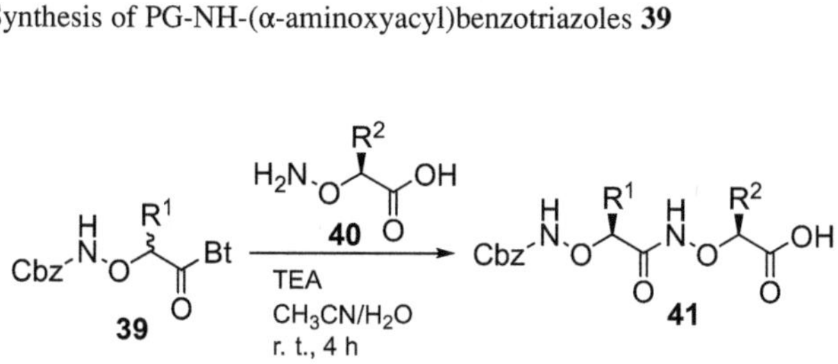

Scheme 12 Synthesis of di-aminoxy acid peptidomimetics

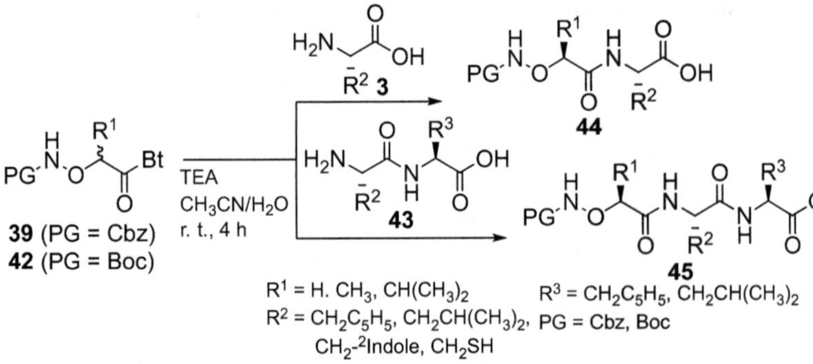

Scheme 13 Synthesis of mono-aminoxyacid di- and tri-peptides

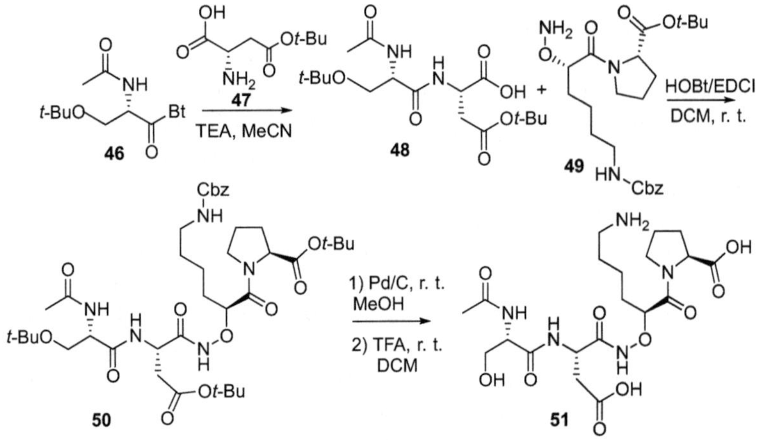

Scheme 14 Synthesis of aminoxy-Lys peptidomimetic of goralatide **51**

solvents (H_2O, MeOH, DMSO) compared to decomposition of goralatide in solutions (2 h in MeOH or DMSO) [53].

2.2.3 Synthesis of Oxyazapeptides Using Benzotriazole Methodology

Biswas et al. recently reported de novo synthesis of oxyazapeptides in which an amino acid is replaced by an aza-hydroxy acid [54]. Conformational analysis of oxyazapeptides revealed that the peptidomimetics of this class should adopt a β-turn secondary structure with greater conformational freedom. This would allow for adaptation to varying steric demands of biological interaction [54]. Oxyazapeptides can be considered as the depsipeptide analogues of azapeptides where the α-amino group of an aza-amino acid is replaced by a hydroxyl group [37].

α-Amino acid ester hydrochloride salts **52** were converted into the acyl imidazoles **54** by their reaction with *N,N′*-carbonyldiimidazole (CDI) **53** in the presence of DIPEA (Hünig's base) in dry DCM. Intermediates **54** were then reacted with *N*-methylhydroxylamine (R^3=Me) or *N*-benzyl hydroxylamine (R^3=Bn) **55** at room temperature for 16 h in dry THF with 1 eq. of base to give oxyazadipeptides **56** in 85–92% yields (Scheme 15).

Various (α-aminoacyl)benzotriazoles **57** were coupled with free oxyazadipeptides **58** in dry THF containing 1 eq. of DIPEA and a catalytic amount of DMAP to give N-protected oxyazatripeptide esters **59** in 85–93% yields (Scheme 16). Chiral HPLC studies of **59** showed that no racemization occurred during coupling of **57** and **58** [54].

Oxyazatetrapeptide esters **62** were prepared in solution by treatment of *N*-PG-(α-dipeptidoyl)benzotriazoles **61** with free oxyazadipeptides **61** in THF containing 1 eq. of DIPEA and a catalytic amount of DMAP for 16 h at room temperature. The yields of the products were within a range of 84–90% (Scheme 17) [54].

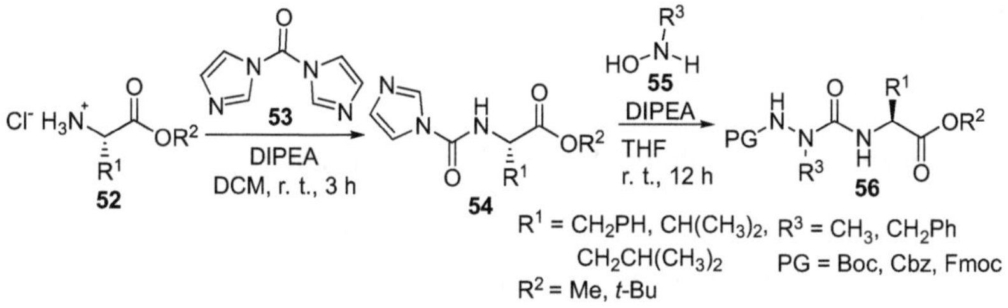

Scheme 15 Synthesis of oxyazadipeptides **56**

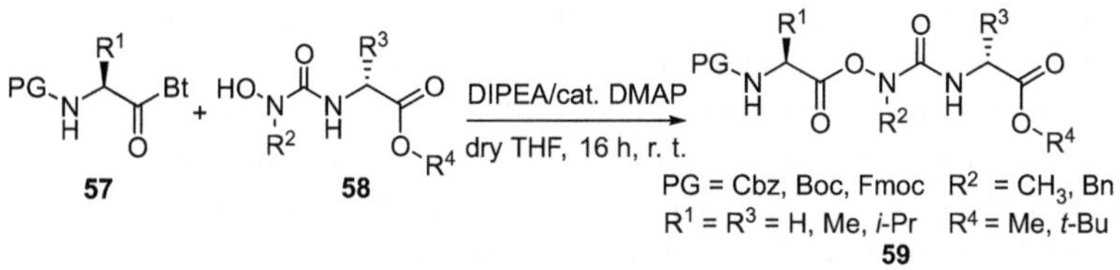

Scheme 16 Synthesis of oxyazatripeptides **59**

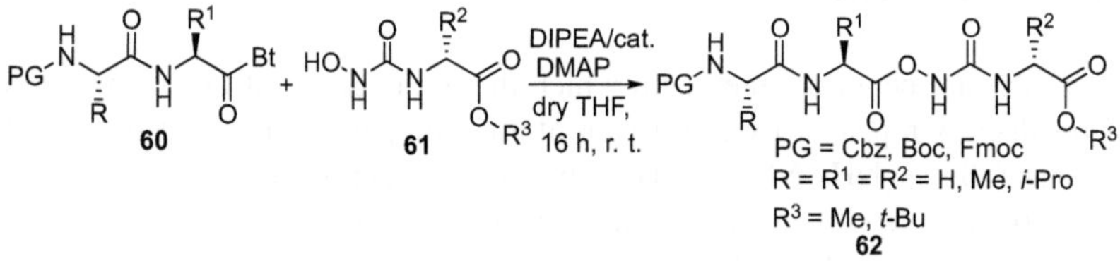

Scheme 17 Coupling of PG-$(AA)_n$-Bt active intermediates **60** with oxyazapeptides **62**

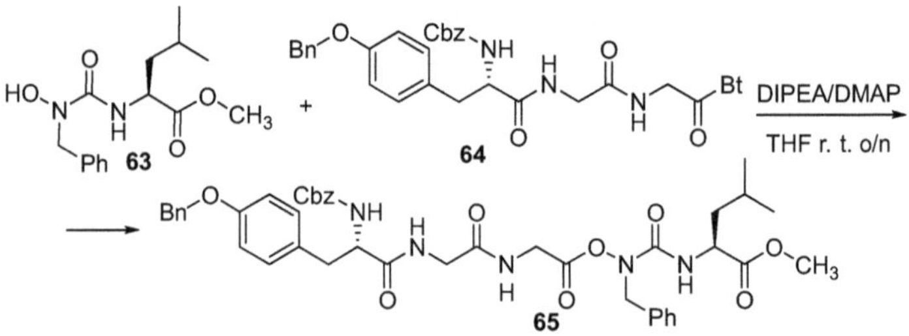

Scheme 18 Construction of the oxyaza analog of leu-enkephalin **65**

The new methodology for the synthesis of oxyazapeptides was validated via the synthesis of oxyaza analogs of leu-enkephalin **65**. Tripeptide **65** was assembled using benzotriazole methodology. After the carbonyl activation with benzotriazole, intermediate **64** was coupled with an oxyaza-Phe-L-Leu-OMe unit **63** to give the peptidomimetic **65** (85%), consisting of five residues, one of which was oxyaza-phenylalanine (Scheme 18). Thus, the new methodology proved to be effective for the solution phase synthesis of chirally pure peptidomimetic sequences in high yields [54].

2.2.4 Benzotriazole-Mediated Synthesis of Hydrazinopeptides

Hydrazinopeptides are those peptides where some C^{β} atoms in a β-amino acid were replaced by a nitrogen atom. The additional hydrazino α-NH group takes part in special structuring effects, such as the tendency toward nanotubular self-organization in solution and solid states [55] and the mimicking of β-turns [56]. Hydrazinopeptides have been used as tags for site-specific protein labeling [57], PNA dimers [58], and have also been considered as potential anticancer drugs [59].

The coupling reaction between the *N*-Cbz-protected α-hydrazino acids **66** and α-aminoacyl benzotriazoles **67** was done with microwave irradiation of the reaction mixture in THF (70°C and 65 W) for 15 min to give new hydrazino hybrid dipeptides **68** (42–71%). No side products were observed after the couplings except minimum traces of unreacted starting materials (Scheme 19) [60].

2.2.5 Benzotriazole-Mediated Synthesis of Azapeptides

Azapeptides are peptidomimetics in which the α-CH group of one or more amino acid residues is replaced by a nitrogen atom. The general formula of aza-amino acids is as follows: NH_2-NR-COOH. In biologically active peptide analogs, the aza substitution has led to enhanced activity and selectivity as well as improved properties, such as prolonged duration of action and metabolic stability [61].

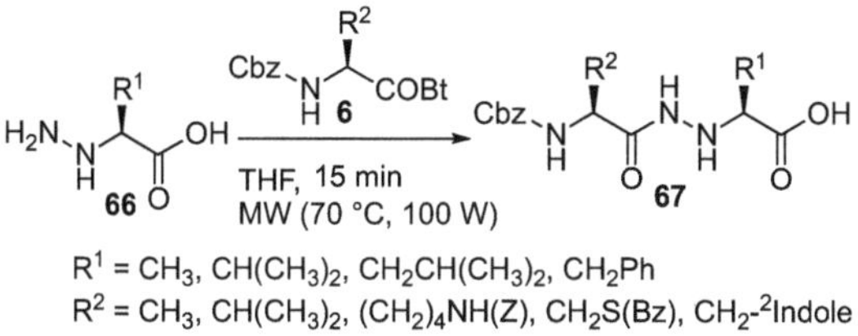

Scheme 19 Synthesis of hydrazino peptides **67** using benzotriazolyl-mediated acylation

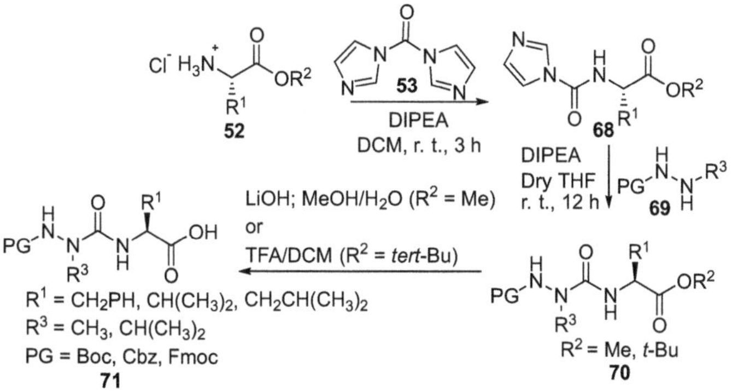

Scheme 20 Synthesis of azadipeptides 71

Azadipeptides were synthesized via activation of the amino acid ester hydrochloride salts **52** by carbonyldiimidazole (CDI) **53** in the presence of 2.5 molar eq. of DIPEA in dry DCM at room temperature for 3 h to obtain the active carbamates **68**. Reaction of imidazolyl intermediates **68** with *N*′-alkyl-*N*-PG-hydrazines **69** at room temperature for 16 h in dry THF in the presence of 1.0 eq. of DIPEA provided protected azadipeptide esters **70**, which were isolated and purified via extraction in the presence of 2N HCl. PG-azadipeptides **71** were either isolated after the hydrolysis of *N*-PG-azadipeptide methyl esters with LiOH in methanol/water mixture or cleavage of the *tert*-butyl ester with trifluoroacetic acid in DCM (Scheme 20) [62].

N-(*N*-PG-azadipeptidoyl) benzotriazoles **77–81** were synthesized in 81–92% yields by treatment of *N*-PG-azadipeptides **72** with 3.0 eq. of 1*H*-benzotriazole, 1.0 eq. of thionyl chloride, and 2.0 eq. of DIPEA in DCM at -30°C (Scheme 21) [62]. Further coupling of active *N*-(*N*-PG-azadipeptidoyl)benzotriazoles **72** with amino acids **3**, dipeptides **73**, aminoxyacetic acid **74**, depsidipeptide **75**, and α-hydroxy-β-phenylpropionic acid **76** led to the formation of aza-peptidomimetics **77** (80–90%), **78** (77–87%), **79** (90%), **80** (85%), and **81** (87%) (Scheme 21).

This methodology allowed for the synthesis of the novel-protected aza analogue of the endogenous opioid peptide neurotransmitter leu-enkephalin, found in animals and humans [63]. A protected analog of aza-leu-enkephalin **84** was

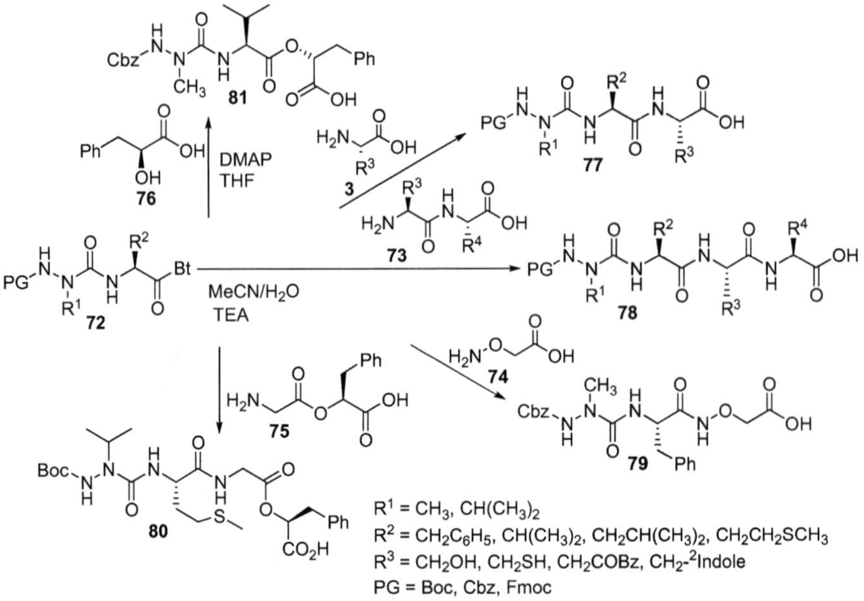

Scheme 21 Coupling of *N*-(*N*-PG-azadipeptidoyl) benzotriazoles **72** with amino acids **3**, dipeptides **73**, aminoxyacetic acid **74**, depsidipeptide **75**, and α-hydroxy-β-phenylpropionic acid **76**

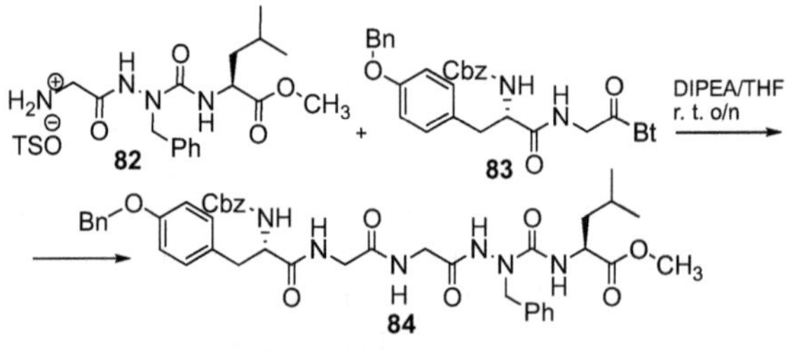

Scheme 22 Synthesis of aza analog of leu-enkephalin **84**

synthesized in 4 steps in 70% yield by coupling of two dipeptide building blocks **82** and **83** to construct the peptidomimetic **84** with aza-phenylalanine linked to the L-leucine methyl ester (Scheme 22) [62].

2.2.6 Benzotriazole-Mediated Synthesis of Azidopeptides

Benzotriazolyl-1-yl sulfonyl azide **86** was prepared from the reaction of sodium azide, sulfuryl chloride, 1*H*-benzotriazole **85** (2 eq.), and pyridine in MeCN (70%) [64]. This diazo transfer reagent was then reacted with L-amino acids and D,L-amino acids in aqueous MeCN in the presence of copper sulfate and amine base to give the

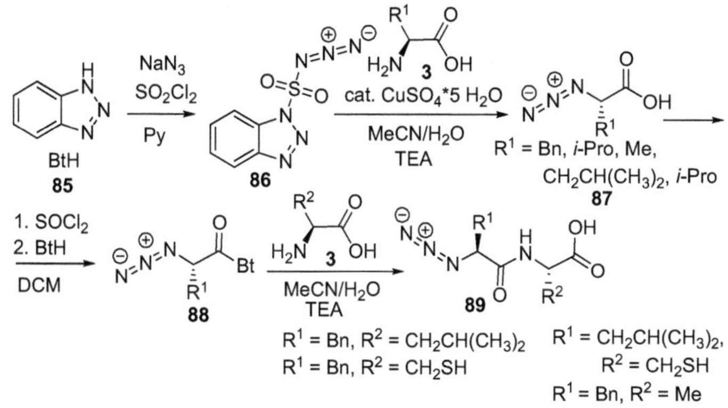

Scheme 23 Synthesis of azidodipeptides **89**

corresponding optically pure α-azido acids **87** in 60–87% yields (Scheme 23). *N*-(α-azidoacyl)benzotriazoles **88** were prepared in 65–98% yields by treatment of corresponding α-azido acids **87** with 1.2 eq. of thionyl chloride and 2 eq. of 1*H*-benzotriazole in DCM. The reaction of *N*-(α-azidoacyl)benzotriazoles **88** with L-amino acids (L-Leu, L-Cys, L-Ala) **3** gave azidodipeptides **89** in 80–87% yields (Scheme 23). Chiral HPLC studies confirmed that no loss of chirality occurred [64].

2.3 Benzotriazole-Mediated Synthesis of Isopeptides

Isopeptides have been found to regulate important metabolic processes in mammalian bodies. Endo-ε(γ-Glu)-Lys isopeptides have been reported to dissolve blood clots [65], and N^{γ}(γ-glutamyl)-lysine isopeptide bonds were detected in histological sections from benign and malignant human breast tissue [66]. An additional amino group, in *N*-, *O*-, or *S*-acyl isopeptides, generally increases the hydrophilicity of the molecule, and this is advantageous in effecting purification of isopeptides by HPLC. The native peptides can then be generated from the corresponding *N*-, *O*, or *S*-acyl isopeptide via an *N* to *N* [67], *O* to *N* [68], or *S* to *N* [40] intramolecular acyl migration reaction to form native peptides [36, 69].

2.3.1 Synthesis of *N*-Acyl Isopeptides

Synthesis of several *N*-acyl isopeptides that include tryptophan residue and *N*- to *N*-acyl migrations via 7-, 10-, 11-, and 12-membered transition states have been recently reported by Katritzky et al. [67].

N-Boc-protected Trp benzyl ester **90** was acylated at the indole nitrogen with Cbz-Ala-Bt **6** in the presence of DBU to afford Boc-protected mono-isopeptide **91** (78%), which after deprotection with HCl in 1,4-dioxane gave the hydrochloride **92**

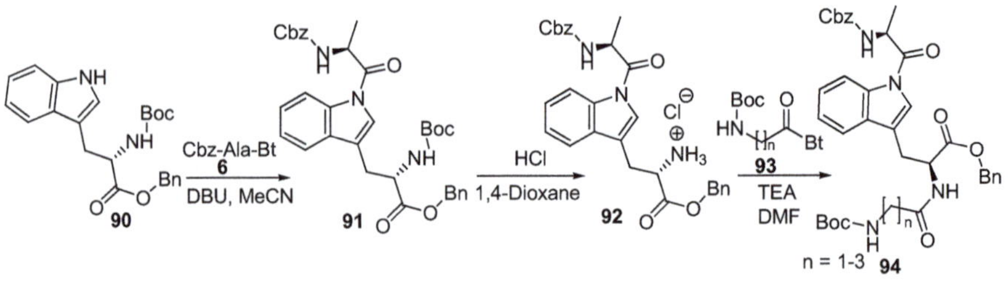

Scheme 24 Synthesis of *N*-acyl isopeptides **94**

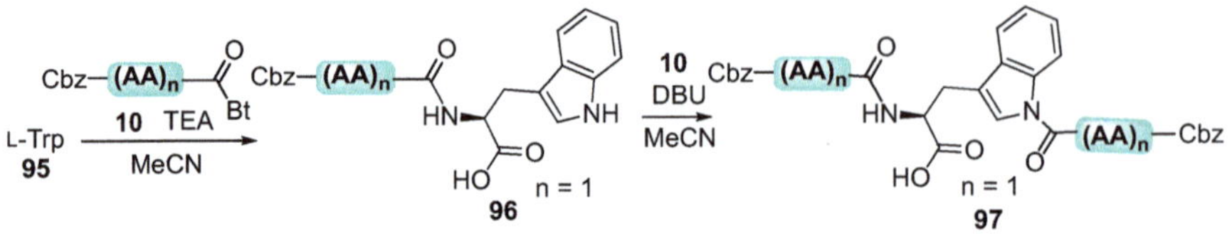

Scheme 25 Synthesis of L-Trp-based monoisopeptides **97**

(90%). Further coupling of the diisopeptide **92** with α-, β-, and γ-amino carbonyl benzotriazolides **93** was carried out in 63–76 % yields (Scheme 24). *N*-indole acylated isopeptides **94** were then studied in native chemical ligation experiments [67].

Katritzky et al. reported solution-phase benzotriazole-based synthesis of *N*-acyl isopeptides **94** [70]. Synthesis of chirally pure isotripeptides, containing an L-Trp residue, was carried out in a two-step benzotriazolyl-mediated acylation of L-Trp **95** at its $^{\alpha}$N position and indole moiety with *N*-Cbz-protected L-amino acids **10** in 75–79% yields [70] (Scheme 25).

2.3.2 Synthesis of *O*-Acyl Isopeptides

O-Acylation of N-protected L-serine and N-protected L-threonine **98** with various *N*-PG-(α-aminoacyl) benzotriazoles **6** in the presence of DIPEA in MeCN at room temperature for 12 h gave *O*-acyl isoserine and *O*-acyl isothreonine dipeptides **99** (Scheme 26) [71].

Katritzky et al. reported a stepwise benzotriazole-mediated synthesis of tyrosine-acylated di- and triisopeptides [68]. First, carbonyl active Boc-protected Gly ($n=1$), β-Ala ($n=2$), and GABA ($n=3$) **93** were coupled with L-Tyr **100** at room temperature in the presence of DBU to give the corresponding Boc-protected dipeptides **101** (69–84%). These intermediates **101** were then *O*-acylated by Cbz-L-Ala-Bt in the presence of TEA to provide N-protected monoisotripeptides **102** (71–84%), which after deprotection by HCl solution in 1,4-dioxane yielded the free monoisotripeptide hydrochlorides **103** (96–98%). *N*-Boc-protected

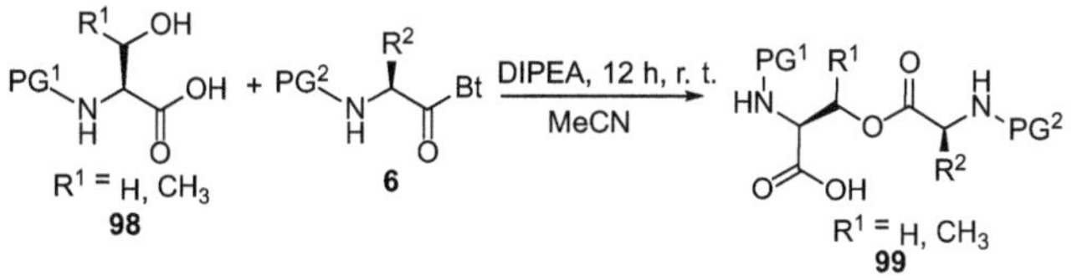

Scheme 26 Synthesis of *O*-acyl isopeptides **99**

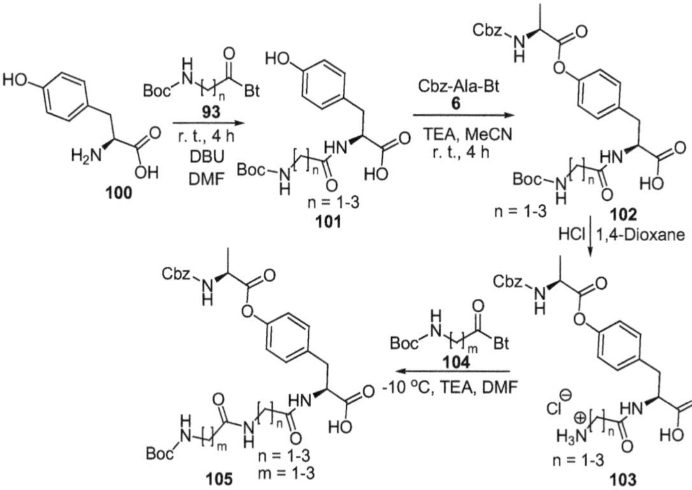

Scheme 27 Synthesis of isodi- and tripeptides **103** and **105**

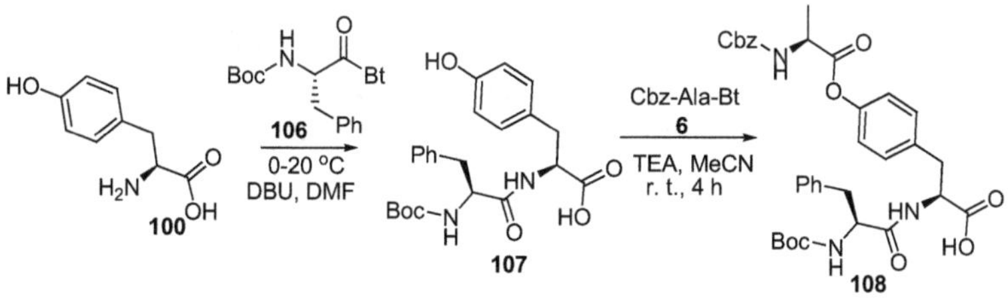

Scheme 28 Synthesis of L-Tyr-based monoisopeptides **108**

monoisotetrapeptides **105** were synthesized in solution phase, by coupling the benzotriazolides **104** with the unprotected monoisotripeptides **103** at −10°C in 96, 98, and 97% yields, respectively (Scheme 27). Synthesized di- and triisopeptides were used for further experiments in 12- and 19-membered native chemical ligation [68].

Katritzky et al. reported a benzotriazole-mediated two-step synthesis of L-tyrosine-based isopeptide **108**, which was synthesized in 65% yield and consisted of *N*-Boc-Phe, L-Tyr, and *N*-Cbz-Ala units [70] (Scheme 28).

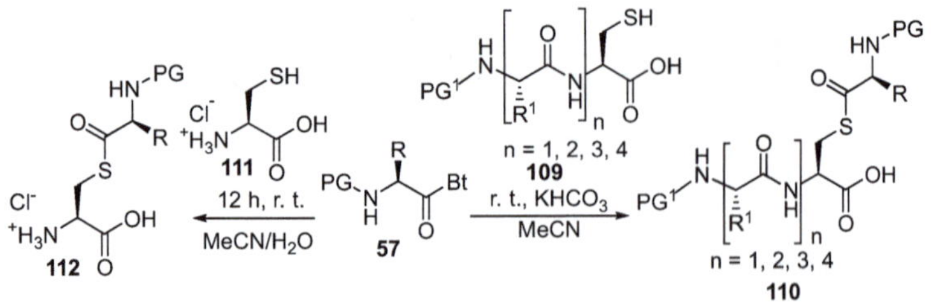

Scheme 29 Synthesis of *S*-acyl isopeptides **110** and **112**

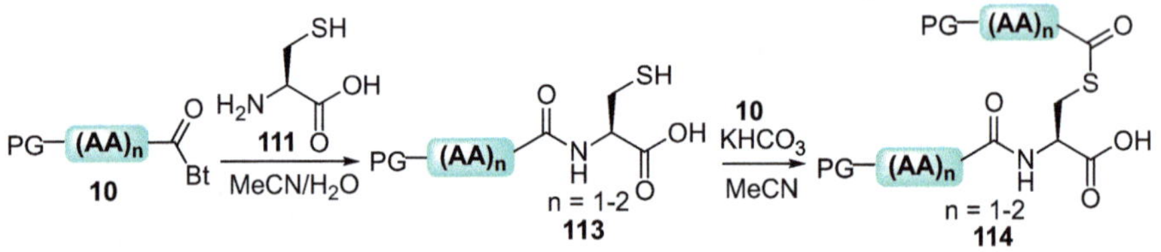

Scheme 30 Synthesis of *S*-acyl isopeptides **114**

2.3.3 Synthesis of *S*-Acyl Isopeptides

S-Acyl isopeptides are usually less prone to aggregate in solution and therefore easier to synthesize and purify relative to the corresponding native peptides [72]. *S*-Acylation of cysteine-containing peptides **109** with **57** was carried out in the presence of $KHCO_3$ at room temperature in MeCN to form **110** (Scheme 29) [40, 72, 73]. Selective *S*-acylation of L-cysteine **111** to yield **112** was also carried out in MeCN/H_2O mixture in the absence of base (Scheme 29) [74].

Katritzky et al. reported the synthesis of *S*-acylated isopeptides **114**. Two-step benzotriazole-mediated acylations of L-cysteine **111** resulted in the synthesis of eight L-cysteine-containing tri-, tetra-, and pentapeptides **114** in 45–97% yields (Scheme 30) [70].

2.4 *Benzotriazole-Mediated Synthesis of Cyclic Peptidomimetics*

2.4.1 Synthesis of Bis-2,5-Diketopiperazines with Symmetrical and Unsymmetrical Linkers

2,5-Diketopiperazines (2,5-DKPs) are found in numerous natural products, often as such, but also embedded in larger, more complex molecular architectures in a variety of natural products from fungi, bacteria, the plant kingdom, and mammals [36, 75].

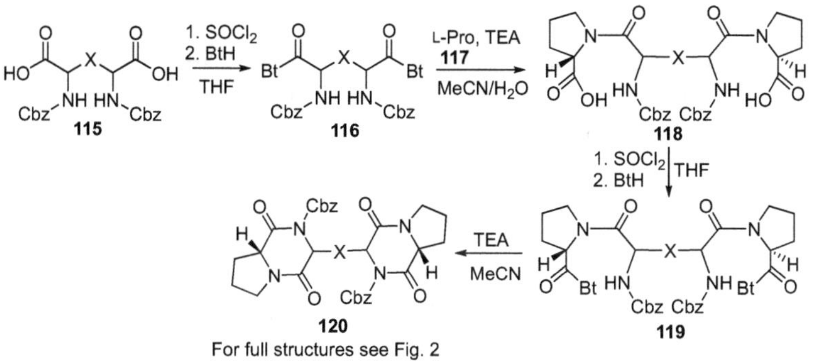

Scheme 31 Synthesis of symmetrical bis-DKPs **120**

A high yield four-step synthesis of bis-2,5-DKPs with symmetrical and unsymmetrical linkers has been reported by Katritzky et al. [76]. Synthesis of 2,5-bis-DKP linked via symmetrical linkers was as follows: (i) reaction of di-COBt linkers with D-proline was complete within 3 h at room temperature and produced dipeptide dimers bis-Cbz-linker-D-Pro-OH **118**; (ii) the reaction of **118** with BtS(O)Bt, generated in situ in dry THF, went without epimerization to give Cbz N-protected dipeptidoyl benzotriazolides **119**; (iii) the final step was intramolecular cyclization of formed Cbz-protected diketopiperazines **120** (Scheme 31) [76]. The authors noted that introduction of D-proline as turn introducer conformationally assisted the processes of cyclization and provided the formation of a library of bis-2,5 DKPs **120** in high yields (72–91%) (Scheme 31).

Introduction of unsymmetrical moieties as linkers for 2,5-diketopiperazines has been carried out in a similar 4-step manner. Reported temperatures for the carbonyl activation with benzotriazole were lower than those used for the activation of intermediates with symmetrical linkers. Step 2 was carried out under –20°C and step 4 under –45°C to give products 120 in 72–91% (Scheme 31) [76]. Structures of the cyclized 2,5-DKP products **120a–j** are shown in the Fig. 2.

2.4.2 Synthesis of Proline- and Hydroxyproline-Containing Hetero-2,5-Diketopiperazines

Hetero-2,5-diketopiperazines (2,5-DKPs) with L-proline and hydroxy-L-proline residues represent a unique class with distinctive structural detail and biological activity [75]. Marine-derived L-proline-containing D,D-diketopiperazines showed strong antibiotic activity against *Vibrio anguillarum* [77].

Katritzky et al. reported a synthesis of a library of 2,5-DKPs – analogs of the natural products and metabolites using the original benzotriazole DKP methodology for both – construction of peptidomimetics and their cyclization into DKP systems [78]. Marine microbial metabolite cyclo-[D-Ala–L-Pro-] **124** was selected as the first target, and it was synthesized in a 4-step procedure (Scheme 32). First,

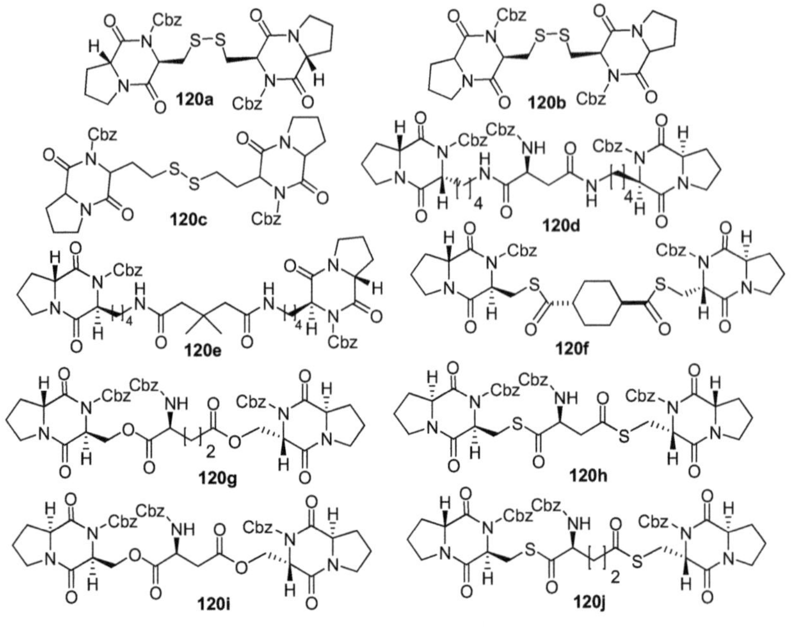

Fig. 2 Structures of symmetrical and unsymmetrical bis-DKPs **120a–j**

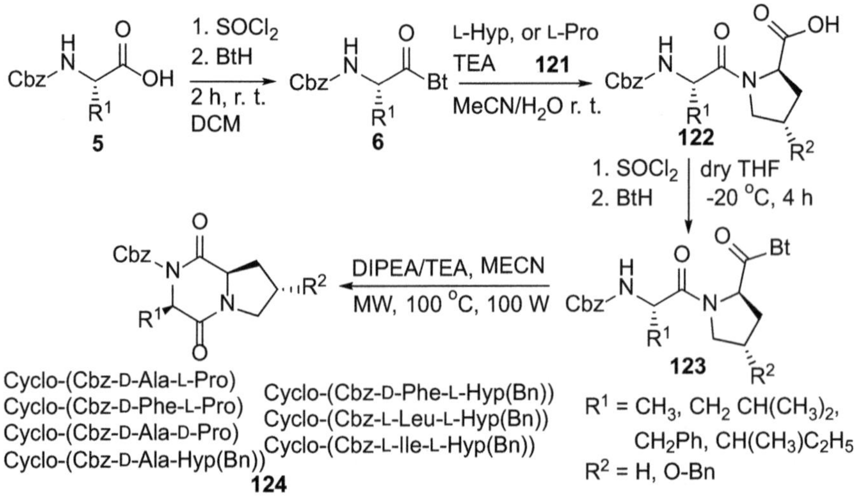

Scheme 32 Synthesis of proline- and hydroxyproline-containing hetero-2,5-diketopiperazines **124**

N-Cbz D-Ala **125** was converted into the benzotriazolide **6** in 83% yield. Reaction of RCOBt intermediate **6** with L-Pro or L-Hyp **121** gave **122**, which was converted into Cbz-dipeptidoyl benzotriazolide **123**. Treatment of **123** under MW irradiation

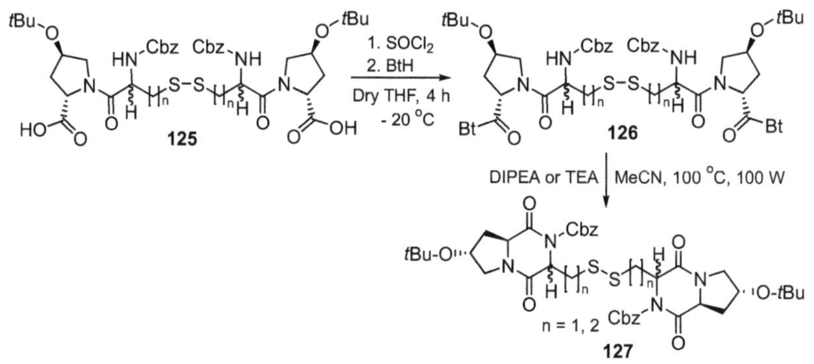

Scheme 33 Synthesis of symmetrically linked hydroxyproline-containing hetero-2,5-diketopiperazines **127**

at 100°C in MeCN in the presence of TEA for 3 h afforded cyclo-(Cbz-D-Ala–L-Pro) **124** in 62% yield (Scheme 32). A library of proline and hydroxyproline-derived cyclic peptidomimetics **124** was synthesized in 62–83% yield to prove the new concept of benzotriazole-mediated DKP cyclization [36, 78].

The lactamization reaction to form dimeric DKPs **127** was a three-step process starting with L-cysteine or homocysteine-linked dipeptides **125** followed by the di-CO activation with 1*H*-benzotriazole into **126** and cyclization under microwave irradiation at 100°C in MeCN to form **127**. Two products **127** were isolated in 84% ($n = 1$) and 88% ($n = 2$) yields. The new methodology made natural DKPs and their dimers easily synthesizable in high yield with retention of chiral purity of the mono- and dimeric products **127** (Scheme 33) [78].

2.5 *Synthesis of Bioactive Peptidomimetics Using Benzotriazole Methodology*

2.5.1 Synthesis of Bioactive Linear Peptidomimetics

The non-stable natural tetrapeptide goralatide (AcSDKP) **130** is a selective inhibitor of primitive hematopoietic cell proliferation [79]. Katritzky et al. reported synthesis of a stable precursor of goralatide **129** that exhibits cytotoxicity toward human myeloid HL-60, HEL, NALM-6 leukemia cells, endothelial HUVEC, and glioblastoma U251, and transformed kidney 293T cells have been reported [80]. N- and O-protected tetrapeptide **129** was synthesized in six steps as a precursor to NAcSDKP **130**. Dipeptide **48** was used to construct the core of protected goralatide tetrapeptide precursor **129**. To obtain the active tetrapeptide **129**, trifluoroacetate dipeptide **128** was reacted with the dipeptide **48** (HOBt/EDCI,

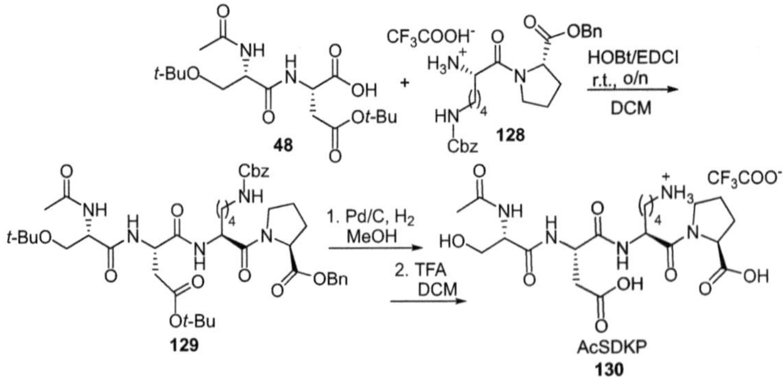

Scheme 34 Synthesis of the bioactive goralatide precursor **129**

r.t. 12 h, 80%) to give O- and N-protected NAcSDKP precursor **129** which, after two deprotection steps, gave the final product AcSDKP **130** (Scheme 34) [80].

In vitro cytotoxicity assays demonstrated that N- and O-diprotected AcSDKP precursor **129** effectively kills cancer cells. It showed highest cytotoxicity toward glioblastoma U251 and kidney 293T cells. Stable protected goralatide precursor was also effective against HL-60, NALM-6 and HEL leukemia cells, and HUVEC endothelial cells at a concentration of 50 μM and toward NALM-6 leukemia cells at a concentration of 20 μM. The high anticancer activity profile of **129** suggests that it could be developed into a novel anticancer and antileukemic drug.

2.5.2 Synthesis of Bioactive Cyclic Peptidomimetics

Katritzky et al. reported the synthesis of macrocyclic peptoids by *S*-acylation of cysteine esters [81]. Optimized selective *S*-acylation of L-cysteine esters gave intermediates **130** for the synthesis of macrocyclic peptoids **132** by a benzotriazole-based method. Cyclization of bis-L-cysteine esters **130** with CO-active di-benzotriazolides **131** under microwave irradiation formed macrocyclic products **132** in 59–92 % yields (Scheme 35) [81].

Mild *S*-acylation of cysteine **111** with diacyl benzotriazoles **133** led to the formation of intermediate **134**, two NH_2 groups of which were acylated with various symmetrical diacyl benzotriazoles **136** (Scheme 36). This led to intramolecular cyclization and the formation of cyclic peptidomimetics **137** in the yields of 64–82%, depending on the R of the final macrocycle (Scheme 36). Screening of two cyclic peptidomimetics **137** ($R=C_6H_5$, 2-pyridine) revealed mild activity against *Bordetella bronchiseptica*, *Micrococcus luteus*, and *Salmonella typhimurium* [82].

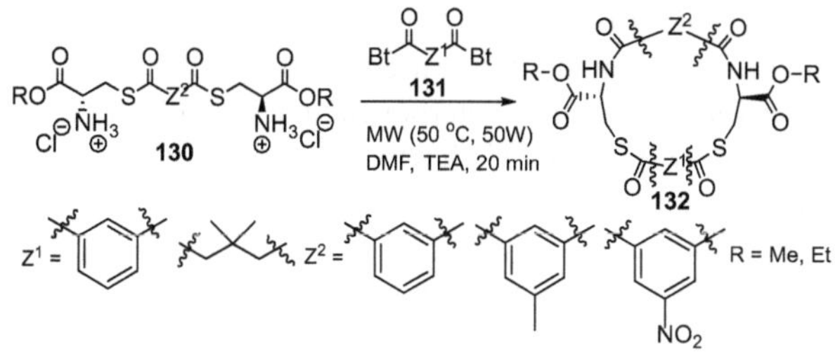

Scheme 35 Synthesis of L-cysteine ester-acylated cyclic peptidomimetics **132**

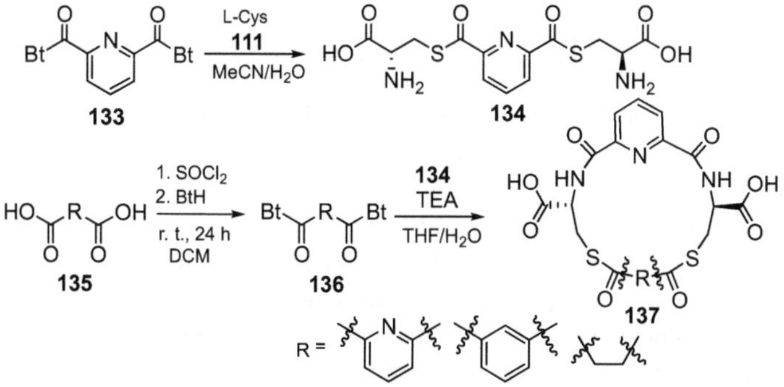

Scheme 36 Synthesis of cyclic peptidomimetics **137**, active against *Bordetella bronchiseptica*, *Micrococcus luteus*, and *Salmonella typhimurium*

3 Conjugation of Aminoacyl/Oligopeptidoyl Benzotriazoles to Nucleophilic Moieties

Irreversible acylation of nucleophilic reagents is crucial for the introduction of RCO moieties into less reactive targets or biologically important nucleophilic targets [83–85]. Katritzky et al. developed an efficient method for the carbonyl activation with benzotriazole, which provides straightforward acylation of various N-, O-, S-, and C-nucleophiles in high yield [39, 86–88]. This methods allows the use of water during the reaction and results in the high-yield formation of the acylated products, which are isolated via an easy workup [39, 86].

3.1 Synthesis of Amino Acid and Peptide N-, O-, S-, *and* C-*Conjugates*

Katritzky's group has established the method of RCOBt activation of the carboxylic acids in amino acids or oligopeptides. This method allowed the high-yield activation of the COOH group with 1*H*-benzotriazole and further couplings with various nucleophiles. This method was used for the activation of carbonyl groups [89–93], regiospecific acylation of heterocyclic nucleophiles [94, 95], synthesis or substituted ureas [96–98], carbonyl diactivation of dicarboxylic acids [99], and synthesis of β-keto esters and β-diketones [100]. Katritzky et al. analyzed in detail the extension of acylation with acylbenzotriazole toward the synthesis of peptide, peptidomimetics, and *N*-, *O*-, *S*-, and *C*-conjugates [101].

3.1.1 Amino Acylation of *N*-Nucleophiles

Katritzky et al. developed a simple and efficient method for the preparation of primary, secondary, and tertiary amides by the treatment of *N*-acylbenzotriazoles with ammonia, primary and secondary amines, respectively [86]. Advantages of this procedure included the following: (1) neutral reaction conditions were useful for ammoniation and amination of compounds possessing acid- or base-sensitive substituents; (2) the use of acyl chlorides was avoided; (3) most *N*-acylbenzotriazoles were purified via recrystallization and were stable to storage at room temperature over months; (4) product isolation and workup was limited to extraction; (5) primary, secondary, and tertiary amides were generally obtained in good to excellent yields; and (6) the method was also applied toward the synthesis of *R*-hydroxyamides and perfluoroalkylated amides. This method was then further extended to the synthesis of peptides and peptidomimetics (Method A, Scheme 37) [37, 39].

Katritzky et al. reported the synthesis of *N*-conjugates of various *N*-nucleophiles (D, L-PhCH(Me)NH_2, L-PhCH(Me)NH_2,L-PhCH(Me)-NH_2, D,L-PhCH(Me)NH_2, *N*-(3-aminopropyl)-imidazole, 2-aminopyridine, 2-amino-6-methoxy-benzothiazole, *N*-methylpiperazine) with tri- and tetrapeptides (Cbz-L-Ala-L-Phe-Gly, Cbz-L-Val-L-Phe-Gly, Cbz-L-Ala-L-Phe-Gly-L-Ala, Cbz-L-Phe-Gly-L-Leu-Gly) (Method **A**, Scheme 37) [88]. The reactions were conducted either at room temperature or under microwave irradiation. Peptide conjugates were isolated in 60–75% yields [88].

The C-terminus conjugation of amino acids to *N*-nucleophiles resulted in the synthesis of amide conjugates (Method **A**, Scheme 37). Since amino acids and peptides are also *N*-nucleophiles, they can be conjugated to the functionalized moieties through the N-terminus side using benzotriazole-mediated N-terminus RCOBt couplings (Method **B**, Scheme 37).

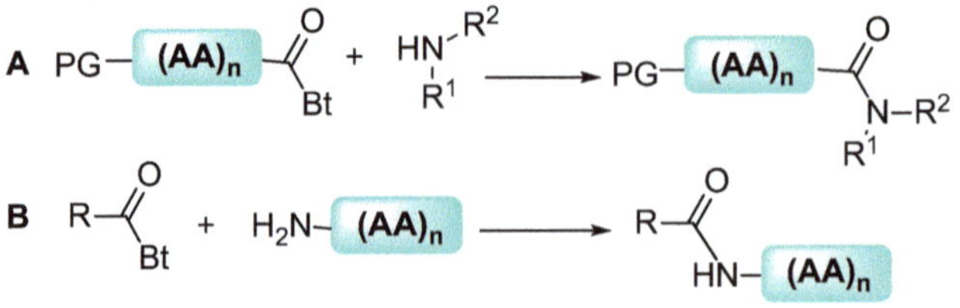

Scheme 37 General scheme of the two possible types of acylations (A and B), between amino acids or peptides and *N*-nucleophiles

Katritzky et al. reported synthesis of *N*-protected amino acid mesalazine conjugates in water using microwave irradiation. 5-Aminosalicylic acid (5-ASA, mesalazine) was acylated at the N-terminus with a variety of *N*-Cbz amino acids, several dipeptides, and one tripeptide (Cbz-Gly, Cbz-L-Ala, Cbz-D,L-Ala, Cbz-L-Phe, Cbz-L-Val, Cbz-L-Trp, *N*-Cbz-*O*-Bz-L-Ser, N^{α}-Cbz-N^{ε}-Cbz-L-Lys, *N*-Cbz-L-Glu-*O*-Bzl, Cbz-L-Ala-L-Phe, Cbz-L-Phe-L-Ala-Gly) with yields of 67–76% (Method **A**) [102]. Katritzky et al. reported a convenient benzotriazole-mediated efficient syntheses for chirally pure quinine conjugates with amino acids, di- and tripeptides (Boc-Gly, Boc-L-Ala, Boc-L-Phe, Boc-L-Ile, Boc-L-His(Tos), Boc-L-Ser (Bzl), Boc-L-Glu(Bz), Cbz-L-Lys(Cbz), Cbz-L-Asp(Bz), Cbz-L-Cys(Bz), Cbz-L-Ala-L-Phe, Cbz-L-Val-L-Leu, Cbz-L-Ileu-Gly, Cbz-Gly-L-Phe-L-Ala) of varying polarity in 52–95% yields. These quinine–peptide conjugates possess an in vitro antimalarial activity similar to that of quinine (Method **A**) [103]. *N*-(Fmoc-α-aminoacyl)benzotriazoles proved to be effective versatile synthetic reagents from proteinogenic amino acids in their reactions with L- or D-PhCH (Me)NH_2 to give Fmoc-L-Tyr(*t*-Bu), Fmoc-L-Val, and Fmoc-L-Tyr(*t*-Bu) conjugates of L-, D-, or D,L-PhCH(Me)NH_2 (Method **A**) [35].

Type **B** couplings (Scheme 37) yielded the conjugates of acylated at N-terminus amino acids. Two glutathione-*S*-CO(4-MeC_6H_4) and glutathione-*S*-CO (4-$NO_2C_6H_4$) conjugates (R=4-MeC_6H_4, 4-$NO_2C_6H_4$) were synthesized in 75 and 80% yields, respectively [104].

Triazoles and tetrazoles are often used for bioisosteric replacement of the amide bond. Katritzky et al. reported the synthesis of amino acid and dipeptide conjugates with 1,2,4-triazoles [105]. 3,5-Diamino-1,2,4-triazole has been coupled to amino acids and peptides; using benzotriazole methodology, both types of couplings took place – the one at the N(1)-endocyclic nitrogen the ring to give **139** and at the exocyclic nitrogen to give peptidomimetics **141** (Schemes 38 and 39, respectively) [105]. The yields for the products **139** (Scheme 38) were 65–95%, and for **141** (Scheme 39) were 57–91%. Microwave irradiation improved the reaction yields and shortened the reaction times to form products **139** and **141** (Schemes 38 and 39).

Katritzky et al. reported the preparation of thiadiazole peptides **144** via a multistep benzotriazole-mediated assembly of a thiadiazole ring, removal of the Cbz-group of **143,** and acylation with Cbz-acyl benzotriazoles **10**. Products **144** were isolated in 57–67% yield (Scheme 40) [106].

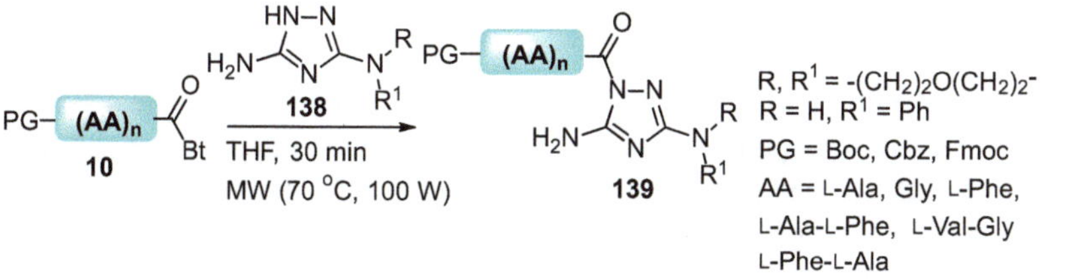

Scheme 38 Acylation of 3,5-diamino-1,2,4-trizoles **138** with N-protected Bt-active amino acids and dipeptides **10**

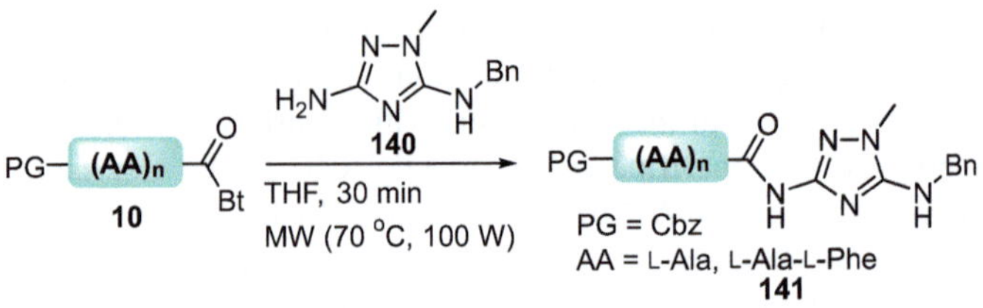

Scheme 39 Acylation of the exocyclic amino group of 3,5-diamino-1,2,4-triazole **140** with *N*-PG Bt-active L-Ala and L-Ala-L-Phe **10**

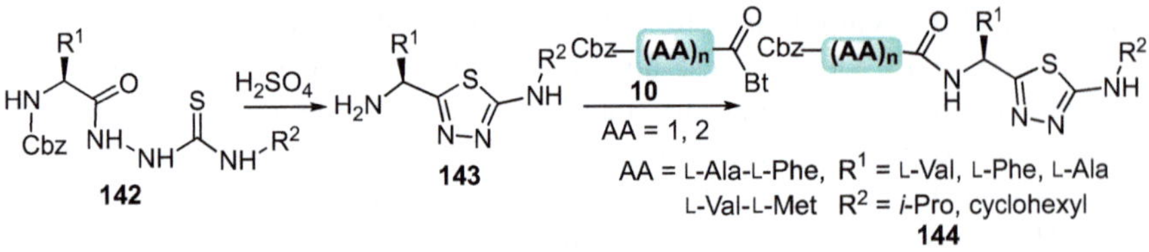

Scheme 40 Synthesis of thiadiazole-containing peptidomimetics **144**

Acylated heterocyclic units of peptides are often formed via biosynthesis and as products of microbial metabolism [107]. For many natural peptidomimetics, which contain heterocyclic units, distinct structural modifications to the nascent peptide chains confer their physiological function [108].

Katritzky et al. reported synthesis of chirally pure (α-aminoacyl) conjugates of heterocycles with weakly nucleophilic heterocyclic amines [109]. Treatment of 2-aminothiazole, 2-amino-6-methoxybenzothiazole, *N*-benzyl-2-aminobenzimidazole, 5-amino-3-methoxy-1,2,4-thiadiazole, 4-amino-1-benzylpyrimidin-2-one **145**, and *N*-Cbz-(α-aminoacyl)benzotriazoles **10** in DMF under microwave irradiation at 70°C for 30–150 min gave N-substituted amides **146** in 50–98% yields (Scheme 41) [105].

Katritzky et al. reported synthesis histidine conjugates of N^{α}-Boc-N^{im}-Ts-L-histidine and N^{α}-Boc-N^{im}-Bn-L-histidine **146**. Di-N-protected histidinyl benzotriazole **10** was conjugated to *N*-nucleophiles **145f–g** to give conjugates **146** in 92% (Het-NH_2 = **145f**) and 91% (Het-NH_2 = **145g**) (Scheme 41) [38].

3.1.2 Acylation of *O*-Nucleophiles

N-Cbz-protected tri- and tetrapeptidoyl benzotriazoles **10** were reacted with a variety of steroids, terpenes, and sugar derivatives **147** in the presence of a catalytic amount of DMAP under microwave irradiation at 70°C and 65 W for 1.5–3 h to afford peptide conjugates **148** in 35–58% yields (Scheme 42) [88].

Cholecalciferol and α-tocopherol were also conjugated to amino acids and oligopeptides via benzotriazole coupling to give conjugates **148** [111]. Amino

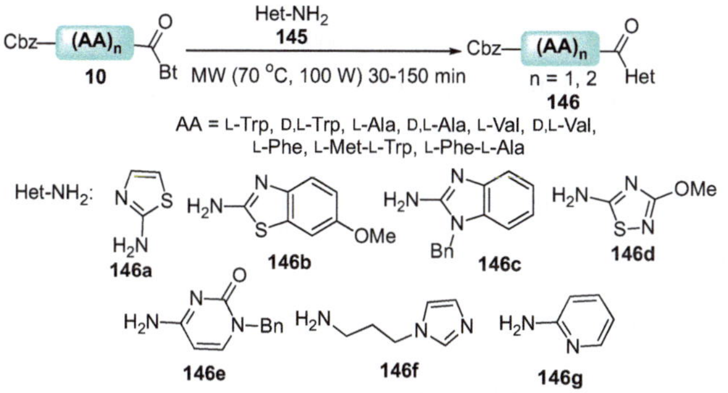

Scheme 41 Synthesis of (α-aminoacyl)- and dipeptidoyl-conjugated amino-heterocycles **146** [110]

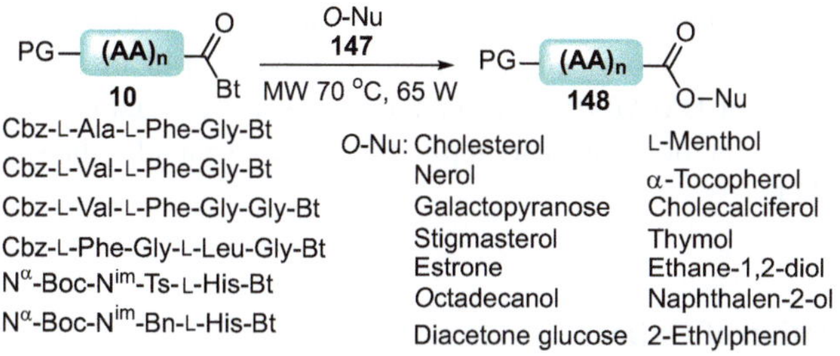

Scheme 42 *O*-acylation of *N*-PG-tri- and tetrapeptidoylbenzotriazoles **148**

acid and peptide conjugates of vitamin D3 were obtained by *O*-acylation of cholecalciferol **147** with Cbz-protected acylbenzotriazoles **10** in the presence of DMAP in THF and under microwave irradiation (50 W, 70°C) for 1–2 h in 52–64% yields (Scheme 42) [111]. *O*-acylation of α-tocopherol **147** with Cbz-protected acylbenzotriazoles **10** under microwave irradiation (20 W, 50°C) in anhydrous DMF in the presence of potassium carbonate resulted in the formation of conjugates **148** in 51–68% yields (Scheme 42) [111].

Katritzky et al. reported synthesis of histidine conjugates using N^{α}-Boc-N^{im}-protected-L-histidine **10**. Di-N-protected histidinyl benzotriazole **10** was conjugated to *O*-nucleophiles such as naphthalene-2-ol and 2-ethylphenol to give conjugates **148** in 55% and 50% yields correspondingly (Scheme 42) [38].

Katritzky and Angrish reported a microwave-assisted benzotriazole-mediated synthesis of *O*-(α-protected-aminoacyl)steroids **151** [112]. Chiral *O*-(α-protected-aminoacyl)steroids and *O*-(α-protected-dipeptidoyl)steroids were conveniently prepared under microwave irradiation in isolated yields of 65–96%, with complete retention of the configuration. The reaction utilized readily available *N*-(Cbz-α-aminoacyl)benzotriazoles and Cbz-L-Met-L-Ala-Bt **10**, with naturally occurring steroidal alcohols such as estrone, cholesterol, and stigmasterol **147**.

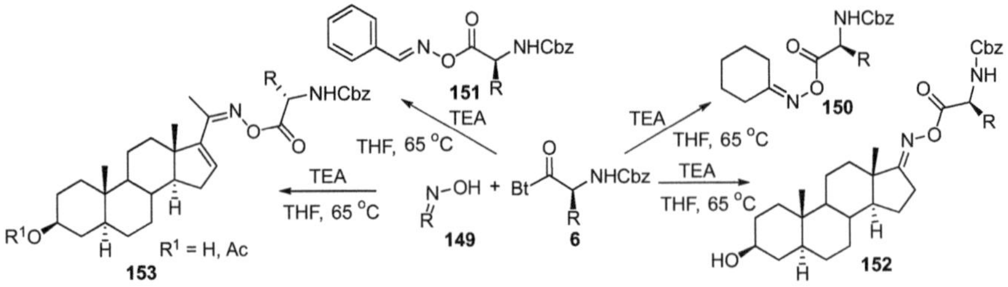

Scheme 43 Synthesis of amino acid–oxime conjugates **150–153**

N-(Cbz-α-aminoacyl)benzotriazoles were synthesized from L-Phe, L-Trp, D-Phe, D, L-Phe, L-Met, and L-Val (Scheme 42) [112].

Katritzky and Angrish also extended the method to the synthesis of various *O*-terpene and *O*-alkane conjugates **148** in 83–97% yields [113]. N-Protected [(α-aminoacyl)oxy]terpenes **148** were prepared by coupling N-protected (α-aminoacyl)benzotriazoles **10** with hydroxyterpenes **148** in the presence of a catalytic amount of DMAP (0.1 eq.) under microwave irradiation at 65°C, 100 W for 15 min. This method was also applied to *O*-octadecanol, ethane-1,2-diol, and thymol **147** [113].

3.1.3 Acylation of Oximes

Katritzky et al. reported acylation of oximes and steroid oximes into amino acid-*O*-N=conjugates **150–153** [114]. First, *N*-Cbz-AA-Bt **6** acylation was studied on commercially available oximes such as cyclohexanone oxime and benzaldehyde oxime to afford products **150** and **151** in 62% and 73% yields. *N*-Cbz-(α-aminoacyl)benzotriazoles **6** were then reacted with 3β-ol and 3β-acetoxy-20-hydroxyimin-5α-pregn-16-en-20-ones **149** at 65°C in THF in the presence of TEA to produce **152** and **153** in the yields of 44–57%. In the case of R^1=H, the acylations proceeded chemoselectively toward the oxime in the presence of free hydroxyl groups. Analogous acylations were conducted with 17-hydroxyimin-5α-androstan-3β-ol with 52–73% yields. *N*-Cbz-L-Ala, *N*-Cbz-L-Val, *N*-Cbz-L-Phe, *N*-Cbz-Gly, and *N*-Cbz-L-Lys(Cbz) were used in the couplings with oximes **152** (Scheme 43) [114].

3.1.4 Acylation of *S*-Nucleophiles

Tri- and tetrapeptidoyl benzotriazoles **10** were reacted with *S*-nucleophiles **154** in the presence of TEA at room temperature for 2 h to give their corresponding *S*-acylated tri- and tetrapeptide conjugates **155** in 68–94% yields (Scheme 44) [88].

Katritzky et al. reported synthesis of various histidine conjugates using N^{α}-Boc-N^{im}-protected-L-histidine. Benzotriazoles **10** were conjugated to *S*-nucleophiles

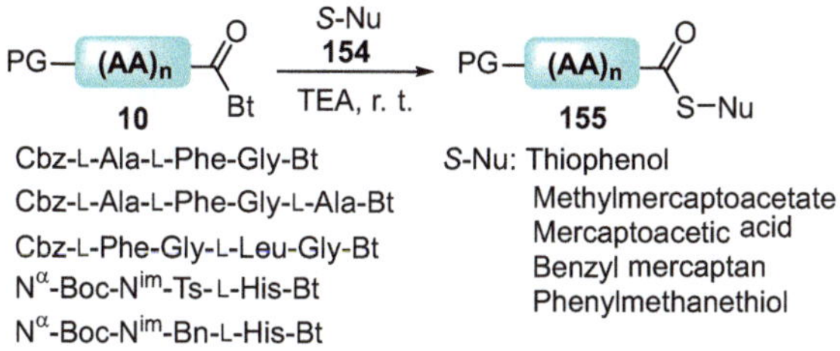

Scheme 44 *S*-acylation of *N*-PG-tri- and tetrapeptidoylbenzotriazoles

such as phenylmethanethiol, thiophenol, and methylmercaptoacetate to give products **155** in 68–80% yields (Scheme 44) [38].

3.1.5 Acylation of *C*-Nucleophiles

Katritzky et al. reported benzotriazole-mediated acylation of nitriles containing active methylene group forming optically pure nitrile-functionalized enol ether conjugates **157** (50–88%) [87]. Two types of reaction conditions for the formation of **157** were studied. Acylation of CH_2-active nitriles **156** with *N*-PG-amino acids and peptides **10** was carried out under microwave irradiation of the reaction mixture for 1 h in THF using DIPEA as base. Analogous reactions, conducted at room temperature overnight, provided longer reaction times and similar yields for the products [87].

Katritzky et al. reported synthesis of various histidine conjugates using N^{α}-Boc-N^{im}-protected-L-histidine. Di-N-protected histidinyl benzotriazole **10** was conjugated to malononitrile (R=CN) using the reaction conditions ii to give product **157** in 55% (Scheme 45) [38].

Katritzky et al. reported synthesis of tri- and tetrapeptides, conjugated to *C*-nucleophiles such as dimedone and 1,3-cyclohexanedione **158** [88]. Two conjugates were synthesized in the presence of 1 eq. of DMAP under MW irradiation at 70°C and 50 W. Tetrapeptide-1,3,-cyclohexanedione **159** was isolated in 61% yield, and the tripeptide–dimedone conjugate **159** was isolated in 69% yield (Scheme 46) [88].

C-Acylation of lithiated 2- and 4-methylpyridine or 2-methylquinolone in THF at −20°C for 1–3 h with *N*-aminoacylbenzotriazolides **6** gave the corresponding (α-aminoacyl) C-linked conjugates **161** in 49–98% yields (Scheme 47) [115].

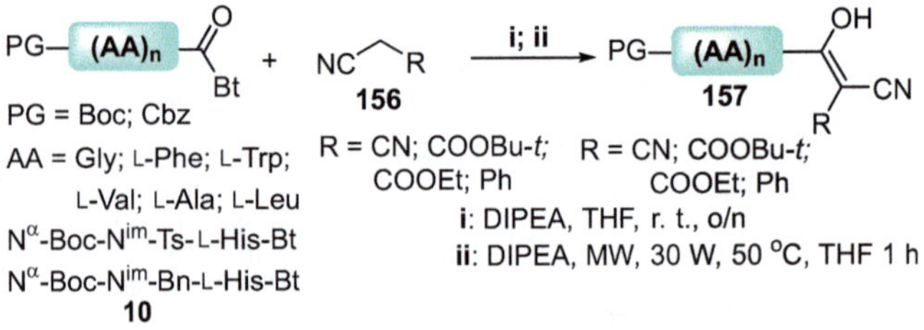

Scheme 45 Synthesis of *C*-acylated enol ether peptidomimetics **157**

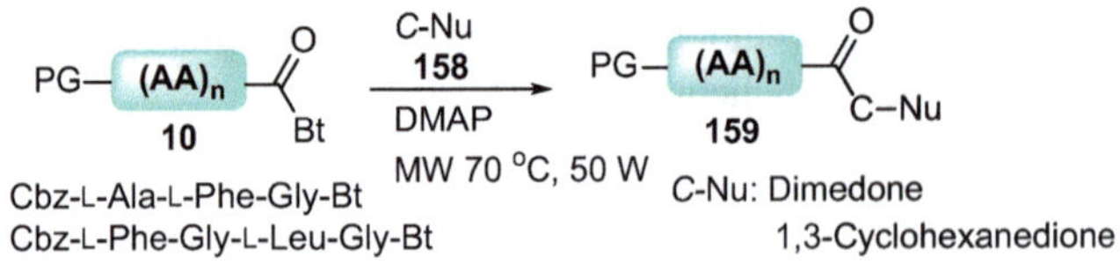

Scheme 46 *C*-acylation of *N*-PG-tri- and tetrapeptidoylbenzotriazoles

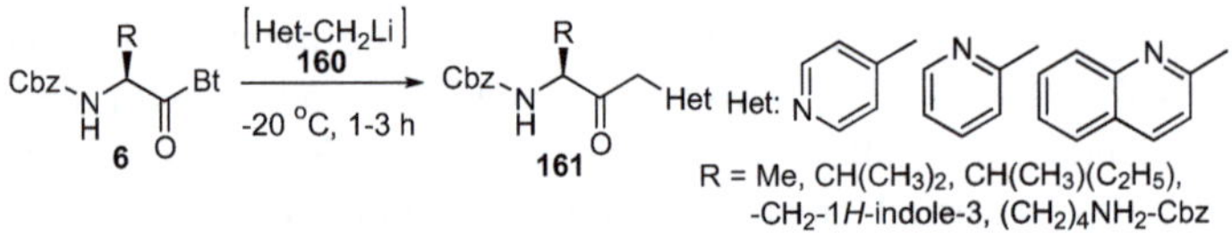

Scheme 47 Synthesis of (α-aminoacyl)-conjugated *N*-heterocycles **161**

3.2 Benzotriazole-Mediated Tagging of Peptides and Peptidomimetics with Dyes and Fluorescent Tags

3.2.1 Synthesis of Azo-Dye-Labeled Peptides

Azo-arene carboxylic acids have found their use as molecular switches in life sciences [116]. Katritzky et al. reported a convenient labeling of amino acids with azo-dye carboxylic acid-labeled amino acid derivatives **164** obtained in high yields (81–99%) by treating N-(4-arylazobenzoyl)-1*H*-benzotriazole **163** with appropriate amino acids **3** in DMF/H_2O (3:1, v/v) mixture in the presence of TEA (Scheme 48). These reactions were completed within 24 h at room temperature (monitored by TLC) to give various *N*-(4-arylazobenzoyl)-amino acid conjugates **164** (Scheme 48) [117].

3.2.2 Labeling of Amino Acids and Peptides with Fluorescent Tags

A fluorescent tag, also known as a label or probe, is a molecule that is attached chemically to aid in the labeling and detection of a biomolecule such as a protein,

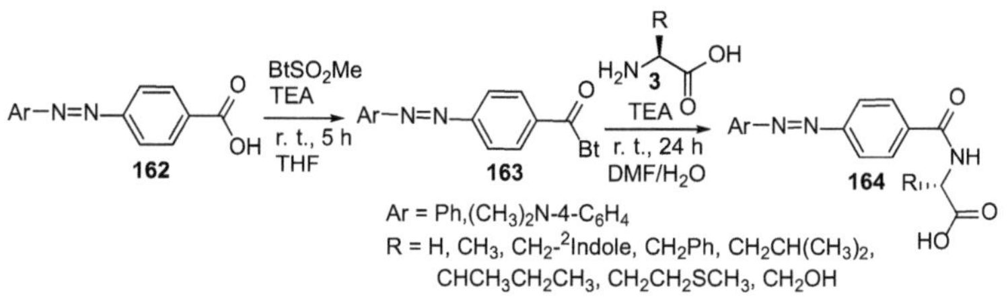

Scheme 48 Synthesis of *N*-acylated peptides, tagged with azo-arene carboxylic acid

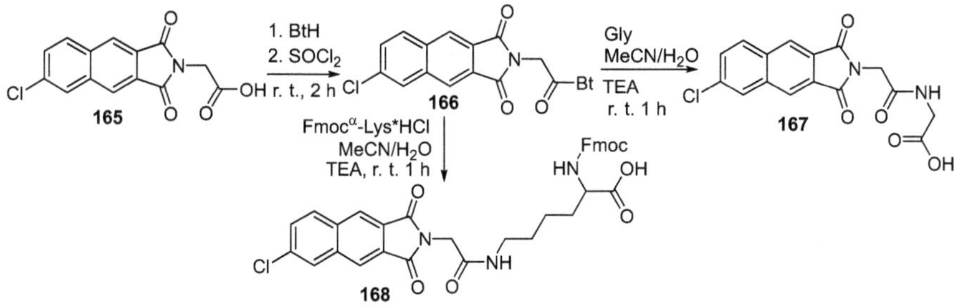

Scheme 49 Synthesis of 6-chloro-2,3-naphthalimide conjugates **167** and **168**

antibody, peptide, or amino acid [118]. Fluorescent tagging has found its increasing use in imaging for monitoring complex cellular processes such as small-molecule-messenger dynamics, enzyme activation, and protein–protein interactions [119].

Naphthalene-based fluorophores are used for the detection of DNA [120] and censoring G-quadruplex ligands [121]. Katritzky et al. reported the synthesis of Gly and Fmoc$^{\alpha}$-Lys-OH conjugates **167** and **168** with the new environment-sensitive fluorophore 6-chloro-2,3-naphthalimide. For that, the naphthalimide carboxylic acid **168** was reacted with 1*H*-benzotriazole and thionyl chloride in THF at room temperature. Intermediate **166** was then coupled with glycine and Fmoc$^{\alpha}$-L-Lys to form products **167** and **168** in 76% and 72% yields, respectively (Scheme 49) [122].

Substituted coumarins are constituents of many commercially important fluorescent dyes since they offer high emission quantum yields, photostability, and good solubility in most solvents [123, 124].

In 2008 Katritzky et al. reported a convenient solid-phase-based methodology on labeling of amino acids and dipeptides with 7-methoxy-substituted coumarin [125]. A model L-Ala-Lys-(N^{ε}-Mca)-NH_2 dipeptide **171** was synthesized using microwave-assisted solid-phase peptide synthesis conditions. After the initial removal of the Fmoc protecting group, free Rink resin-NH_2 was coupled with **170** in DMF under microwave irradiation at 70°C for 10 min. The second step was performed with the (N^{α}-Fmoc-aminoacyl)benzotriazole reagent **3** derived from Fmoc-L-Ala, and the desired peptide was then cleaved from the resin to give labeled dipeptide amide **171** in a total yield of 26% (Scheme 50) [125].

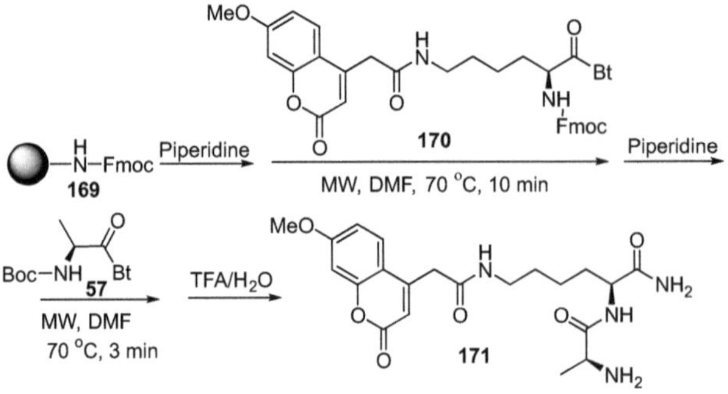

Scheme 50 Synthesis of 7-methoxycoumarin-labeled dipeptide amide **171**

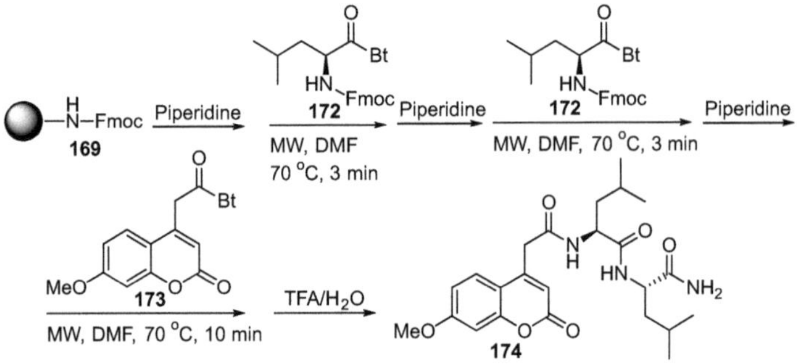

Scheme 51 Synthesis of 7-methoxycoumarin-labeled dipeptide-NH_2

Synthesis of 7-methoxycoumarin-labeled L-Leu-L-Leu amide **177** was carried out in a similar manner. After *N*-Fmoc-L-Leu was added to the free Rink resin, the Fmoc group was removed and the amino group of L-Leu was acylated by *N*-Fmoc-Leu-Bt **172**. After another deprotection step, the intermediate dipeptide was acylated by the Bt-active 7-methoxycoumarin **173**, which, after cleaving off the resin, gave the final coumarin-labeled dipeptide amide **174** in 26% overall yield (Scheme 51) [125].

3.3 Benzotriazole-Mediated Conjugation of Amino Acids and Peptides to Bioactive Molecules

3.3.1 Mono-Conjugation of Amino Acids and Oligopeptides to Bioactive Molecules

Benzotriazole methodology is advantageous for solution-phase synthesis of chirally pure *O*-(α-aminoacyl) sugar conjugates [126] and *N*-(α-aminoacyl)sugar conjugates [127]. Katritzky et al. reported the acylation of tetra-*O*-pivaloyl-β-D-galactopyranosylamine **175c** by *N*-(Cbz or Fmoc-α-aminoacyl)benzotriazoles **10** under microwave irradiation (100 W, 60°C, 75 min). The reaction proceeded diastereoselectively to give β-*N*-glycoaminoacids and glycosylated asparagine building block **176** (65%). 1,2:3,4-Di-*O*-isopropylidene-R-D-galactopyranose, 1,2:5,6-di-*O*-isopropylidene-D-glucose, and 2,3:5,6-di-*O*-isopropylidene-α-D-mannofuranose **175** were efficiently *O*-acylated with *N*-(Cbz-α-aminoacyl) benzotriazoles **10** under microwave irradiation to give chirally pure and diastereomeric mixtures of conjugates **176** in 78–93% yields [126]. *N*-Cbz-protected oligopeptidoylbenzotriazoles (*AA* = 2,3) **10** at their reaction with various sugars **175** afforded β-N-glycodipeptides **176** (76–81%) in a similar manner (Scheme 52) [127].

Katritzky et al. reported synthesis of biotin and niacin benzotriazolides coupled with free amino acids, dipeptides, and tripeptides under microwave irradiation (50 W) at 70°C in the presence of TEA to give the expected bioconjugates **179** (Scheme 53) [111]. Niacin was coupled with L-Phe, L-Lys(Cbz), L-Met, L-Leu-L-Leu, and Gly-L-Ile in 43–81% yield. Biotin was conjugated to L-Ile, L-Phe, L-Leu (Cbz), L-Met, Gly-L-Val, L-Leu-L-Leu, and Gly-L-Ile in 35–82% yield [111].

Conjugation of amino acids and a dipeptide **178** to plant hormone benzotriazolides **177a,b** resulted in **179** and prepared in 86–90% yields. Indole-3-acetic acid and 1,2-indole-3-propanoic acid benzotriazolides **177c,d** were coupled

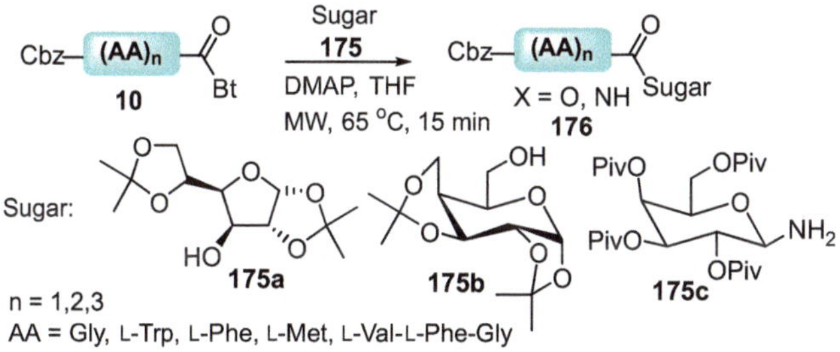

Scheme 52 Preparation of (α-aminoacyl)sugar conjugates **176**

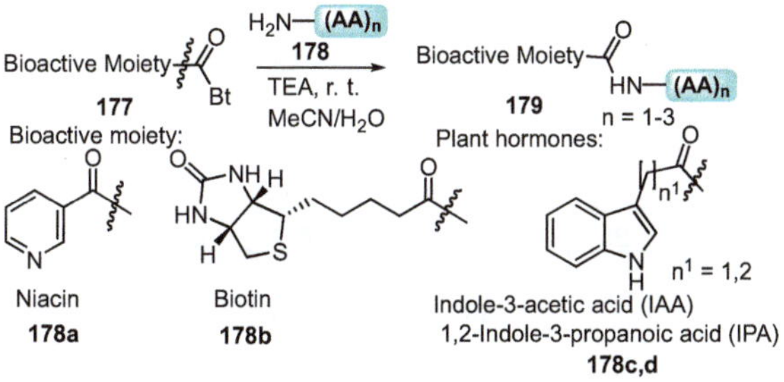

Scheme 53 Conjugation of RCOBt biologically active moieties **177** to the amino acids or peptides **178**

with Gly, L-Ala, L-Phe, D,L-Ala, D,L-Phe, L-Met, L-Lys, L-Arg, L-Trp, and Gly-D-Phe **178** to give plant hormone–amino acid conjugates **179** in 32–70% yields (Scheme 53) [128].

Katritzky et al. reported the synthesis of tetrapeptide-17-ethinyl estradiol (EE) conjugate, linked via a 1,2,3-triazole ring [129]. Conjugate **186** was synthesized by solution-phase coupling–deprotection methodology, starting from the conjugation of 2-azidoethanamine to Boc-protected L-tryptophan employing BtH as a coupling additive (ii) followed by deprotection of the Boc group (iii). A sequence of successive steps of Boc-L-amino acid couplings and the Boc group deprotections resulted in adding L-Val, L-Phe, and L-Ala units to the peptide chain to afford azidopeptide **185**. It was then labeled with 17EE in the presence of Cu(I) and deprotected into conjugate **186** (Scheme 54) [129].

Utilization of prodrugs, which temporarily mask the carboxyl groups of NSAIDs, may increase the uptake and reduce stomach irritation caused by direct contact [130]. Katritzky et al. reported synthesis of benzotriazole-mediated conjugation of D,L-ibuprofen and L-naproxen **190** with L-amino acids and peptides **181** in 76–84% yields into prodrugs or potential drug candidates. L-Ala, L-Val, L-Phe, L-Ser, L-Leu, L-Trp, Gly-L-Ala, and Gly-Gly **181** were coupled with D, L-ibuprofen and L-naproxen using stepwise solution-phase benzotriazole-mediated methodology (Scheme 55) [131].

Katritzky et al. reported conjugation of NSAID benzotriazolides **188** with amino acids and a bioactive dipeptide carnosine (Gly-L-His) in aqueous MeCN at room temperature (Scheme 55). Ibuprofen, mefenamic acid, diclofenac, naproxen, and indomethacin were CO activated with benzotriazole to form **188** followed by their

Scheme 54 Synthesis of tetrapeptide–17EE conjugate **186**, linked via a 1,2,3-triazole ring

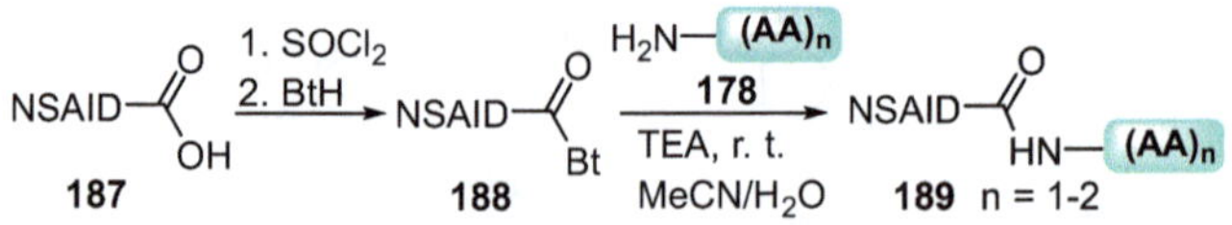

Scheme 55 Synthesis of NSAID–amino acid and peptide conjugates **192**

reaction with carnosine, L-His, L-Thr, or L-Ala **178** to give NSAID–amino acid or peptide conjugates **189** in 50–97% yields (Scheme 55) [132].

Prodrugs formed by linking quinolone acids with amino acid esters are more lipophilic than the parent drugs and show enhanced in vivo antibacterial properties with pronounced therapeutic effects [133].

Katritzky et al. developed a benzotriazole-mediated methodology toward N-protected amino acid–antibiotic conjugates **191** [134]. The coupling of ciprofloxacin, pipemidic acid, and norfloxacin **190** with Cbz-*N*-(aminoacyl) benzotriazoles **6** in aqueous MeCN in the presence of TEA for 3 h at room temperature resulted in the formation of conjugates **191**: amino acid–ciprofloxacin (68–77%), amino acid–pipemidic acid (51–82%), and amino acid–norfloxacin (75–86%). The following amino acids were used in couplings with quinolone antibiotics (conditions i): Cbz-Gly, Cbz-L-Ala, Cbz-L-Lys(Cbz), Cbz-L-Val, Cbz-L-Phe, and Cbz-D,L-Ala **5** (Scheme 56) [134]. Couplings of Cbz-*N*-(aminoacyl)benzotriazoles **6** (conditions ii) with metronidazole **190a** resulted in the formation of *N*-Cbz-amino acid–metronidazole conjugates **191** of Cbz-Gly, Cbz-L-Ala, Cbz-L-Lys (Cbz), and Cbz-L-Phe in 72–85% yield (Scheme 57). Coupling of Cbz-L-Ala, Cbz-L-Asp(Cbz), and Cbz-L-Phe **5** with sulfadiazine **190b** using conditions iii resulted in the formation of *N*-Cbz-amino acid–sulfadiazine conjugates **191** in

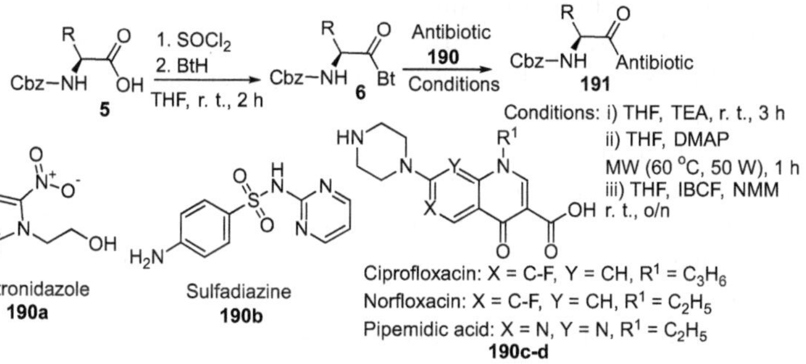

Scheme 56 Reaction of Cbz-*N*-(aminoacyl)benzotriazoles **6** with antibiotics **191**

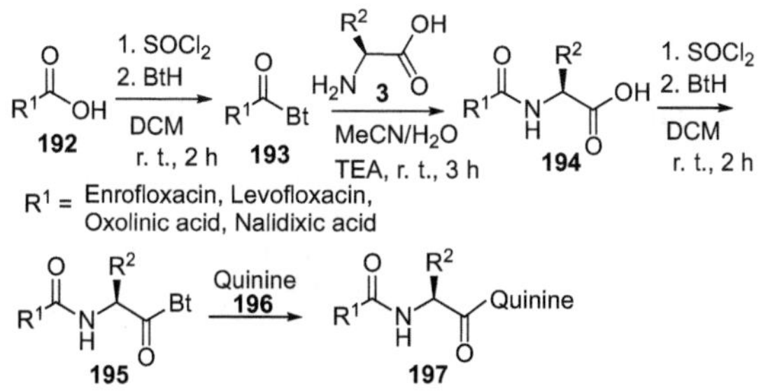

Scheme 57 Synthesis of bis-conjugates of quinine, quinolone antibiotics, linked via an amino acid **197**

65–72% (Scheme 56). Antibiotic conjugates were tested for activity in four medically relevant organisms: *Staphylococcus aureus* (RN4220), *Escherichia coli* (DH5α), *Pseudomonas aeruginosa* (PAO1), and *Bacillus subtilis* (168). Several antibiotic conjugates show promising results against several of the strains screened [134].

3.3.2 Bis-Conjugation of Amino Acids and Oligopeptides to Bioactive Molecules

Katritzky et al. reported a benzotriazole-mediated efficient synthesis of quinine bis-conjugates **197** incorporated with quinolone (or fluoroquinolone) antibiotic and peptide fragments via two alternative routes. Some of these bis-conjugates display antimalarial activity similar to quinine (IC_{50} values ranging from 12 to 207 nM) [135]. L-amino acids **3** were first acylated with benzotriazolyl derivatives of antibiotics (oxolinic acid-Bt, nalidixic acid-Bt, levofloxacin-Bt, enrofloxacin-Bt) **193** [136], followed by carbonyl activation of the amino acid with 1*H*-benzotriazole to form **194**. The final step was microwave-assisted coupling with the quinine **196** to give bis-conjugates of quinine, quinolone antibiotics, and L-amino acids **197** in 68–72% yields. L-Phe, L-Ala, L-Val, Gly, L-Leu, and L-Ile amino acids were used to connect quinolone antibiotics to quinine in 71–90% yields (Scheme 57) [135].

Katritzky et al. developed an effective method of conjugation of NSAIDs to glucosamine hydrochloride **200** via an amino acid linker. Indomethacin, diclofenac, and mefenamic acid benzotriazolides **198** were synthesized, coupled with amino acids **3** at the *N*-terminus and then conjugated to the glucosamine moiety **200** in 36–70% yields (Scheme 58) [137]. The presence of rotamers in products **201** was observed due to a lack of free rotation around amide bonds caused by the close proximity of the amino acid R^2 group to both the NSAID and the carbohydrate moiety [137].

Katritzky et al. reported microwave-assisted benzotriazole-mediated synthesis of biotin–quinolone conjugates **204** linked via amino acid residues [138]. The Boc-protected aminoacyl benzotriazoles **6** were reacted with antibiotics **190** in the presence of TEA in DMF under microwave irradiation (20 W, 50°C) for 1 h

Scheme 58 Synthesis of glucosamine–amino acid–NSAID bis-conjugates **201**

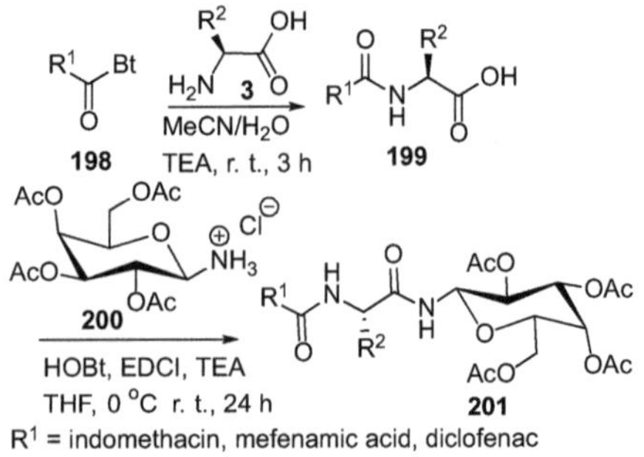

R¹ = indomethacin, mefenamic acid, diclofenac
R² = H, CH₃, CH(CH₃)₂, Bn

to give conjugates **204** in 58–91% yields (Scheme 59). Amino acid–quinolone antibiotic intermediates **202** were synthesized using benzotriazole-mediated coupling in 68–89% yield [138].

Katritzky et al. developed a benzotriazole-based methodology toward fluorescently-labeled amino acid–antibiotic conjugates, which allows double conjugation of amino acids at both C- and N-termini to produce bis-conjugates [139]. This method provides novel fluorescent-labeled amino acid–quinolone antibiotic conjugates **209** in good yields with the retention of the absolute configuration. The photophysical properties of all the fluorescent-labeled antibiotic conjugates were also determined and reported [139]. Coumarin–amino acid antibiotic bis-conjugates were synthesized with the yields of 65–88% for the final conjugation step to form **209**. In such a way, ciprofloxacin (R^3=cyclopropyl, X=C-F, Y=CH) was conjugated to Gly (R^2=H), L-Ala (R^2=CH_3), and L-Val (R^2=$CHMe_2$) at the C-terminus and to coumarin carboxylic acids (R^1=H, OMe) at the N-terminus. Pipemidic acid (R^3=ethyl, X=N, Y=N) was conjugated to Gly (R^2=H), L-Ala (R^2=CH_3), and L-Val (R^2=$CHMe_2$) at the C-terminus and to coumarin carboxylic acids (R^1=H, OMe), and norfloxacin (R^3=ethyl, X=C-F, Y=CH) was conjugated to the analogous amino acids and coumarins. All the conjugations were carried out via a stepwise activation of different carboxyl groups with benzotriazole (Scheme 60) [139].

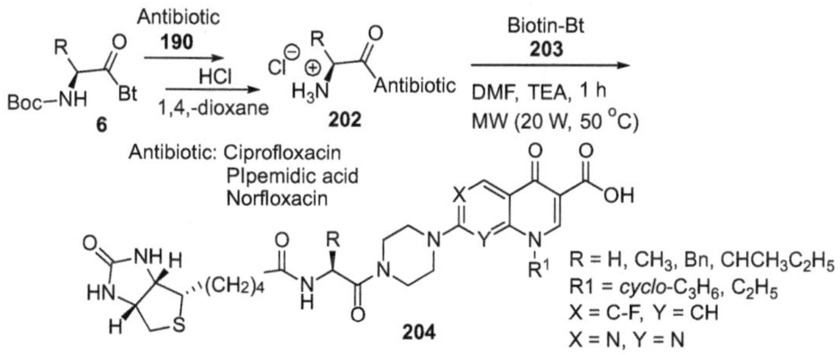

Scheme 59 Synthesis of antibiotics conjugate with amino acids and biotin into bis-conjugates **204**

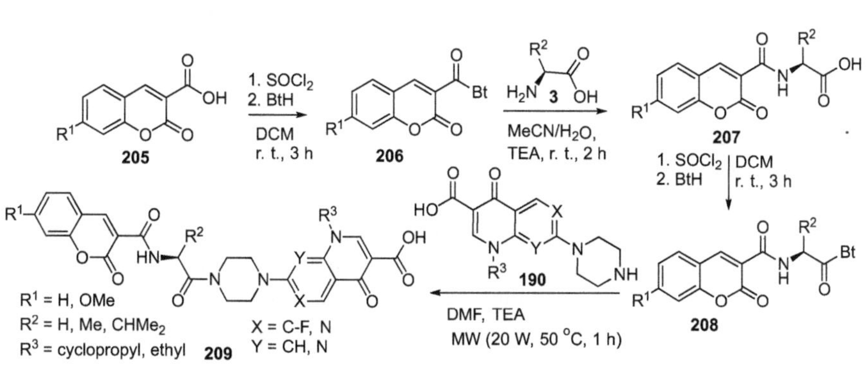

Scheme 60 Synthesis of coumarin–amino acid–quinolone bis-conjugates **209**

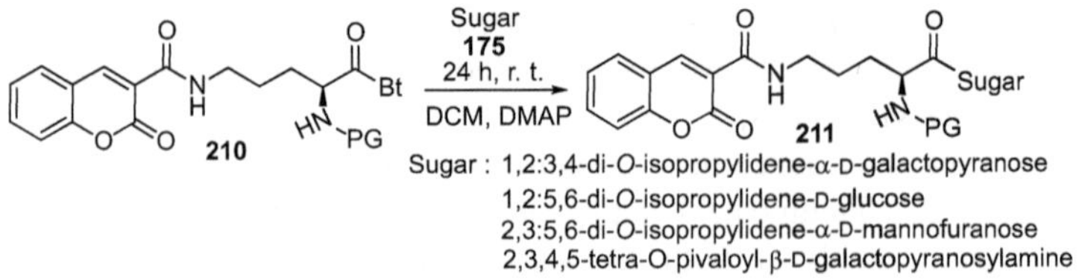

Scheme 61 Synthesis of coumarin–L-Lys (PG)–sugar bis-conjugates **211**

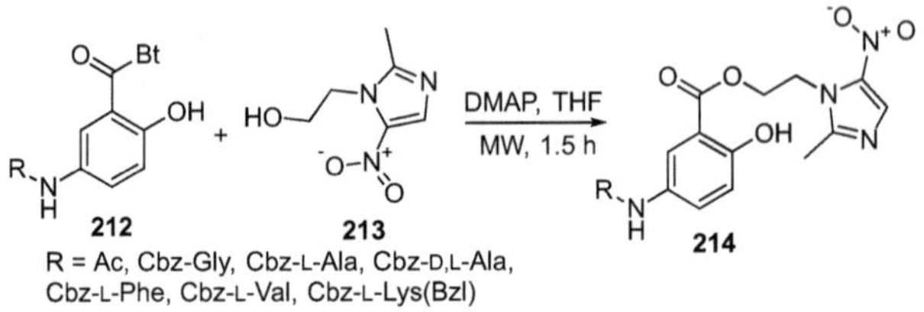

Scheme 62 Synthesis of *N*-acyl or amino acid–mesalamine–metronidazole conjugates **214**

Katritzky et al. reported the development of L-lysine-based monosaccharide water-soluble fluorescent tags [140]. Monosaccharides **175** were labeled with coumarin fluorescent tags via an L-Lys linker **210**. 1,2:3,4-Di-*O*-isopropylidene-*R*-D-galactopyranose, 1,2:5,6-di-*O*-isopropylidene-D-glucose, and 2,3:5,6-di-*O*-isopropylidene-*R*-D-mannofuranose **175** were linked to the N^{ε}-coumarin tagged-L-Lys **210** via benzotriazole-mediated *O*-acylation. The amino group of 2,3,4,5-tetra-*O*-pivaloyl-β-D-galactopyranosylamine was acylated with N^{ε}-coumarin tagged-L-Lys to give the final conjugate **211**. Coumarin–lysine–sugar scaffolds **211** were isolated in 67–85% yields (Scheme 61). After removal of the diisopropylidene groups, water-soluble fluorescent derivatives were obtained [140, 141].

Naumov et al. reported benzotriazole-mediated synthesis of mesalazine–metronidazole bis-conjugates **214**. Mesalazine was used either as the N-acetyl derivative or was acylated at the 5-amino group with various amino acid benzotriazolides (Cbz-Gly, Cbz-L-Ala, Cbz-L-Phe, Cbz-L-Val, Cbz-L-Lyz(Bz)) using microwave irradiation (50 W, 70°C, 1–1.5 h) and H_2O or DMF as solvent. The carbonyl group of the intermediate was then activated with BtH ($SOCl_2$, r. t., DCM), and the RCOBt intermediates **212** were reacted with metronidazole **213** using microwave irradiation (50 W, 70°C, 1.5 h) to give bis-conjugated products **214** in 48–82% yields (Scheme 62) [142].

Katritzky et al. developed a benzotriazole-mediated microwave-assisted methodology toward acetaminophen–amino acid–NSAID conjugates [143]. In addition to the QSAR studies, biological data acquired for the novel bis-conjugates **217** showed that they exhibit a more potent anti-inflammatory activity than their parent drug. Final bis-conjugates **217** were synthesized in 31–76% yields using solution-phase microwave-assisted coupling of amino acid–acetaminophen conjugates **216** with RCOBt NSAIDs such as ibuprofen, naproxen, and indomethacin **188** (Scheme 63) [143].

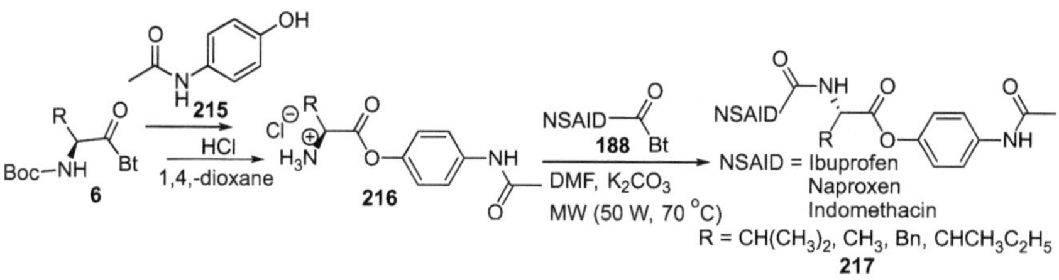

Scheme 63 Synthesis of acetaminophen–amino acid–NSAID conjugates **217**

3.4 Benzotriazole-Mediated Acylation of Various Nucleophiles with Modified Amino Acids or Peptidomimetics

Chiral *N*-Boc and *N*-Cbz-protected α-aminoacyl benzotriazoles **6** were reacted under microwave irradiation for 10 min with ethyl (triphenylphosphoranylidene) acetate and produced chirally pure phosphorus ylides **218** conjugated to N-protected L-Ala, L-Val, L-Phe, L-Asp(OMe), and L-Trp in 65–90% yields (Scheme 64) [144].

Efficient conversions of carboxylic acids into Weinreb amides **220** were achieved by treatment of *N*-acylbenzotriazoles **6** with *N,O*-dimethylhydroxylamine hydrochloride **219** under mild conditions as reported. Various carboxylic acids and amino acids such as *N*-Boc-L-Ala, *N*-Boc-L-Val, and *N*-Boc-L-Phe were used to synthesize the corresponding Weinreb amides **220** in 75–99% yields with retained chiral purity (Scheme 64) [145].

3.4.1 Glutamic and Aspartic Acid β- and γ-Derivatives

Katritzky et al. reported novel β- and γ-amino acid derivatives **225** synthesized from aspartic and glutamic acid through their corresponding benzotriazole intermediates **222** in 35–89% yields (Scheme 65). Complete retention of the absolute configuration in these sequences was confirmed by chiral HPLC [146].

3.4.2 Conjugates of L-Arginine with *N*-Nucleophiles

Katritzky et al. reported synthesis of conjugates of L-$^{\omega}NO_2$-Arg-Bt **226** with several *N*-nucleophiles **227** under different reaction conditions to produce products of two types: conjugates **228** (mainly with primary amines) and cyclic lactam **229** (mainly during the reaction with secondary amines). The ratio of the conjugate versus cyclic by-product varied on the reaction conditions and the nucleophile. Reaction with *n*-propylamine in THF gave products **228** and **229** in 85% and 11%, respectively.

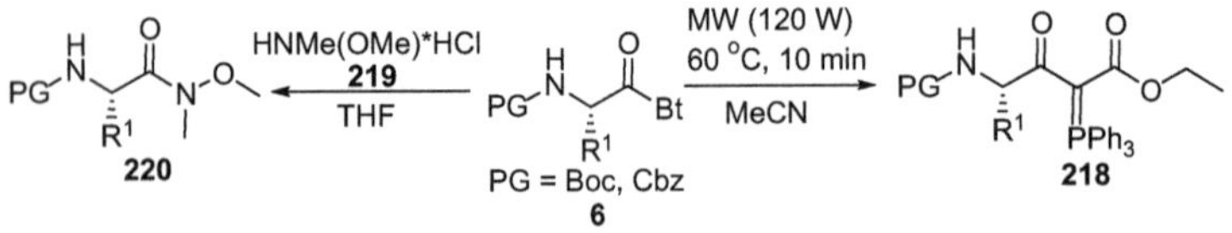

Scheme 64 Synthesis of peptidyl phosphorus ylides **218** and Weinreb amides **220**

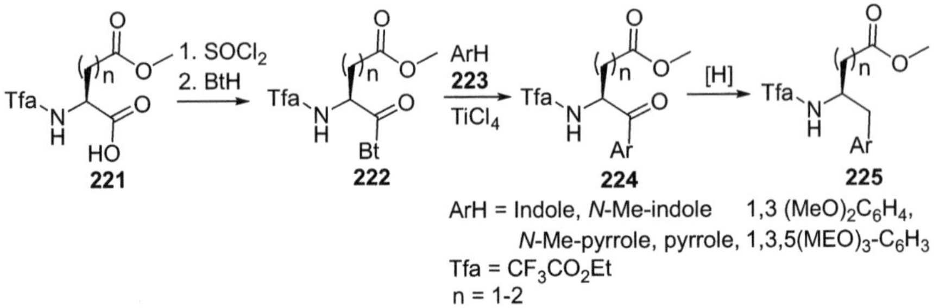

Scheme 65 Synthesis of novel β- and γ-amino acid derivatives of aspartic and glutamic acid **225**

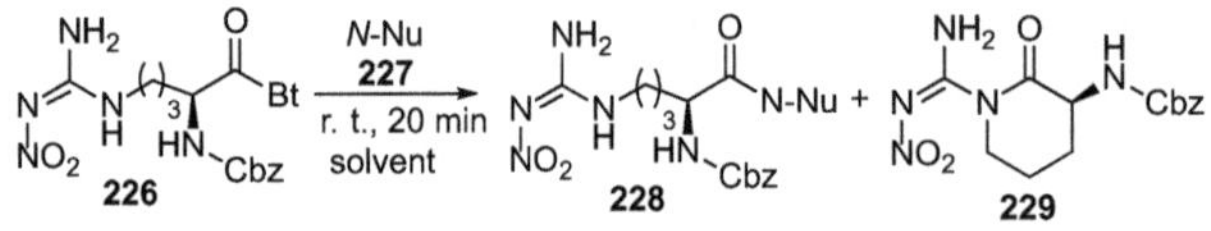

Scheme 66 Synthesis of L-arginine conjugates with *N*-Nu **228**

Reaction of *N*-(3-aminopropyl)imidazole in THF gave products **228** and **229** in 83% and 12% yields. Dipropylamine produced traces of the conjugate **228** and 90% of the cyclic by-product **229**, and the reaction with glycine resulted in 80% of **228** and 15% of **229**, respectively (Scheme 66) [147].

3.4.3 Azido Acid Conjugates

Azido acids **88**, synthesized via the reaction of amino acids with benzotriazol-1-yl-sulfoyl azide [64], were activated to give benzotriazolides **89** under regular reaction conditions (r. t., 1.2 eq. thionyl chloride, 2 eq. 1*H*-benzotriazole, Scheme 67). The reaction of azidobenzotriazolide intermediates **89** with *N*-, *O*-, *S*-, and *C*-nucleophiles led to the formation of products **231** in 57–95%, yields depending on the nature of the nucleophile. Also, various reaction conditions have been optimized to improve the reaction yields [64].

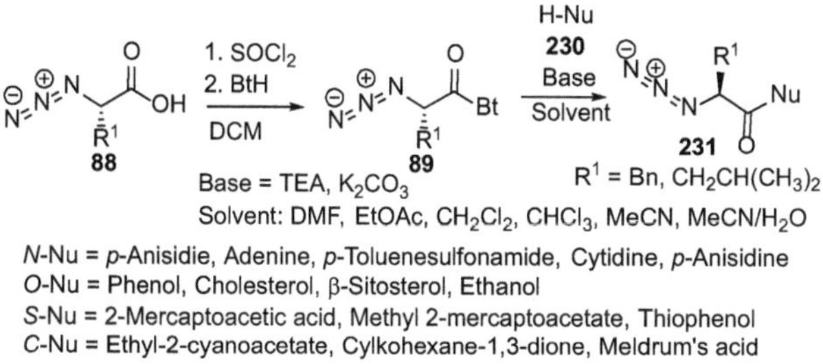

Scheme 67 Benzotriazole-mediated synthesis of azido acid *N*-, *O*-, *S*-, *C*-conjugates **231**

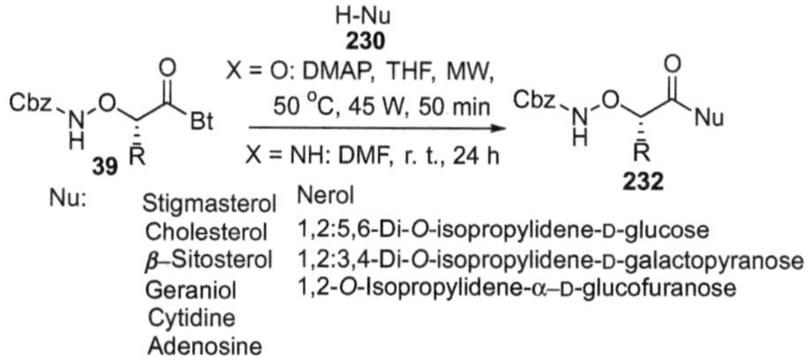

Scheme 68 Synthesis of *O*- and *N*-conjugates of *N*-Cbz α-aminoxy acids **232**

3.4.4 Aminoxy Acid Conjugates of Peptidomimetics

Katritzky et al. reported an efficient microwave-assisted synthesis of aminoxy acid conjugates with O- and *N*-nucleophiles [148]. *N*-Cbz-protected (α-aminoxyacyl) benzotriazoles **39** were prepared in yields ranging from 76 to 86% from the corresponding *N*-Cbz-protected α-aminoxy acids. *N*-Cbz-(α-aminoxyacyl)benzotriazoles **39** were then coupled with *O*-nucleophiles **230** to give **232** in 22–60% yields. Coupling of Bt intermediates **39** with the polynucleophilic systems **230**, such as cytidine and adenosine, resulted in the isolation of the *N*-acylated products **232** in 21% (cytidine) and 54% (adenosine) yields (Scheme 68) [148].

Katritzky et al. reported the incorporation of coumarin into aminoxy acids in aqueous acetonitrile at room temperature to give coumarin-labeled aminoxy acids **40**. Coumarin-labeled aminoxy hybrid peptides were obtained in two steps from the respective labeled aminoxy-acid benzotriazoles **206** (Scheme 69). Coumarin-aminoxy acids **233** and aminoxy dipeptides **235** were isolated in 76–94% yields. The 7-methoxycoumarin derivatives have quantum yields of 0.35–0.71 and may therefore be useful in peptide assays [149].

Katritzky et al. developed an efficient benzotriazole-mediated synthesis of azo-dye-labeled aminoxy acids **237** and peptides **239** [150]. For the synthesis of aminoxy acid–azo-dye conjugates **237**, *N*-(4-arylazobenzoyl)-1*H*-benzotriazoles **236** were treated with aminoxy acids **40** in THF/H_2O in the presence of TEA for 4–8 h at room temperature to afford the conjugated aminoxy products **237** in 65–

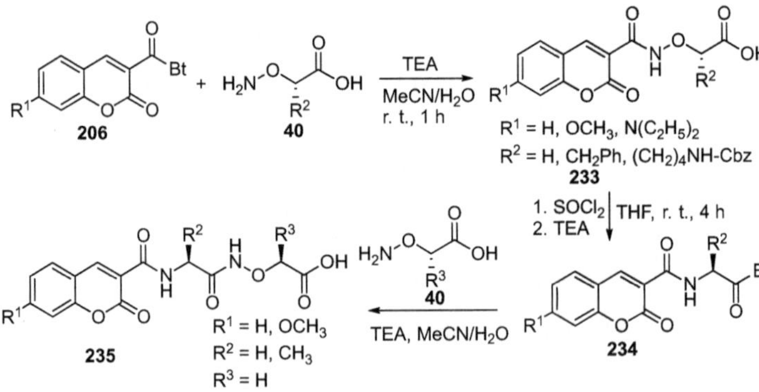

Scheme 69 Synthesis of coumarin-labeled aminoxy hybrid peptides **235**

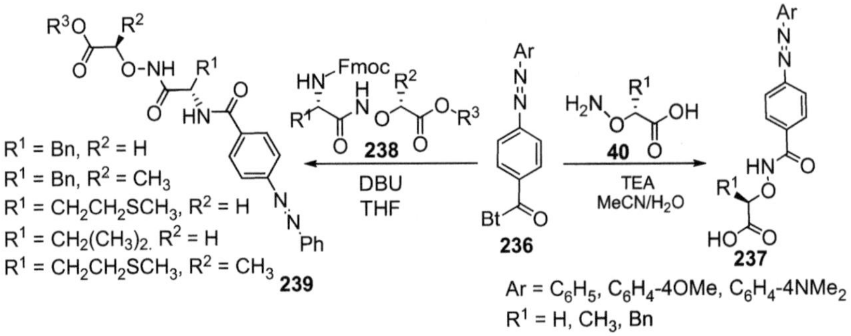

Scheme 70 Synthesis of azo-dye-labeled aminoxy acids **237** and peptides **239**

80% yields. In the synthesis of dipeptide azo-dye conjugates **239**, *N*-Fmoc-protected aminoxy dipeptides **238** were treated with a solution of 1 eq. of DBU in dry THF for 2 h at 0–5 °C, followed by the treatment with 4-phenylazobenzoyl-1*H*-benzotriazole **236**, which gave the dye-labeled aminoxy peptides **239** in 55–65% yields (Scheme 70) [150].

3.4.5 Depsipeptide Conjugates of Peptidomimetics

Biswas et al. reported the synthesis of coumarin-labeled depsidi- and tripeptides **242** and **243** [50]. L-Depsidipeptides **242** were synthesized using benzotriazole methodology [49, 50]. Coumarinoyl-labeled depsidipeptides **242** were obtained by treatment of coumarinoyl benzotriazoles **206** with unprotected depsipeptides **240** in the presence of TEA and MeCN/H_2O at room temperature in 69–92% yields (Scheme 73). An analogous approach was used in the couplings of coumarinoyl benzotriazoles **206** with depsitripeptides **241** to form coumarin-tagged L-depsitripeptides **243** in 83–92% yields (Scheme 71). Variations of quantum yields in

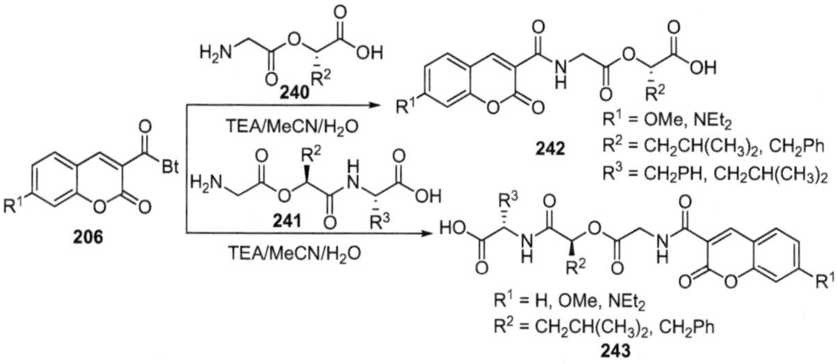

Scheme 71 Synthesis of coumarin-labeled depsidi- and tripeptides **242** and **243**

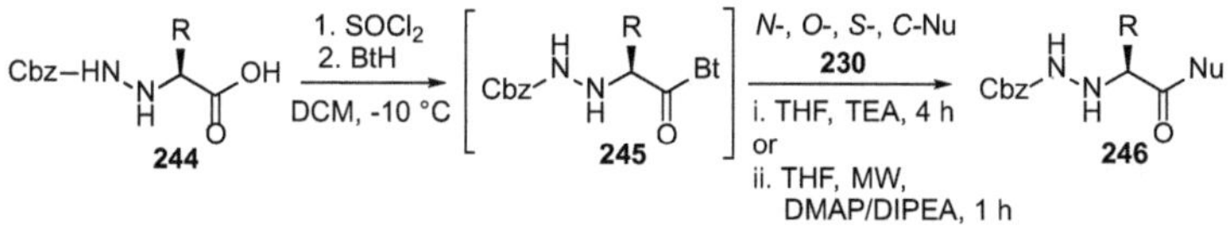

Scheme 72 Synthesis of *N*-L-Cbz-hydrazino acid conjugates with *N*-, *O*-, *S*-, *C*-nucleophiles

different solvents were studied and reported, and it was suggested that coumarin-labeled depsipeptides could be good candidates for the real-time monitoring of physiological processes [50].

3.4.6 Hydrazino Acid Conjugates

N-protected hydrazino acids **244** were converted into the carbonyl active benzotriazolides **245**, which were highly hygroscopic and used without further purification for the reaction with various *N*-, *O*-, *S*-, and *C*-nucleophiles **230**. Under the appropriate conditions (solvent: THF, base: TEA or DIPEA), *N*-Cbz-L-hydrazino benzotriazolides **245** underwent a series of acylations with *N*-, *O*-, *S*-, and *C*-nucleophiles **233** to give products **246** with complete retention of configuration (confirmed by chiral HPLC) with 49–88% yields (Scheme 72) [60].

References

1. Du AW, Stenzel MH (2014) Biomacromolecules 15:1097
2. Albericio F, Kruger HG (2012) Future Med Chem 4:1527
3. Vagner J, Qu H, Hruby VJ (2008) Curr Opin Chem Biol 12:292
4. Fear G, Komarnytsky S, Raskin I (2007) Pharm Therapeut 113:354
5. Schwenk RW, Vogel H, Schürmann A (2013) Mol Metab 2:337

6. Kanasaki K, Koya D (2011) J Biomed Biotechnol 2011:197636
7. Randolph JT, DeGoey DA (2004) Curr Top Med Chem 4:1079
8. Kazmierski WM, Kenakin TP, Gudmundsson KS (2006) Chem Biol Drug Des 67:13
9. Ahrens VM, Bellmann-Sickert K, Beck-Sickinger AG (2012) Future Med Chem 4:1567
10. Latacz G, Pekala E, Ciopinska A, Kiec-Kononowicz K (2006) Acta Pol Pharm 63:430
11. Walensky LD, Kung AL, Escher I, Malia TJ, Barbuto S, Wright RD, Wagner G, Verdine GL, Korsmeyer SJ (2004) Science 305:1466
12. Li L, Thomas RM, Suzuki H, De Brabander JK, Wang X, Harran PG (2004) Science 305: 1471
13. Wu G, Chai J, Suber TL, Wu J-W, Du C, Wang X, Shi Y (2000) Nature 408:1008
14. González-López M, Welsh K, Finlay D, Ardecky RJ, Ganji SR, Su Y, yuan H, Teriete P, Mace PD, Riedl SJ, Vuori K, Reed JC, Cosford NDP (2011) Bioorg Med Chem Lett 21:4332
15. Patick AK, Binford SL, Brothers MA, Jackson RL, Ford CE, Diem MD, Maldonado F, Dragovich PS, Zhou R, Prins TJ, Fuhrman SA, Meador JW, Zalman LS, Matthews DA, Worland ST (1999) Antimicrob Agents Chemother 43:2444
16. Kostis JB, Packer M, Black HR, Schmieder R, Henry D, Levy E (2004) Am J Hypertens 17: 103
17. Gokhale AS, Satyanarayanajois S (2014) Immunotherapy 6:755
18. Giuliani A, Rinaldi AC (2011) Cell Mol Life Sci 68:2255
19. Vaishampayan U, Glode M, Du W, Kraft A, Hudes G, Wright J, Hussain M (2000) Clin Cancer Res 6:4205
20. Lhospice F, Bregeon D, Belmant C, Dennler P, Chiotellis A, Fischer E, Gauthier L, Boedec A, Rispaud H, Savard-Chambard S, Represa A, Schneider N, Paturel C, Sapet M, Delcambre C, Ingoure S, Viaud N, Bonnafous C, Schibli R, Romagne F (2015) Mol Pharmaceut 12:1863
21. Gupta S, Jain A, Chakraborty M, Sahni JK, Ali J, Dang S (2013) Drug Deliv 20:237
22. Chugh A, Eudes F, Shim YS (2010) IUBMB life 62:183
23. Lee S, Xie J, Chen X (2010) Biochemistry 49:1364
24. Hansen PR, Flyge H, Holm A, Lauritzen E, Larsen BD (1996) Int J Pept Prot Res 47:419
25. Zhang P, Cheetham AG, Lock LL, Cui H (2013) Bioconj Chem 24:604
26. Srinivasan D, Muthukrishnan N, Johnson GA, Erazo-Oliveras A, Lim J, Simanek EE, Pellois JP (2011) PloS one 6, e17732
27. Dupont E, Prochiantz A, Joliot A (2011) Methods Mol Biol 683:21
28. Futaki S (2006) Biopolymers 84:241
29. El Khafagy S, Morishita M (2012) Adv Drug Deliv Rev 64:531
30. Chen F, Gerion D (2004) Fluorescent CdSe/ZnS Nano Lett 4:1827
31. Marmé N, Knemeyer J-P, Wolfrum J, Sauer M (2004) Angew Chem Int Ed 43:3798
32. McGuire MJ, Gray BP, Li S, Cupka D, Byers LA, Wu L, Rezaie S, Liu Y-H, Pattisapu N, Issac J, Oyama T, Diao L, Heymach JV, Xie X-J, Minna JD, Brown KC (2014) Sci Rep 27: 4480
33. Cai W, Chen X (2008) Nat Prot 3:89
34. Newman DJ (2008) J Med Chem 51:2589
35. Katritzky AR, Singh A, Haase DN, Yoshioka M (2009) Arkivoc 8:47
36. Monbaliu JCM, Hansen FK, Beagle LK, Panzner MJ, Steel PJ, Todadze E, Stevens CV, Katritzky AR (2012) Chem Eur J 18:2632
37. Avan I, Hall CD, Katritzky AR (2014) Chem Soc Rev 43:3575
38. Bajaj K, Panda SS, Ibrahim MA, El-Feky SA, Katritzky AR (2013) J Pept Sci 19:110
39. Panda SS, Hall CD, Scriven E, Katritzky AR (2013) Aldrichim Acta 46:43
40. Ha K, Chahar M, Monbaliu J-CM, Todadze E, Hansen FK, Oliferenko AA, Ocampo CE, Leino D, Lillicotch A, Stevens CV (2012) J Org Chem 77:2637
41. Katritzky AR, Todadze E, Angrish P, Draghici B (2007) J Org Chem 72:5794
42. Potschka H, Feuerstein TJ, Löscher W (2000) Naunyn Schmiedeberg's Arch Pharmacol 361: 200

43. Lebedyeva IO, Ostrov DA, Neubert J, Steel PJ, Patel K, Sileno SM, Goncalves K, Ibrahim MA, Alamry KA, Katritzky AR (2014) Bioorg Med Chem 22:1479
44. Ripps H, Shen W (2012) Molecular Vis 18:2673
45. Vertesaljai P, Biswas S, Lebedyeva I, Broggi E, Asiri AM, Katritzky AR (2014) J Org Chem 79:2688
46. Duclohier H (2007) Chem Biodivers 4:1023
47. Suarez-Jimenez G-M, Burgos-Hernandez A, Ezquerra-Brauer J-M (2012) Mar drugs 10:963
48. Zampella A, D'Auria MV, Paloma LG, Casapullo A, Minale L, Debitus C, Henin Y (1996) J Am Chem Soc 118:6202
49. Avan I, Tala SR, Steel PJ, Katritzky AR (2011) J Org Chem 76:4884
50. Biswas S, Avan I, Basak AK, Abo-Dya NE, Asiri A, Katritzky AR (2013) Amino acids 45: 159
51. Chen F, Song K-S, Wu Y-D, Yang D (2008) J Am Chem So 130:743
52. Katritzky AR, Avan I, Tala SR (2009) J Org Chem 74:8690
53. Li Z, Lebedyeva I, Zhao D, Myers L, Pillai GG, Hall CD, Katritzky AR (2014) J Pept Sci 20: 923
54. Biswas S, Abo-Dya NE, Oliferenko A, Khiabani A, Steel PJ, Alamry KA, Katritzky AR (2013) J Org Chem 78:8502
55. Moussodia R-O, Acherar S, Romero E, Didierjean C, Jamart-Grégoire B (2015) J Org Chem 80:3022
56. Aubry A, Mangeot JP, Vidal J, Collet A, Zerkout S, Marraud M (1994) Int J Pept Prot Res 43: 305
57. Eldridge GM, Weiss GA (2011) Bioconj Chem 22:2143
58. Cerea P, Giannini C, Dall'Angelo S, Licandro E, Maiorana S, Marchelli R (2007) Tetrahedron 63:4108
59. Bouget K, Aubin S, Delcros J-G, Arlot-Bonnemains Y, Baudy-Floc'h M (2003) Bioorg Med Chem 11:4881
60. Panda SS, El-Nachef C, Bajaj K, Katritzky AR (2013) Eur J Org Chem 2013:4156
61. Proulx C, Sabatino D, Hopewell R, Spiegel J, García Ramos Y, Lubell WD (2011) Future Med Chem 3:1139
62. Abo-Dya NE, Biswas S, Basak A, Avan I, Alamry KA, Katritzky AR (2013) J Org Chem 78: 3541
63. Henderson G, Hughes J, Kosterlitz H (1978) Nature 271:677
64. Katritzky AR, El Khatib M, Bol'shakov O, Khelashvili L, Steel PJ (2010) J Org Chem 75: 6532
65. Zavalova L, Lukyanov S, Baskova I, Snezhkov E, Akopov S, Berezhnoy S, Bogdanova E, Barsova E, Sverdlov ED (1996) Mol Gen Genet 253:20
66. Roch A-M, Noel P, Alaoui SE, Charlot C, Quash G (1991) Int J Cancer 48:215
67. Popov V, Panda SS, Katritzky AR (2013) Org Biomol Chem 11:1594
68. Popov V, Panda SS, Katritzky AR (2013) J Org Chem 78:7455
69. Monbaliu J-CM, Dive G, Stevens CV, Katritzky AR (2013) J Chem Theory Comput 9:927
70. Liaqat S, Panda SS, Rauf A, Al-Youbi AO, Katritzky AR (2014) Synthesis 46:0067
71. El Khatib M, Jauregui L, Tala SR, Khelashvili L, Katritzky AR (2011) Med Chem Comm 2: 1087
72. Katritzky AR, Tala SR, Abo-Dya NE, Ibrahim TS, El-Feky SA, Gyanda K, Pandya KM (2010) J Org Chem 76:85
73. Panda SS, El-Nachef C, Bajaj K, Al-Youbi AO, Oliferenko A, Katritzky AR (2012) Chem Biol Drug Des 80:821
74. Katritzky AR, Tala SR, Abo-Dya NE, Gyanda K, El-Gendy BE-DM, Abdel-Samii ZK, Steel PJ (2009) J Org Chem 74:7165
75. Borthwick AD (2012) Chem Rev 112:3641
76. Ha K, Lebedyeva I, Li Z, Martin K, Williams B, Faby E, Nasajpour A, Pillai GG, Al-Youbi AO, Katritzky AR (2013) J Org Chem 78:8510

77. Avilov SA, Antonov AS, Silchenko AS, Kalinin VI, Kalinovsky AI, Dmitrenok PS, Stonik VA, Riguera R, Jimenez C (2003) J Nat Prod 66:910
78. Nsengiyumva O, Hamedzadeh S, McDaniel J, Macho J, Simpson G, Panda SS, Ha K, Lebedyeva I, Faidallah HM, Al-Mohammadi MM, Hall CD, Katritzky AR (2015) Org Biomol Chem 13:4399
79. Masse A, Ramirez LH, Bindoula G, Grillon C, Wdzieczak-Bakala J, Raddassi K, Deschamps de Paillette E, Mencia-Huerta JM, Koscielny S, Potier P, Sainteny F, Carde P (1998) Blood 91:441
80. Li Z, Lebedyeva IO, Golubovskaya VM, Cance WG, Alamry KA, Faidallah HM, Dennis Hall C, Katritzky AR (2015) Bioorg Med Chem 23:5056
81. Ibrahim MA, Panda SS, Nhon L, Hamed A, El-Feky SA, Katritzky AR (2013) Synthesis 45: 0767
82. Berhanu WM, Ibrahim MA, Pillai GG, Oliferenko AA, Khelashvili L, Jabeen F, Mirza B, Ansari FL, El-Feky SA, Katritzky AR (2012) Beilstein J Org Chem 8:1146
83. Shaul PW, Smart EJ, Robinson LJ, German Z, Yuhanna IS, Ying Y, Anderson RG, Michel T (1996) J Biol Chem 27:6518
84. Hess DT, Patterson SI, Smith DS, Skene JP (1993) Nature 366:562
85. Guo L, Lim KB, Poduje CM, Daniel M, Gunn JS, Hackett M, Miller SI (1998) Cell 95:189
86. Katritzky AR, He HY, Suzuki K (2000) J Org Chem 65:8210
87. Sahu S, Lebedyeva IO, Panda SS, Katritzky AR (2013) Synthesis 45:1256
88. Bajaj K, Panda SS, El-Nachef C, Katritzky AR (2012) Chem Biol Drug Des 80:17
89. Katritzky AR, Zhang Y, Singh SK (2003) Synthesis 18:2795
90. Katritzky AR, Kirichenko N, Rogovoy BV (2003) Synthesis 18:2777
91. Katritzky AR, Rogovoy BV, Kirichenko N, Vvedensky V (2002) Bioorg Med Chem Lett 12: 1809
92. Katritzky AR, Suzuki K, Zuoquan W (2005) Synlett 11:1656
93. Katritzky AR, Singh SK, Cai C, Bobrov S (2006) J Org Chem 71:3364
94. Katritzky AR, Khelashvili L, Mohapatra PP, Steel PJ (2007) Synthesis 23:3673
95. Katritzky AR, Suzuki K, Singh SK, He H-Y (2003) J Org Chem 68:5720
96. Katritzky AR, Rogovoy BV (2003) Chem Eur J 9:4586
97. Katritzky AR, Rogovoy BV, Cai X, Kirichenko N, Kovalenko KV (2004) J Org Chem 69:309
98. Katritzky AR, Kirichenko N, Rogovoy BV (2003) Arkivoc 8:8
99. Katritzky AR, Meher NK, Cai C, Singh SK (2004) J Mex Chem Soc 48:275
100. Katritzky AR, Wang Z, Wang M, Wilkerson CR, Hall CD, Akhmedov NG (2004) J Org Chem 69:6617
101. Katritzky AR, Angrish P, Todadze E (2009) Synlett 2009:2392
102. Ibrahim MA, Panda SS, Alamry KA, Katritzky AR (2013) Synthesis 45:3255
103. Panda SS, Ibrahim MA, Küçükbay H, Meyers MJ, Sverdrup FM, El-Feky SA, Katritzky AR (2013) Chem Biol Drug Des 82:361
104. El-Fekyb SA (2010) Synlett 9:1337
105. Hansen FK, Beagle LK, Todadze E, Katritzky AR (2011) Heterocycles 84:515
106. Katritzky AR, El-Nachef C, Bajaj K, Kubik J, Haase DN (2010) J Org Chem 75:6009
107. Kelly WL (2012) Nat Chem Biol 8:505
108. Walsh CT, Nolan EM (2008) Proc Nat Acad Sci 105:5655
109. Katritzky AR, El-Gendy BE-DM, Todadze E, Abdel-Fattah AAA (2008) J Org Chem 73: 5442
110. Katritzky AR, Todadze E (2009) Arkivoc 9:248
111. El-Nachef C, Bajaj K, Koblick J, Katritzky AR (2012) Eur J Org Chem 23:4412
112. Katritzky AR, Angrish P (2006) Steroids 71:660
113. Katritzky AR, Angrish P (2006) Synthesis 24:4135
114. Barbakadze NN, Jones RA, Rosario NR, Nadaraia NS, Kakhabrishvili ML, Hall CD, Katritzky AR (2014) Tetrahedron 70:7181
115. Katritzky AR, Bajaj K, Charpentier M, Zadeh EG (2010) J Org Chem 75:3938

116. Pace G, Ferri V, Grave C, Elbing M, von Hänisch C, Zharnikov M, Mayor M, Rampi MA, Samorì P (2007) Proc Nat Acad Sci 104:9937
117. Katritzky AR, Chen Q-Y, Tala SR (2008) Org Biomol Chem 6:2400
118. Sahoo H (2012) RSC Advances 2:7017
119. Jung D, Min K, Jung J, Jang W, Kwon Y (2013) Molecular BioSyst 9:862
120. Suzuki A, Takahashi N, Okada Y, Saito I, Nemoto N, Saito Y (2013) Bioorg Med Chem Lett 23:886
121. Doria F, Nadai M, Sattin G, Pasotti L, Richter SN, Freccero M (2012) Org Biomol Chem 10: 3830
122. Katritzky AR, Ozcan S, Todadze E (2010) Org Biomol Chem 8:1296
123. Signore G, Nifosì R, Albertazzi L, Bizzarri R (2009) J Biomed Nanotechnol 5:722
124. Lavis LD, Raines RT (2008) ACS Chem Biol 3:142
125. Katritzky AR, Yoshioka M, Narindoshvili T, Chung A, Johnson JV (2008) Org Biomol Chem 6:4582
126. Katritzky AR, Angrish P, Narindoshvili T (2007) Bioconj Chem 18:994
127. Katritzky AR, Narindoshvili T, Draghici B, Angrish P (2008) Regioselective J Org Chem 73: 511
128. Katritzky AR, Khelashvili L, Munawar MA (2008) J Org Chem 73:9171
129. Bol'shakov OI, Lebedyeva IO, Katritzky AR (2012) Synthesis 18:2926
130. Wallace JL, McKnight W, Reuter BK, Vergnolle N (2000) Gastroenterology 119:706
131. Katritzky AR, Jishkariani D, Narindoshvili T (2009) Chem Biol Drug Des 73:618
132. Sahu S, Panda SS, Asiri AM, Katritzky AR (2013) Synthesis 45:3326
133. Aboul-Fadl T, Fouad E (1996) Pharmazie 51:30
134. Ibrahim MA, Panda SS, Birs AS, Serrano JC, Gonzalez CF, Alamry KA, Katritzky AR (2014) Bioor Med Chem Lett 24:1856
135. Panda SS, Bajaj K, Meyers MJ, Sverdrup FM, Katritzky AR (2012) Org Biomol Chem 10: 8985
136. Katritzky AR, Munawar MA, Kovacs J, Khelashvili L (2009) Org Biomol Chem 7:2359
137. Jones RA, Thillier Y, Panda SS, Rosario NR, Hall CD, Katritzky AR (2014) Org Biomol Chem 12:8325
138. Panda SS, Naumov RN, Asiri AM, Katritzky AR (2014) Synthesis 46:1511
139. Tiwari AD, Panda SS, Asiri AM, Hall CD, Katritzky AR (2014) Synthesis 46:2430
140. Katritzky AR, Cusido J, Narindoshvili T (2008) Bioconj Chem 19 7:1471
141. Katritzky AR, Narindoshvili T (2009) Org Biomol Chem 7:627
142. Naumov RN, Panda SS, Girgis AS, George RF, Farhat M, Katritzky AR (2015) Bioorg Med Chem Lett 25:2314
143. Tiwari AD, Panda SS, Girgis AS, Sahu S, George RF, Srour AM, La Starza B, Asiri AM, Hall CD, Katritzky AR (2014) Org Biomol Chem 12:7238
144. Katritzky AR, Vincek AS, Suzuki K (2005) Arkivoc 116:126
145. Katritzky AR, Yang H, Zhang S, Wang M (2002) Arkivoc 11:39
146. Katritzky AR, Tao H, Jiang R, Suzuki K, Kirichenko K (2007) J Org Chem 72:407
147. Khaybullin RN, Panda SS, Mirzai S, Toneff E, Asiri AM, Hall CD, Katritzky AR (2014) RSC Advances 4:55210
148. Draghici B, Hansen FK, Buciumas A-M, El-Gendy BE-DM, Todadze E, Katritzky AR (2011) RSC Advances 1:602
149. Katritzky AR, Abdelmajeid A, Tala SR, Amine M, Steel PJ (2011) Synthesis 1:83
150. Biswas S, Tala SR, Kalatozishvili A, Katritzky AR (2011) Arkivoc 3:212

Top Heterocycl Chem (2016) 43: 143–176
DOI: 10.1007/7081_2015_180

Published online: 14 October 2015

Benzotriazole-Mediated Synthesis of Oxygen-Containing Heterocycles

Oleg I. Bolshakov

Abstract This review aims at delivering an overview of the synthetic utility of benzotriazole and its derivatives for the preparation of oxygen-containing heterocycles. This review serves both as a synthetic guide and as a theoretical handbook of benzotriazole-assisted preparations of oxygen heterocycles and offers a complete overview of existing applications for benzotriazole and its derivatives in heterocyclic synthesis. Multiple examples reveal the superiority of benzotriazole-based approaches over conventional ones.

Keywords 1*H*-1,2,3-Benzotriazole · Carbanions · Cyclization · Cycloaddition · Furans · Oxadiazoles · Oxazepines · Oxazines · Oxazoles · Oxazolidines · Oxiranes · Pyrans · β-Lactones

Contents

O.I. Bolshakov (✉)
South Ural State University, 76, Lenin prospekt, Chelyabinsk 454080, Russia
e-mail: bolshakovoi@susu.ru

Abbreviations

$(P(C_6H_5)_3)_2PdCl_2$	Bis(triphenylphosphine)palladium(II) dichloride
BtH	1*H*-1,2,3-Benzotriazole
BuLi	Butyllithium
CH_3CN	Acetonitrile
DBU	1,8-Diazabicyclo[5.4.0]undec-7-ene
DCM	Dichloromethane
DMD	3,3-Dimethyldioxirane
DME	Dimethoxyethane
DMF	Dimethylformamide
DMSO	Dimethylsulfoxide
E	Electrophile
h	Hour
Hal	Halogen
HMPA	Hexamethylphosphoramide
i-PrOH	2-Propanol
LDA	Lithium diisopropylamide
LiHMDS	Lithium hexamethyldisilazide
m-CPBA	3-Chloroperbenzoic acid
NIS	*N*-Iodosuccinimide
NMP	*N*-Methyl-2-pyrrolidone
Py	Pyridine
rt	Room temperature
t-BuOH	*tert*-Butyl alcohol
t-BuOK	Potassium *tert*-butoxide
TFA	Trifluoroacetic acid
THF	Tetrahydrofuran
TMSCl	Trimethylchlorosilane
TsOH	4-Toluene sulfonic acid

1 Introduction

There is no doubt that oxygen-containing heterocycles are one of the most prominent organic heterocyclic compounds; they have widespread occurrence in organic life and found versatile industrial, nutritional, and medicinal applications. For classical reviews on the preparation and chemistry of oxygen heterocycles, author refers to the volumes of “Comprehensive Heterocyclic Chemistry” of 1984 [1], 1996 [2], and 2008 [3] edited by Alan Roy Katritzky.

This review gives an overview of a significant part of “benzotriazole methodology” [4], i.e., the synthesis of oxygen-containing heterocycles in which benzotriazole enables cycle formation. This review is focused on the cyclization process and excludes benzotriazole-assisted ring functionalization and derivatization [5, 6].

Benzotriazole-mediated synthesis of oxygen-containing heterocycles benefits from the major assets of benzotriazole chemistry, such as benzotriazole good leaving group ability, benzotriazole α-C–H activation to proton loss, electron-donor properties of benzotriazole, and carbanion stabilization with benzotriazolyl substituent, as described in earlier reviews from the Katritzky group [6–15]. In most cases, the benzotriazole methodology provides convenient synthetic alternatives to the previously established protocols. For instance, most of these transformations are done in one pot and use easily prepared and large-scale compatible precursors. Besides, the resulting heterocyclic products bear a residual benzotriazole moiety, which opens up tremendous opportunities for further derivatization.

The synthetic versatility of benzotriazole has allowed for the preparation of a diverse range of its active derivatives, which can be readily functionalized and incorporated into an almost infinite number of heterocyclic systems. Significant advances have been achieved in the entrapment of benzotriazol-1-ylalkyl carbanions with various electrophiles to give oxiranes, furans, and pyrans. Most of the reactions covered in this chapter proceed regioselectively and with satisfactory yields.

Multiple synthetic approaches to diverse heterocycles allow systematization according to the type of either the heterocycle or the synthetic methodology. In order to avoid confusion, this review is organized in a classical manner and presents a systematic structural approach according to the size and the nature of the heterocycle formed, as well as the presence of another heteroelement. Thus, this review is divided into two parts: (1) oxygen only and (2) oxygen- and nitrogen-containing heterocycles. Some represented polycyclic heterocycles are included as well. Established or suggested mechanisms are provided for most of the reactions in order to get a deeper understanding of benzotriazole chemical polyvalence.

As such, it is expected that the present review will make it clear to the reader that there is much done and left to do within benzotriazole-mediated synthesis of oxygen-containing heterocycles.

2 Synthesis of Oxygen-Containing Heterocycles

2.1 Three-Membered Rings

The first benzotriazole-assisted synthesis of oxiranes dates back to 1995 [16]. α-Benzotriazol-1-ylallyl ether **3**, readily available from benzotriazole **1** and acrolein acetal **2**, was deprotonated with BuLi to give anion **4**, which reacted with a variety of ketones to give alkoxyanions **6**. Following the cyclization of adducts **6** to epoxides **7** was accomplished with the assistance of $ZnBr_2$. The reaction of anion

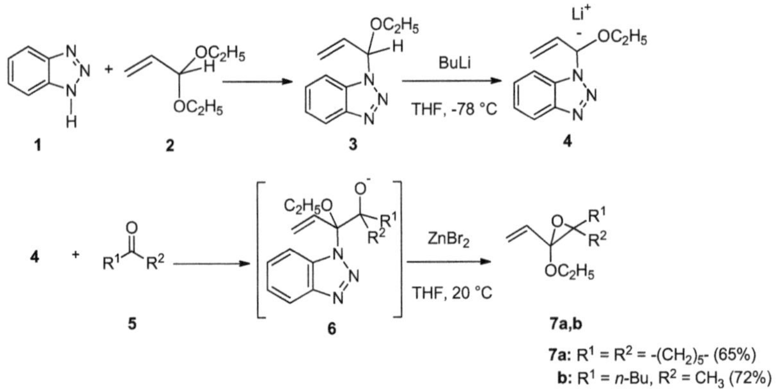

Scheme 1 Synthesis of ethoxy oxiranes **7**

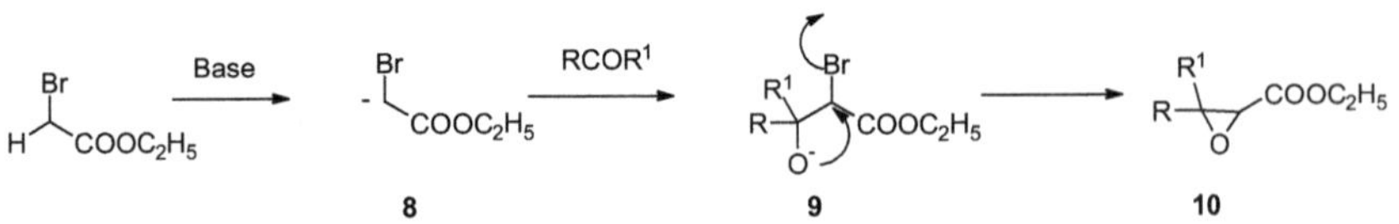

Scheme 2 Darzens glycidic ester cyclization

4 with aliphatic ketones or alicyclic ketones followed by stirring with $ZnBr_2$ in THF gave epoxides **7** in good yields (65–72%) (Scheme 1). The excellent leaving group ability of the benzotriazolyl moiety in compound **6** facilitated the formation of the oxirane ring via an intramolecular S_N2 nucleophilic displacement.

Such an addition-substitution sequence to form oxiranes is closely related to the Darzens glycidic ester condensation (Scheme 2), in which haloester carbanion **8** adds to an aldehyde or a ketone to give diastereomeric aldolates **9**. The last step involves an internal S_N2-displacement that provides α,β-epoxy ester **10** [16].

This methodology was later applied on variety of benzotriazolyl-stabilized anions **12**, obtained by the lithiation of 1-(1-alkoxyalkyl)- or 1-(1-aryloxyalkyl)-benzotriazoles **11**. Anion **12** was then quenched with both aliphatic and aromatic ketones **13** and treated with zinc bromide at elevated temperatures to trigger oxirane formation. 2-Alkoxyoxiranes **15** were obtained in moderate to good yields. The method was adapted to a one-pot procedure, providing a convenient preparation of polysubstituted 2-alkoxyoxiranes or 2-aryloxyoxiranes **15** (Scheme 3) [17].

Another benzotriazole-stabilized carbanion 1-benzotriazol-1-ylchloro-methyllithium **17**, which is generated from the corresponding chloride **16** by reaction with LDA, reacted with both enolizable and nonenolizable ketones **18** to give benzotriazolyloxiranes **19** in good yields (Scheme 4) [18].

The allyllithium generated by lithiation of 1-(3-chloroprop-2-enyl)-1*H*-1,2,3-benzotriazole **20** with LDA at −78°C in the presence of HMPA reacted with

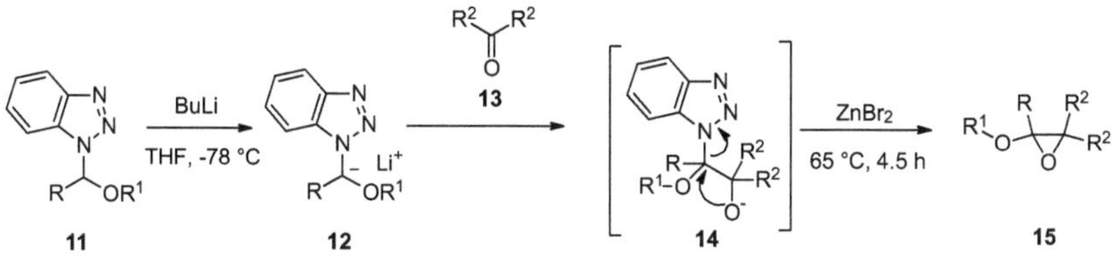

15a: R = n-C_5H_{11}, R^1 = C_6H_5, R^2 = -$(CH_2)_5$- (56 %)
b: R = n-C_5H_{11}, R^1 = C_6H_5, R^2 = C_6H_5 (43 %)
c: R = 4-ClC_6H_4, R^1 = CH_3, R^2 = -$(CH_2)_5$- (71 %)

Scheme 3 Synthesis of alkoxy- and aryloxyoxiranes **15**

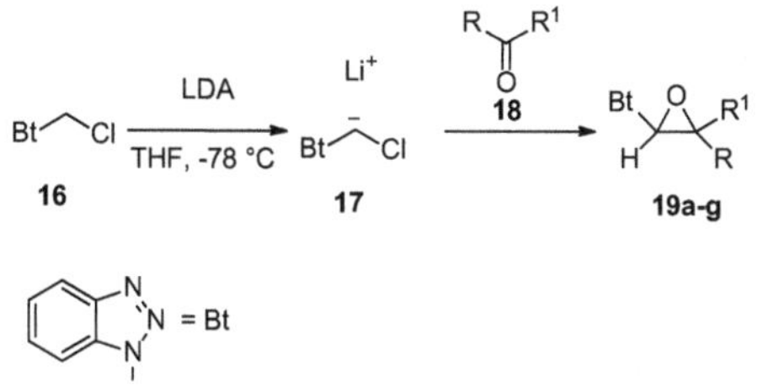

	R	R^1	Yield (%)
19a	C_6H_5	C_6H_5	68
b	C_2H_5	C_2H_5	72
c	-$(CH_2)_5$-		70
d	C_6H_5	C_2H_5	68
e	C_6H_5	CH_3	62
f	C_6H_5	4-ClC_6H_5	75
g	C_6H_5	4-$CH_3C_6H_5$	69

Scheme 4 Synthesis of benzotriazolyloxiranes **19**

enolizable and nonenolizable ketones and aldehyde regioselectively at the CCl terminus to give vinyloxiranes **22** in good yields (Scheme 5) [19].

Benzotriazole derivatives could be utilized not only as a stabilized carbanion source for the synthesis of oxiranes but also as carbonyl synthetic equivalent. For instance, benzotriazol-1-yl-methanol ($BtCH_2OH$) was used as an alternative source of formaldehyde in the reaction with lithiated chloroalkyloxazole **23** giving terminal oxazolinyloxiranes **25**, through chlorohydrin **24**, in good yields (70–90%) (Scheme 6) [20].

Xiao and coworkers reported that 1-benzyl-3-methylbenzotriazolium iodide **26** reacted with *t*-BuOK in a 1:1 mixture of *t*-BuOH/DMSO at room temperature to form 1-benzyl-3-methylbenzotriazolium ylide. 1-Benzyl-3-methylbenzotriazolium ylide then reacted with aromatic aldehydes **27** as electrophiles, yielding nonsymmetrical diaryl oxiranes **28** in moderate to good yields. The resulting oxiranes were obtained mostly as *trans*-isomers (Scheme 7). The formation of 1-aryl-2-(benzotriazol-1-yl)ethanol **29** as major by-product was reported along with oxiranes **28**.

The mechanism for the formation of **28** and **29** is described in Scheme 7. Firstly, iodide **26** reacted with *t*-BuOK to form ylides **30** and **31**. Since ylide **31** is a more reactive species than ylide **30**, it reacted preferentially with an aryl aldehyde to give intermediate **32**. The latter occurred in equilibrium with more nucleophilic ylide **33**,

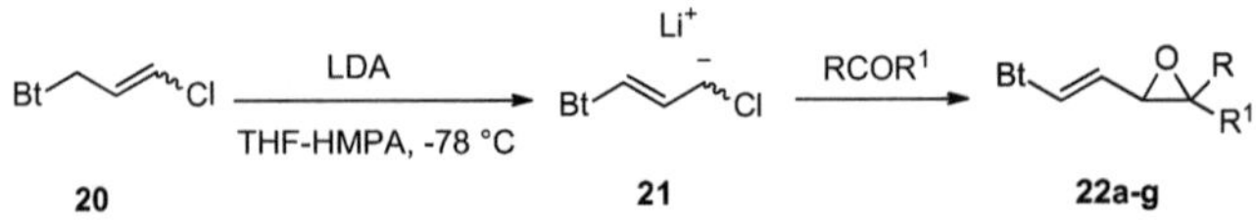

	R	R^1	Yield (%)
22a:	C_6H_5	C_6H_5	78
b:	C_6H_5	C_6H_5	73
c:	-$(CH_2)_5$-		72
d:	C_6H_5	H	69
e:	C_6H_5	4-ClC_6H_5	76
f:	C_6H_5	4-$CH_3OC_6H_5$	82
g:	C_6H_5	4-$CH_3C_6H_5$	61

Scheme 5 Synthesis of vinyloxiranes **22**

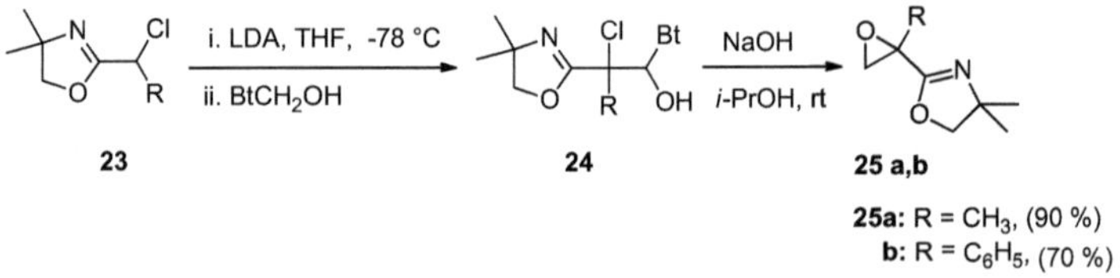

Scheme 6 Synthesis of oxazolinyloxiranes **25**

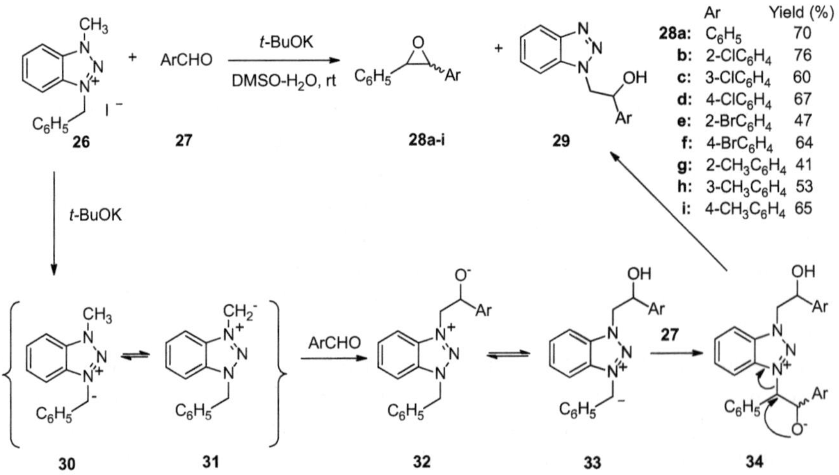

	Ar	Yield (%)
28a:	C_6H_5	70
b:	2-ClC_6H_4	76
c:	3-ClC_6H_4	60
d:	4-ClC_6H_4	67
e:	2-BrC_6H_4	47
f:	4-BrC_6H_4	64
g:	2-$CH_3C_6H_4$	41
h:	3-$CH_3C_6H_4$	53
i:	4-$CH_3C_6H_4$	65

Scheme 7 Synthesis of symmetric diaryloxiranes **28**

which reacted with another equivalent of aromatic aldehyde to form intermediate **34**. Finally, intermediate **34** underwent epoxide ring closure to give **28** and **29** [21].

Katritzky reported the preparation of 1-alkenylbenzotriazoles **36** by *t*-BuOK catalyzed isomerization of the corresponding allyl derivatives **35** [22]. Lithiation of **36**, followed by reaction with various electrophiles, provided α-substituted 1-vinylbenzotriazoles **37** (Scheme 8). In the last step, the epoxidation of the double

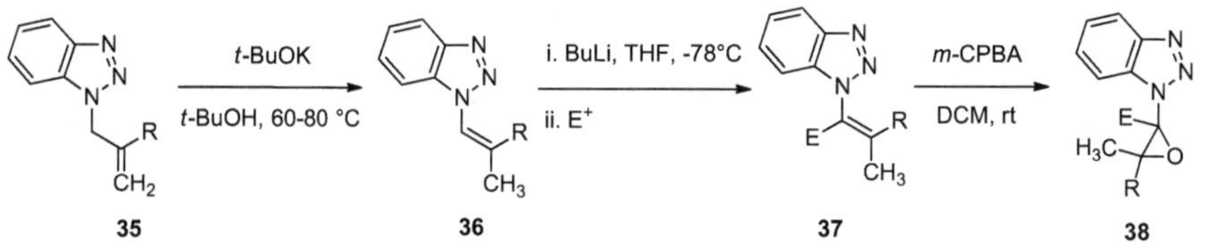

	R	E	Yield (%)
38a:	CH_3	Octyl	72
b:	CH_3	$4\text{-BrC}_6H_4CH_2$	61
c:	CH_3	$(C_6H_5)_2COH$,	43
d:	H	$(CH_2)_5COH$	80
e:	CH_3	C_6H_5CHOH	53

Scheme 8 Synthesis of benzotriazolyloxiranes **38**

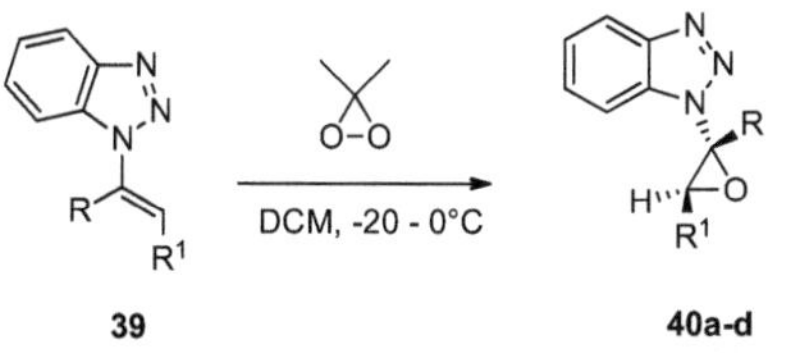

	R	R^1	Yield (%)
40a:	H	H	100
b:	CH_3	H	97
c:	H	CH_3	99
d:	H	C_6H_5	96

Scheme 9 Synthesis of benzotriazol-1-yloxiranes **40**

bond with 2 equiv. of *m*-chloroperbenzoic acid converted **37** into oxiranes **38** in 43–80% yields. The reaction conditions required a Na_2HPO_4 buffer since oxiranes are sensitive to acidic conditions [22].

Alternatively, benzotriazole-1-yloxiranes were obtained in practically quantitative yields by the epoxidation of various benzotriazol-1-ylalkenes **39** with dimethyldioxirane (DMD) in dichloromethane at temperatures ranging from −20°C to 0°C (Scheme 9). This procedure is very convenient since epoxides **40** were isolated in excellent yields after solvent evaporation [23]. No further purification was required.

2.2 Four-Membered Rings

H. Shick and colleagues reported an interesting procedure for the preparation of β-lactones from *N*-acylbenzotriazoles **41** with one hydrogen atom in α-position to the carboxamide group [24]. *N*-Acylbenzotriazoles **41** were deprotonated with lithium diisopropylamide (LDA) or lithium hexamethyldisilazide (LiHMDS) to furnish enolates **42**. Enolates **42** reacted with ketones or aldehydes at temperatures ranging from −95 to −90°C to furnish β-lithiated β-hydroxyacyl benzotriazoles **43**. Carboxamides **43** underwent cyclization with concomitant elimination of lithium benzotriazolide towards the corresponding di- and trisubstituted β-lactones **44** (Scheme 10). This procedure emphasized the advantages of benzotriazolides **41**

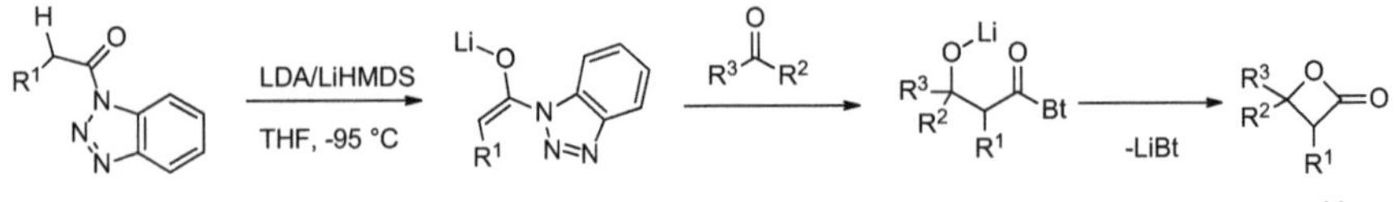

41 42 43 44

R^1	R^2	R^3	Yield of **44** (%)
n-C_2H_5	-(C_2H_5)-		78
n-C_6H_{13}	-(C_2H_5)-		89
$C_6H_5CH_2$	-(C_2H_5)-		87
H	-(C_2H_5)-		80
n-C_2H_5	CH_3	CH_3	53
n-C_6H_{13}	CH_3	CH_3	54
$C_6H_5CH_2$	CH_3	CH_3	65
n-C_2H_5	t-Bu	H	67
n-C_6H_{13}	t-Bu	H	90
$C_6H_5CH_2$	t-Bu	H	84

R^1	R^2	R^3	Yield of **44** (%)
n-C_2H_5	n-Pr	H	44
n-C_6H_{13}	n-Pr	H	46
$C_6H_5CH_2$	n-Pr	H	50
n-C_2H_5	n-C_7H_{15}	H	50
n-C_6H_{13}	n-C_7H_{15}	H	56
$C_6H_5CH_2$	n-C_7H_{15}	H	51
n-C_2H_5	$C_6H_5(CH_2)_2$	H	42
n-C_6H_{13}	$C_6H_5(CH_2)_2$	H	51
$C_6H_5CH_2$	$C_6H_5(CH_2)_2$	H	58
n-C_2H_5	n-C_6H_{11}	H	64
n-C_6H_{13}	n-C_6H_{11}	H	55
$C_6H_5CH_2$	n-C_6H_{11}	H	76

Scheme 10 Conversion of 1-acylbenzotriazoles **41** into β-lactones **44**

over the corresponding classical *S*-phenyl [25] or *O*-phenyl esters [26] for the formation of α-monosubstituted β-lactones **44** from aldehydes.

In all cases involving aldehydes and their reactions with the amide enolates **42**, both diastereoisomers of the corresponding α,β-disubstituted β-lactones **44** were formed. The *trans*-configurated diastereoisomers predominated in all cases [24].

Tetrahydrolipstatin **45** is a lipase inhibitor and is used as an antiobesity agent [27]. It consists of an (3*S*,4*S*)-oxetan-2-one core with *trans*-orientated alkyl groups and a (*S*)-configured hydroxy group at the 2nd position of the 4-alkyl chain. Shick et al. disclosed an efficient three-step procedure for the synthesis of (3*S*,4*S*)-3-hexyl-4-[(2*S*)-2-hydroxytridecyl]oxetan-2-one **50a**, which is a common intermediate for the preparation of **45**. This procedure featured a one-pot tandem aldol lactonization of lithium enolate **49** with aldehyde **48** (Scheme 11) [28].

The photolysis of vinyl methyl ethers **51** was reported for the construction of highly sterically hindered spirooxindoles **53**. The authors also reported the presence of 2-iminooxetane **52a** in the reaction mixture. For instance, the photolysis of 1-(benzotriazole-1-yl)-methoxymethylene adamantane **51a** afforded a 1:1 separable mixture of 2-iminooxetane **52a** and **53a** in a combined yield of 80%, demonstrating the ability of this reaction to form a C–C bond within a sterically congested environment. Even more steric congestion had no perceptible effect on the reaction outcome in case of substrate **51b**: irradiation of **51b** by high pressure mercury lamp yielded a separable 1:1 mixture of photoadducts **52b** and **53b** in a combined yield of 70% (Scheme 12) [29, 30].

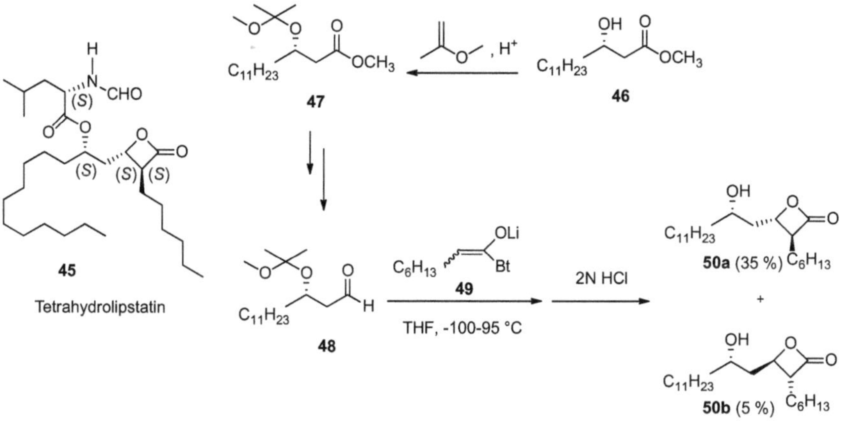

Scheme 11 Synthesis of a precursor **50a** towards tetrahydrolipstatin **45**

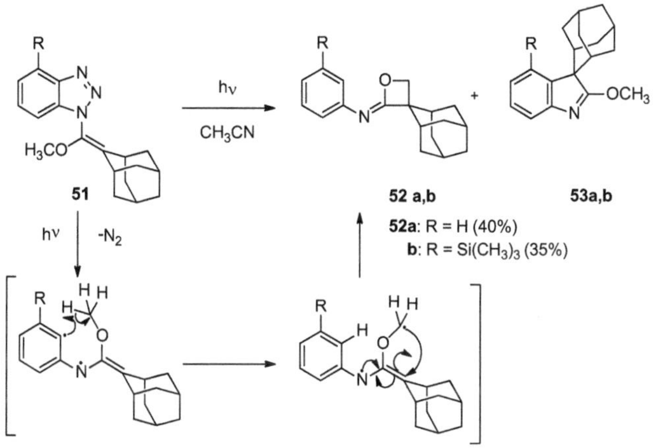

Scheme 12 Synthesis of 2-iminooxetanes **52**

2.3 *Five-Membered Rings*

This part follows the narrative of benzotriazole-stabilized carbanions in the synthesis of oxygen heterocycles. In this regard, the synthesis of 5-membered ring oxygen-containing heterocycles could be accomplished with elongated carbanions. For instance, 1-(*o*-silyloxybenzyl)benzotriazole **55** served as a precursor for the formation of furan **58**. This example illustrates two important properties of the benzotriazole group, namely, (1) α-negative charge stabilization and (2) leaving group ability.

Lithiation of **56** and its subsequent reaction with benzaldehyde afforded benzofuran precursor **57**. Heating of **57** in the presence of *p*-toluenesulfonic acid

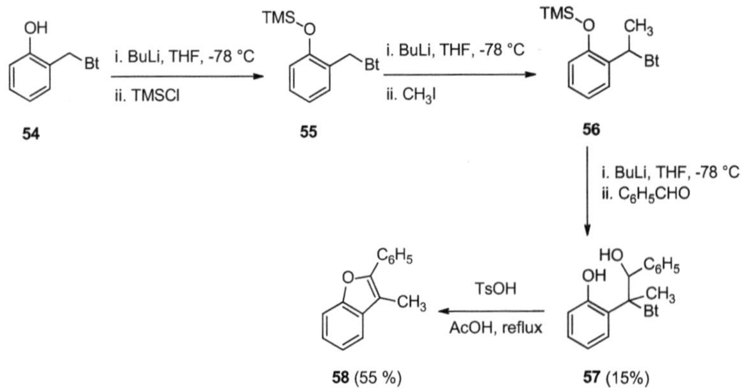

Scheme 13 Synthesis of 2-phenyl-3-methylbenzofuran **51**

promoted a sequential intramolecular etherification/benzotriazole elimination to give 2-phenyl-3-methylbenzofuran **58** in 55% yield (Scheme 13) [31].

2,5-Dihydrofurans were obtained via an intramolecular allene-hydroxyl cyclization. [32]. While allenes are quite unstable, they can be conveniently generated in situ from alkynes. Katritzky reported that the deprotonation of readily available 1-propargylbenzotriazole **59** and its subsequent coupling with aromatic aldehydes or ketones provided propargyl alcohols **60** in high yields (Scheme 14). Compounds **60** (R^1=H) were converted into the corresponding 2-substituted furans **63** in 53–81% yield upon heating with ethanolic NaOH via a sequential cyclization/elimination of benzotriazole. The mechanism was believed to proceed through the formation of allenyl alkoxides **61** (derived from the isomerization of the acetylenes **60**), which then underwent, under the reaction conditions, an intramolecular cyclization into 2,5-dihydrofurans **62**. Further aromatization under the reaction conditions with elimination of the benzotriazole yielded furans **63**.

An analogous reaction of **59** with aromatic ketones afforded l-(5,5-diaryl-2,5-dihydrofuran-2-yl)benzotriazoles **64** in 68–90% yield (Scheme 14). In contrast to their analogs **62**, 2,5-dihydrofurans **64** were stable under basic conditions [32].

1-Propargylbenzotriazole **59** also found application in the synthesis of furans as C_3-annulation unit: deprotonated 1-propargylbenzotriazole **59** reacted with α-bromoketones to give 4-substituted and 4,5-disubstituted 2-(benzotriazol-l-ylmethyl)furans **71**. This sequence of reactions was accomplished in one pot and gave the product in good yields (Scheme 15) [33].

The following mechanism was proposed: anion **65** reacted at the carbonyl group of an α-bromo ketone to yield adduct **66**, which led to epoxide **67** through an intramolecular Williamson reaction. Epoxide **61** could be isolated in some cases. In the next step, deprotonation in the presence of potassium *tert*-butoxide at the propargylic position triggered a rearrangement from alkyne **67** to cumulene alkoxide **68**. Compound **68** then underwent an *5-endo-dig* intramolecular cyclization to give anion **69**. Protonation of **69** led to the formation of compound **70**, which was rearranged to form 2-(benzotriazol-1-ylmethyl)furans **71**.

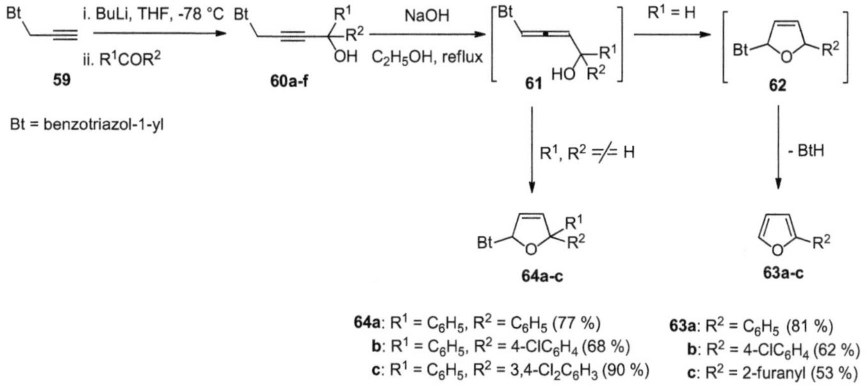

Scheme 14 Conversion of propargyl benzotriazole **59** into furans **63** and 2,5-dihydrofurans **64**

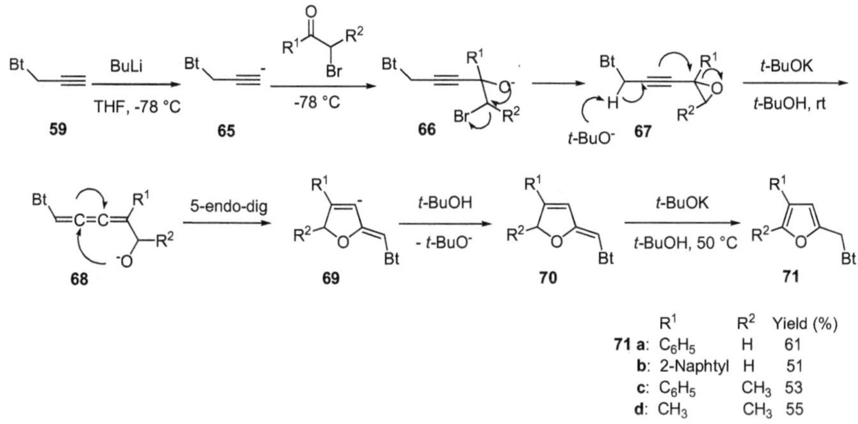

Scheme 15 Synthesis of 2-(benzotriazol-1-ylmethyl)furans **71**

Further exploration of the synthetic utility of propargyl benzotriazole **59** resulted in the preparation of 2-substituted benzo[*b*]furan **73** via a palladium-catalyzed heteroannulation with *o*-iodophenol **72** in the presence of $(P(C_6H_5)_3)_2PdCl_2$ and CuI, using DMF as solvent (Scheme 16) [34].

Scheme 17 depicts an example of allene-alcohol cyclization to furans using benzotriazole-substituted allene **75**, which was obtained in 90% yield by the reaction of **74** with two equivalents of LDA followed by 4-methoxybenzaldehyde. In contrast to allene **61** and cumulene **68**, allene **75** was isolated and could be stored. Allene **75** was further cyclized into the 2-benzotriazolyl-2,5-dihydrofuran **76** by treatment with an alkali (Scheme 17) [35].

Katritzky also reported an alternative preparation of novel dihydrofurans **79** by reacting benzotriazolium *N*-ylides **78** with aldehydes [36]. 3-Methyl-l-phenacyl-benzotriazolium bromide **77**, prepared by refluxing of 1-methyl-benzotriazole and α-bromoacetophenone in toluene, was conveniently converted to ylide **78** upon treatment with triethylamine (Scheme 18). When **78** was refluxed with the

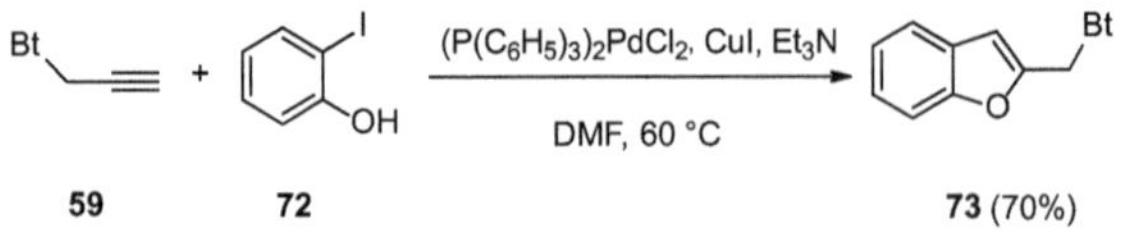

Scheme 16 Synthesis of 2-(benzotriazol-1-ylmethyl)benzo[*b*]furan **73**

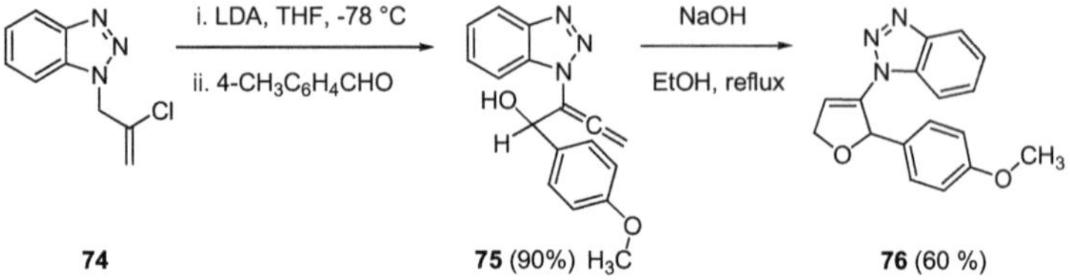

Scheme 17 Synthesis of 2-benzotriazolyl-2,5-dihydrofuran **76**

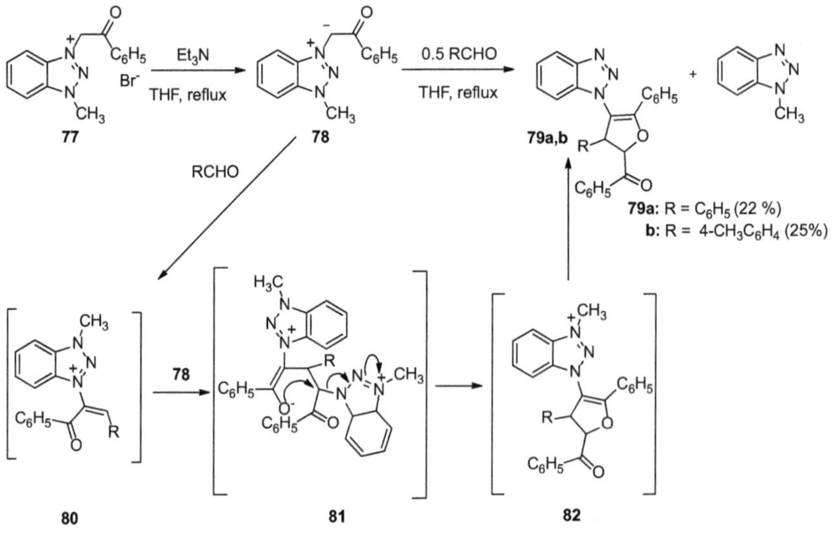

Scheme 18 Synthesis of 4,5-dihydrofurans **79**

appropriate aldehydes in THF, 4,5-dihydrofurans **79a,b** were obtained in low yields (22–25%). *N*-Methylbenzotriazole was observed as a side product.

The mechanism for the conversion of **78** to **79** is proposed in Scheme 18. Ylide **78** reacted with various aldehydes to give α,β-unsaturated ketones **80**. Then, in the next step, **80** served as a Michael acceptor for the condensation with another molecule of ylide **78**. A subsequent intramolecular nucleophilic substitution in **81** led to compound **82** together with the elimination of 1-methylbenzotriazole. In the final step, cation **82** eliminated 1 equiv. of bromomethane to give 4,5-dihydrofurans **79** [36].

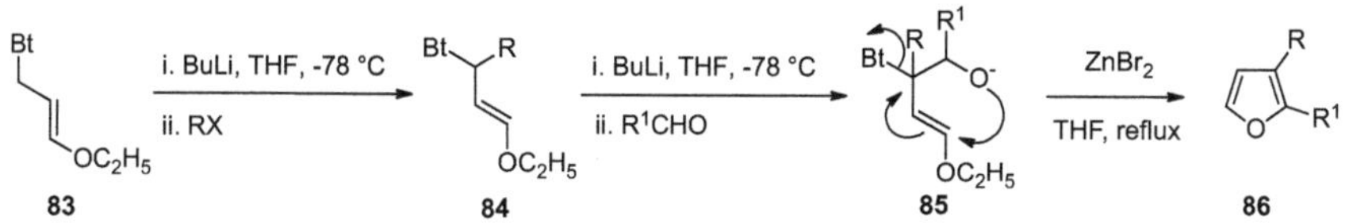

	R	R^1	Yield (%)
86a:	*n*-Bu	4-$CH_3C_6H_4$	46
b:	*n*-Bu	*n*-Pr	49
c:	C_2H_5	C_6H_5	52

Scheme 19 Synthesis 2,3-disubsustituted furans **86**

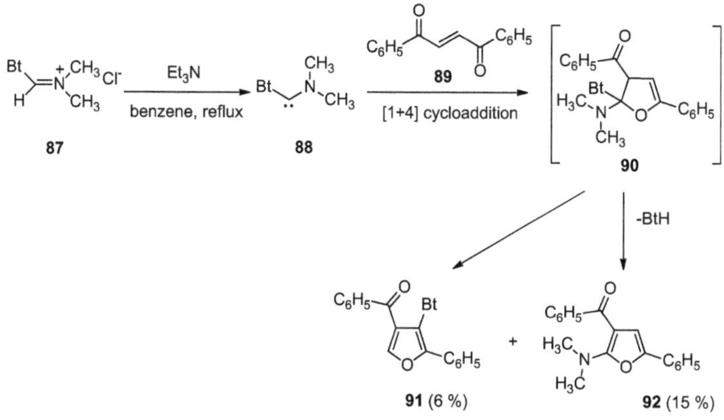

Scheme 20 Synthesis of 2-phenyl-4-benzoyl furans **91** and **92**

2,3-Disubsustituted furans were prepared according to a one-pot procedure starting from 1-(3-ethoxyallyl)-benzotriazoles **83** (Scheme 19). Substituents at the 3-position of the furan were introduced via a lithiation/alkylation sequence from **83** at the allylic position. The resulting substituted allyl intermediates **84** were deprotonated with BuLi followed by the addition of various aldehydes, to yield **85**. Intermediate **85** underwent 5-*endo-trig* cyclization upon exposure to zinc bromide and gave 1,2-disubstituted furans **86a–c** in fair yields [4].

A quite exotic and rare [1+4] cycloaddition was reported by Katritzky for the synthesis of furans. It involved an acyclic aminocarbene **88**, generated by deprotonation of *N*,*N*-dimethylaminobenzotriazolylmethyleniminium chloride **87** under mild conditions. Trapping of the carbene with *trans*-dibenzoylethylene **89** afforded 2-phenyl-3-[benzotriazol-l-yl]-4-benzoylfuran **91** and 2-dimethylamino-3-benzoyl-5-phenylfuran **92** via a [1+4] cycloaddition (Scheme 20) [37].

1-(Phenoxymethyl)benzotriazoles **94** served as substrates for the preparation of benzofurans by insertion of their anions into alkyl and aryl aldehydes followed by a Lewis acid-promoted intramolecular cyclization of the α-aryloxy ketones **97** (Scheme 21).

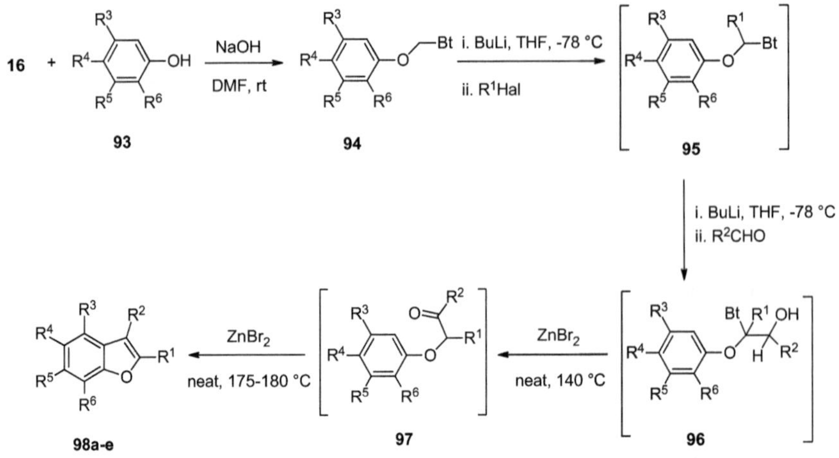

	R^1	R^2	R^3	R^4	R^5	R^6	Yield (%)
98a:	$C_6H_5CH_2$	$C_6H_5(CH_2)_2$	H	H	H	H	80
b:	n-C_8H_{17}	4-$CH_3C_6H_4$	H	H	H	H	79
c:	CH_3	4-$CH_3C_6H_4$	H	H	H	H	80
d:	n-C_4H_9	2,4-$Cl_2C_6H_3$	H	H	H	H	59
e:	n-C_3H_7	C_6H_5	H	H	H	C_2H_5	72

Scheme 21 Synthesis of polysubstituted benzofurans **98**

1-(Phenoxymethyl)benzotriazoles **94** were prepared from polysubstituted phenols **93** and 1-chloromethylbenzotriazole **16** (Scheme 21). Further lithiation of phenyl ethers **94** with BuLi, followed by quenching with an electrophile, gave the substituted benzotriazole derivatives **95**. An additional equivalent of BuLi, followed by the addition of various aldehydes, yielded **96**. Subsequent heating in the presence of $ZnBr_2$ afforded the corresponding α-aryloxy ketones **97** as a result of a pinacol-type rearrangement. Prolonged heating gave the desired polysubstituted benzo[*b*]furans **98** in good overall yields [38].

Katritzky drastically extended the scope of his benzotriazole-based strategy for the preparation of substituted benzofurans starting with 1-(benzotriazol-1-yl)alkyl chlorides **100** and developed a convenient multistep preparation procedure thereof (Scheme 22).

1-Chloroalkylbenzotriazoles **100**, which were prepared from aldehydes **99**, benzotriazole, and thionyl chloride, reacted with the sodium salts of various *o*-hydroxybenzophenones **101** to give the corresponding intermediates **102** (Scheme 22). Intermediates **102** were then lithiated with an equivalent amount of LDA in THF at low temperatures to give the corresponding 2-(benzotriazol-1-yl)-3-substituted-2,3-dihydro-1-benzofuran-3-ols of type **103**. Compounds **103** were reacted with an equivalent amount of a low valent titanium reagent to provide benzofurans **104** in moderate to good yields [39].

The deprotonation of 2-(1-benzotriazolylalkoxy)-benzophenones **107** with LDA triggered a cyclization towards 2,3-dihydrobenzofuran-3-ols **108**, which

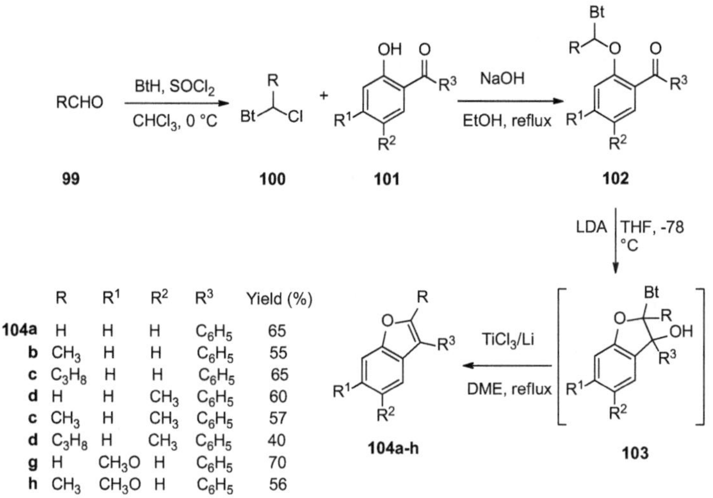

	R	R1	R2	R3	Yield (%)
104a	H	H	H	C_6H_5	65
b	CH_3	H	H	C_6H_5	55
c	C_3H_8	H	H	C_6H_5	65
d	H	H	CH_3	C_6H_5	60
c	CH_3	H	CH_3	C_6H_5	57
d	C_3H_8	H	CH_3	C_6H_5	40
g	H	CH_3O	H	C_6H_5	70
h	CH_3	CH_3O	H	C_6H_5	56

Scheme 22 Synthesis of polysubstituted benzofurans **104**

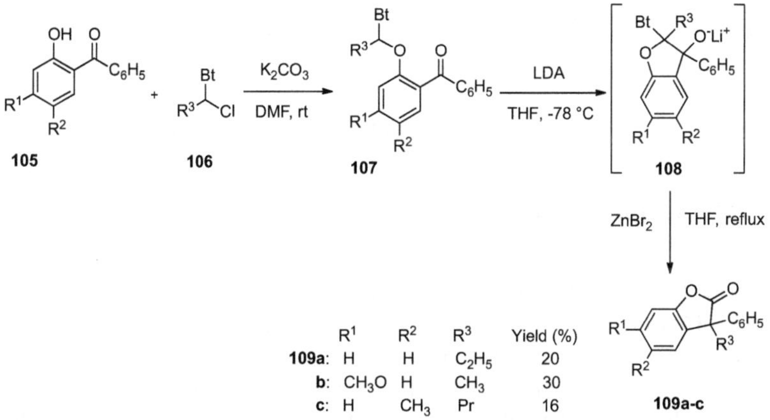

	R1	R2	R3	Yield (%)
109a:	H	H	C_2H_5	20
b:	CH_3O	H	CH_3	30
c:	H	CH_3	Pr	16

Scheme 23 Synthesis of 2,3-dihydrobenzofuran-2-ones **109**

subsequently underwent Lewis acid-promoted pinacol-type rearrangement to 3-alkyl-3-aryl-substituted 2,3-dihydrobenzofuran-2-ones **109** (Scheme 23) [40].

Synthesis of non-annulated furan-2-one **111** concluded this series of acid-promoted intramolecular cyclizations to furans. The benzotriazole-stabilized anion **4** derived from the lithiation of 1-(α-ethoxyallyl)-benzotriazole **3** (Scheme 24) reacted with sterically hindered 2,4-dimethyl-3-pentanone to give **110**. Alcohol **110** then underwent an acid-promoted intramolecular 5-*endo-trig* cyclization to γ-lactone **111** (Scheme 24) [41].

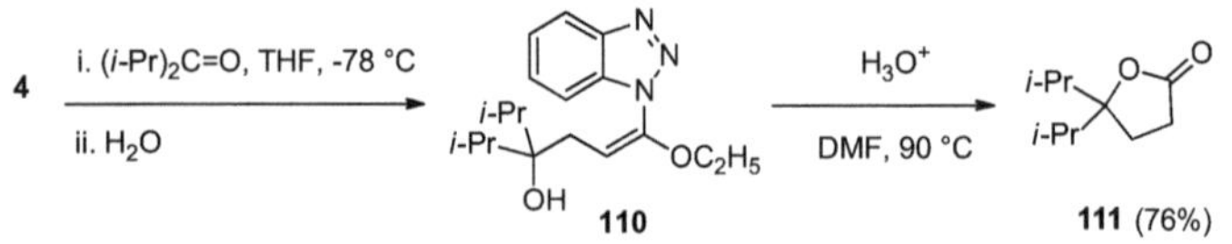

Scheme 24 Synthesis of substituted γ-butyrolactone **111**

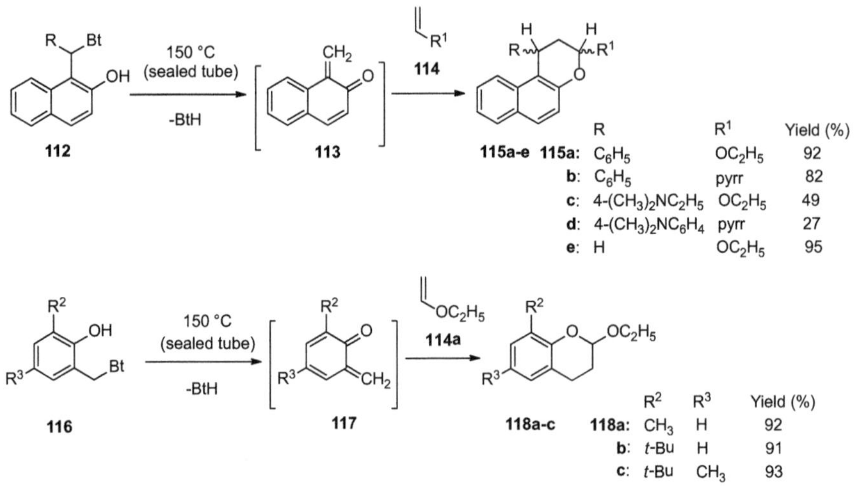

Scheme 25 Synthesis of benzo[*f*]chromanes **115** and 2-ethoxychromanes **118**

2.4 *Six-Membered Rings*

l-[α-(Benzotriazol-l-yl)alkyl]-2-napthols **112** and 2-[(benzotriazol-l-yl)-methyl] phenols **116** were reported as precursors of *o*-quinone methide derivatives **113** and **117**, obtained by elimination of the benzotriazolyl fragment. *o*-Quinone methides are reactive intermediates, and their annulation with electron-rich olefins **114** gave chromane derivatives. For instance, l-[α-(benzotriazol-l-yl)alkyl]-2-napthols **112** and 2-[(benzotriazol-l-yl)-methyl]phenols **116** gave benzo[*f*] chromanes **115** and 2-ethoxychromanes **118**, respectively, in high yields (Scheme 25) [42].

Katritzky reported the preparation of chromanes **123a–c** and **126** starting from (benzotriazol-1-yl)methanes **119** [43]. This example illustrates the interplay of two fundamental properties of the benzotriazole moiety, i.e., α-negative charge stabilization and excellent leaving group ability. The anions **120** were easily generated from **119** and could be trapped with various electrophiles. Ethers **122** and **124** were prepared accordingly in excellent yields using the 1-(2-bromoethoxy)benzene **121**, as an electrophile. Subsequently, **122** and **125** were converted into and chromanes **123a–c** and **126** in low to good yields by six-membered ring annulations (Scheme 26). These cyclizations were carried out in the presence of $ZnBr_2$ [44].

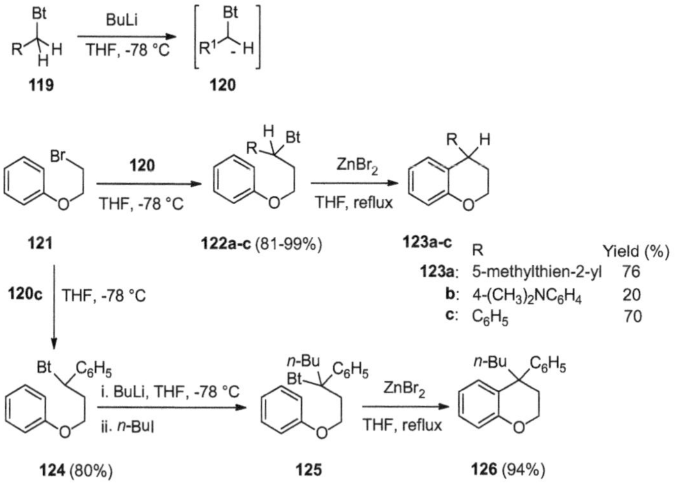

Scheme 26 Synthesis of the chromanes **123a–c** and **126**

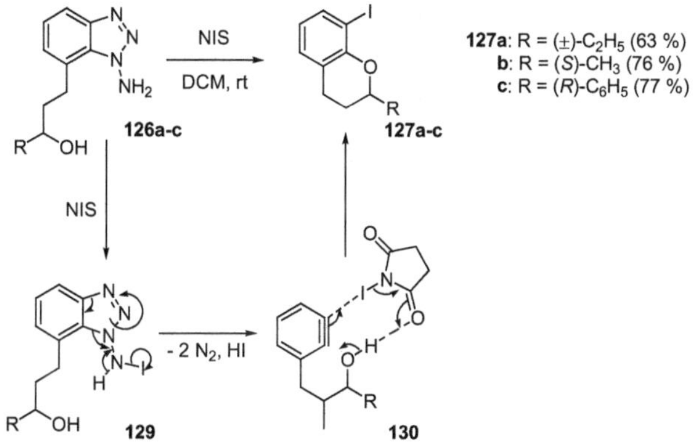

Scheme 27 Synthesis of 8-iododihydrobenzopyrans **127**

1-Aminobenzotriazole was widely used as a convenient benzyne precursor, especially as it can be converted into highly reactive intermediate under nonbasic conditions by treatment with lead(IV) acetate [45]. Knight and coworkers reported that 1-aminobenzotriazole derivatives **126** bearing 7-hydroxyalkyl substituents were efficiently converted into the corresponding benzynes when treated with *N*-iodosuccinimide. The authors proposed that the intermediate benzyne reagents were intramolecularly trapped by the leading iododihydrobenzopyrans (chromans) **127** in good yields (Scheme 27) [46, 47].

This 1-aminobenzotriazole methodology was extended for the production of diverse 8-iodochromanes and 8-iodochromenes. For this purpose, the *ortho*-iodo derivative **131** was reacted with a range of alk-1-yn-3-ols via an efficient Sonogashira coupling to provide the arylalkynes **132**. In the next step, total or

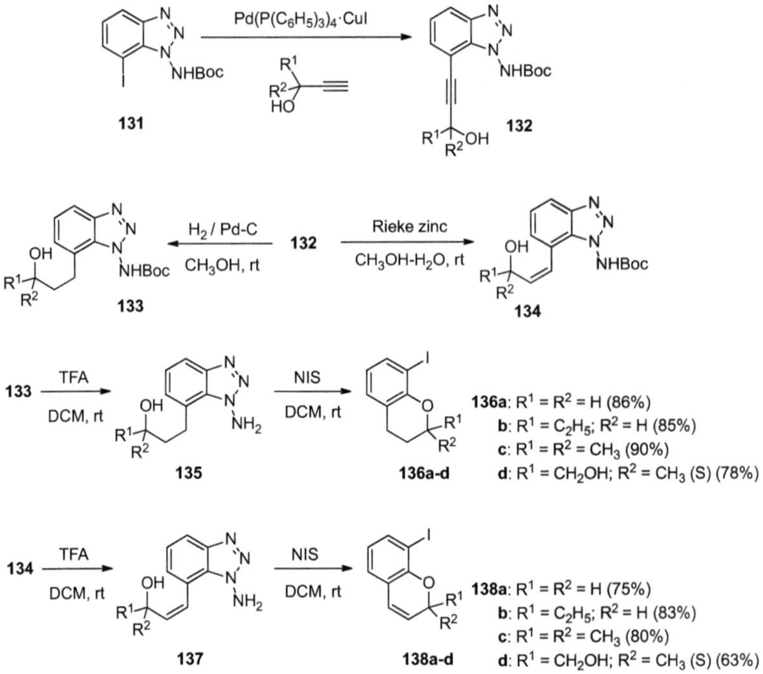

Scheme 28 Synthesis of 8-iodochromanes **136** and 8-iodochromenes **138** using 1-aminobenzotriazole methodology

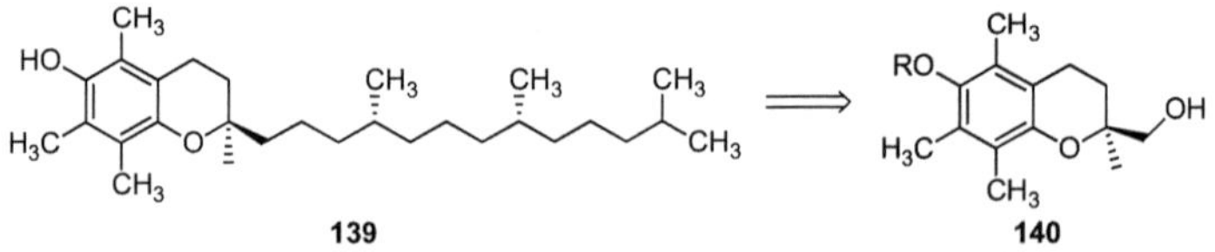

Fig. 1 α-Tocopherol **139** and its suggested synthon **140**

partial reduction of **132** led to arylpropanols **133** or the (*Z*)-allylic alcohols **134**, respectively. *N*-Deprotection and subsequent exposure to 2 equiv. of *N*-iodosuccinimide then led to smooth benzyne generation and intramolecular trapping by the hydroxyl functions together with iodine incorporation to give the iodochromanes **136** and iodochromenes **138**, respectively, in good overall yields (Scheme 28) [48].

A formal total synthesis of α-tocopherol **139**, the main component of vitamin E (Fig. 1), was reported by Knight; it involved the intramolecular trapping of a highly substituted benzyne by an alcohol group to establish the pyran ring [49]. Following the modified Stille-type coupling employing tetramethyltin as the methyl source delivered the targeted α-tocopherol precursor **140** in the mixture with 8-H (**143a**) and 8-iodo (**143b**) derivatives in a combined yield of 67% (Scheme 29).

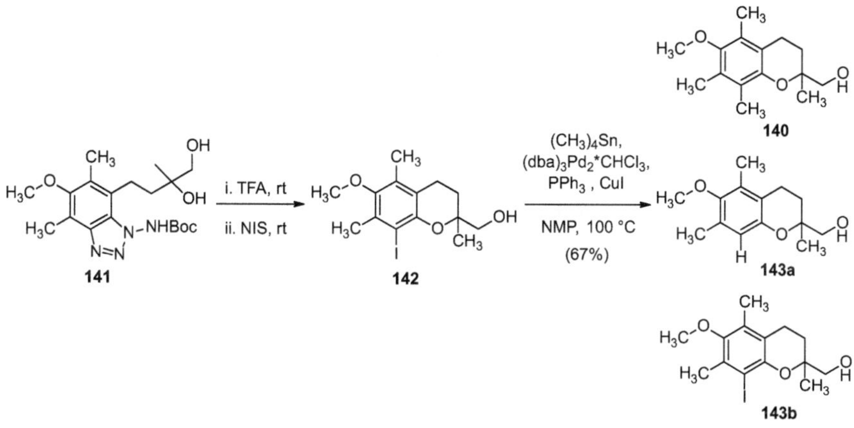

Scheme 29 Synthesis of α-tocopherol precursor **140**

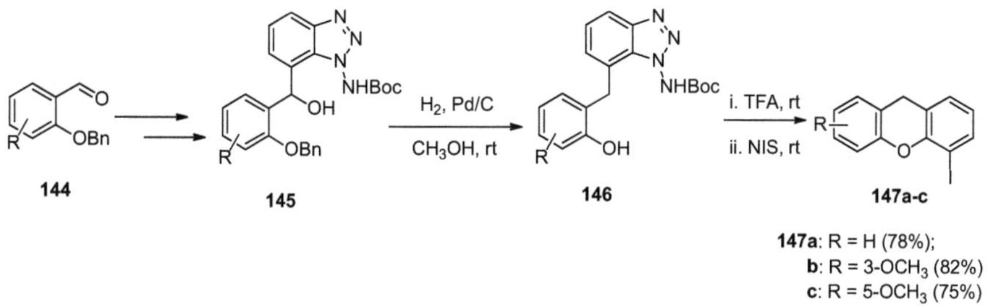

Scheme 30 1-Aminobenzotriazole-mediated synthesis of iodoxanthenes **147**

Such a methodology involving an intramolecular benzyne capture was further extended by Knight and coworkers. For instance, phenols **146**, which were prepared via double hydrogenolysis of intermediate alcohols **145**, were Boc-deprotected and treated with *N*-iodosuccinimide. The corresponding benzyne derivative underwent an intramolecular cyclization to give the iodoxanthenes **147** in excellent overall yields. This work demonstrated for the first time a viable method for the intramolecular trapping of benzyne derivatives by phenolic groups (Scheme 30) [50, 51].

Similarly to the formation of oxetan-1-ones **44** from enolates **42** (Scheme 10), the preparation of pyran-2-ones was achieved by the reaction of α,β-unsaturated ketones with the anions derived from *N*-acylbenzotriazoles **148**. Thus, the anions obtained from the lithiation of aliphatic 1-acylbenzotriazoles **148** underwent a Michael addition with α,β-unsaturated ketones **149** to afford 3,4,6-trisubstituted 3,4-dihydropyran-2-ones **151**, **152** (Scheme 31). Mechanistically, it was assumed that the 1,4-addition of lithiated acylbenzotriazoles **148** to chalcones **149** generated a reactive enolate anion **150**, which then reacted intramolecularly with the acylbenzotriazole to form the 3,4-dihydropyran-2-ones as a mixture of diastereoisomers **151** and/or **152** [52].

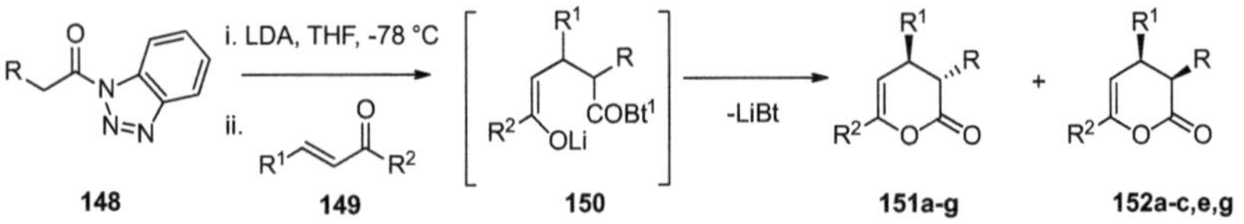

	R	R[1]	R[2]	Yield (%)
151a/152a	C_2H_5	C_6H_5	C_6H_5	70
151b/152b	C_2H_5	4-ClC_6H_4	C_6H_5	74
151c/152c	C_2H_5	4-$NO_2C_6H_4$	C_6H_5	81
151d	C_2H_5	C_6H_5CO	C_6H_5	72
151e/152e	C_2H_5	C_6H_5	$C_6H_5CH{=}CH$	53
151f	n-C_6H_{13}	C_6H_5	C_6H_5	79
151g/152g	$(CH_3)_2CH$	C_6H_5	C_6H_5	24

Scheme 31 Synthesis of 3,4-dihydropyran-2-ones **151** and/or **152**

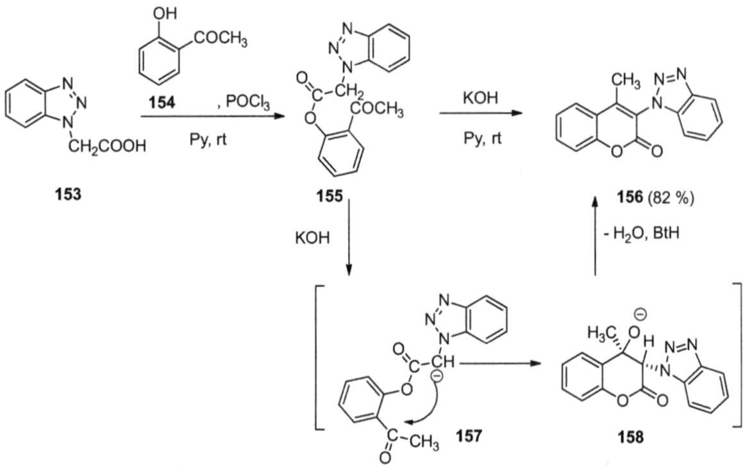

Scheme 32 Synthesis of 3-(benzotriazol-1-yl)-4-methyl-coumarin **156**

3-(Benzotriazol-1-yl)coumarins **156** were prepared from benzotriazole acetic acid **153** [53]. The esterification of **153** with 2-hydroxyacetophenone **154** was followed by a base-catalyzed cyclodehydration of 2-acetylphenyl-2-(1-benzotriazolyl)acetate **155**. A novel coumarin derivative, 3-(benzotriazole-1-yl)-4-methyl-coumarin **156**, was obtained accordingly (Scheme 32).

The mechanism of the aforementioned cyclodehydration is shown in Scheme 32. The benzotriazol-1yl moiety acted as an electron-withdrawing group and activated the adjacent methylene group for proton loss upon treatment with a potassium hydroxide. The stabilized anion **157** then reacted intramolecularly with the adjacent ester to form intermediate **158**. Subsequent protonation and dehydration led to the formation of 3-(benzotriazole-1-yl)-4-methyl-coumarin **156**.

Three novel coumarin derivatives containing benzotriazole moieties, namely, MBMBC, BMBC, and BMC (Fig. 2), were prepared according to this method. They

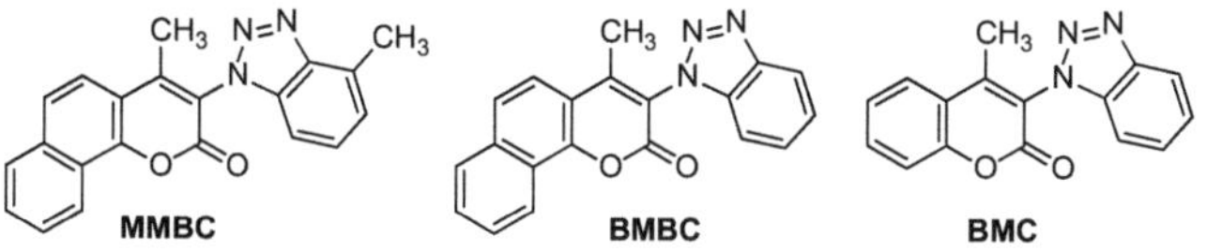

Fig. 2 3-(Benzotriazol-1-yl)-4-methyl coumarins – efficient HepG-2 growth inhibitors

all showed efficient in vitro inhibition of the human hepatocellular carcinoma cell lines (HepG-2) [54].

3 Synthesis of Oxygen- and Nitrogen-Containing Heterocycles

3.1 Five-Membered Rings with One Oxygen and One or Two Nitrogens

3.1.1 Oxazoles and Oxazolidines

N-(α-Aminoalkyl)benzotriazoles **159** provided an easy access to substituted α-aminocarbanions **160** that could be captured with various electrophiles [55, 56]. This allowed the preparation of 1,3-oxazolidines from ketones as electrophiles. *N*,*N*-Bis(benzotriazol-1-ylmethyl)alkyl amines **161**, which were conveniently prepared from primary amines [57], reacted with diethylketone in the presence of samarium diiodide in THF-HMPA to give oxazolidines **162** as the major [3+2] cyclization product (Scheme 33) [58].

Besides, the reaction of *N*-acylbenzotriazoles **163** with 2-amino-2-methyl-1-propanol under mild microwave conditions afforded the corresponding 2-substituted 2-oxazolines (Scheme 34). A variety of readily available *N*-acylbenzotriazoles **163** produced 2-substituted 1,3-oxazolines **164** in high (84–98%) yields under mild conditions and short reaction times [59].

Benzotriazol-1-yl methylisocyanide (BetMIC) **166** was prepared from formamide **165** [60] and was reported as a convenient entry towards oxazolines **167** and oxazoles **168** (Scheme 35). It reacted with various ketones and aldehydes under mild conditions, for 2 h to afford 4-ethoxy-2-oxazolines **167** and oxazoles **168**, respectively. This example illustrated the interplay of the unique properties of benzotriazole: electronegative nature, facilitating deprotonation at α-position and a good leaving group ability.

The preparation of BetMIC **166** was amenable on large scale from readily available and inexpensive starting materials and was conveniently implemented as stable and nontoxic alternative to tosylmethyl isocyanate [61].

Katritzky also reported the preparation 5-acylaminooxazoles **171** from 1,2-diacylamino-l,2-di(benzotriazol-l-yl)ethanes **170**. Compounds **170** were

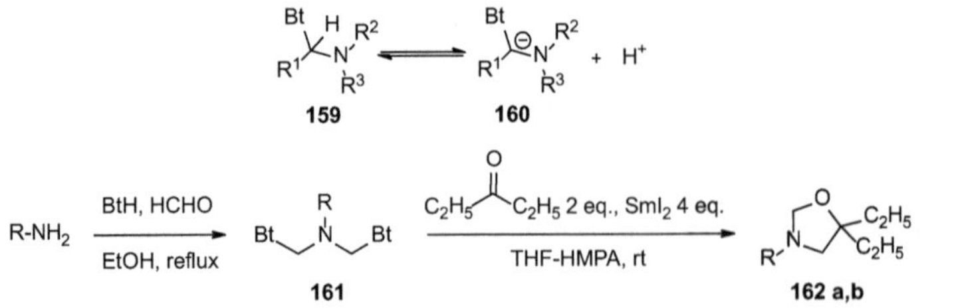

Scheme 33 Synthesis of oxazolidines **162**

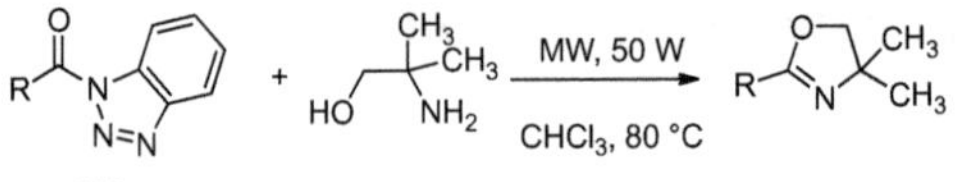

	R	Yield (%)
164a	4-$CH_3C_6H_4$	98
b	4-$CH_3OC_6H_4$	95
c	4-$NO_2C_6H_4$	90
d	4-ClC_6H_4	90
e	2-ClC_6H_4	98
f	C_6H_5	86
g	1-naphtyl	95
h	2-furyl	95
i	2-phenylethenyl	84
j	1-(6-methoxy-2-naphthyl)ethyl	91

Scheme 34 Synthesis of 2-substituted 1,3-oxazolines **164**

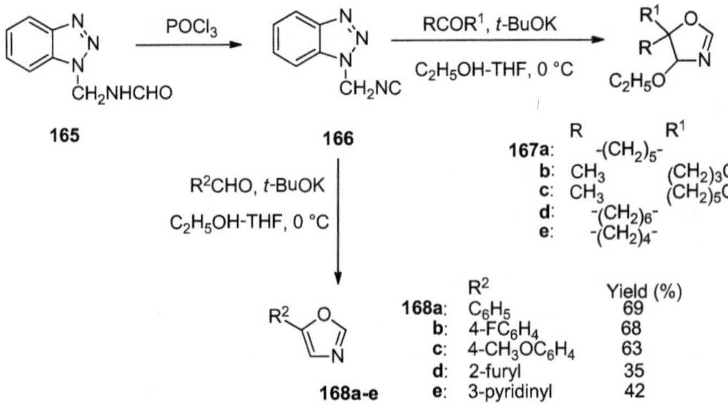

	R	R^1	Yield (%)
167a:	-$(CH_2)_5$-		65
b:	CH_3	$(CH_2)_3CH_3$	93
c:	CH_3	$(CH_2)_5CH_3$	92
d:	-$(CH_2)_6$-		63
e:	-$(CH_2)_4$-		59

	R^2	Yield (%)
168a:	C_6H_5	69
b:	4-FC_6H_4	68
c:	4-$CH_3OC_6H_4$	63
d:	2-furyl	35
e:	3-pyridinyl	42

Scheme 35 Synthesis of oxazolines **167** and oxazoles **168** from BetMIC **166**

conveniently prepared from the condensation of 1,2-di(benzotriazol-1-yl)ethane-1,2-diol **169** and primary amides. Its reaction with sodium hydride in dry DMF at 100°C led to 5-acylaminooxazoles **171** in good to moderate yields (Scheme 36). As illustrated in Scheme 36, the formation of oxazoles proceeded via the intramolecular nucleophilic displacement of one benzotriazolyl group, followed by the elimination of the benzotriazole to complete aromatization [62].

Novel chiral pyrido[2,1-*b*][1,3]oxazole **174** [63] and oxazolo[2,1-*b*]pyrrolidine **175** [64] were prepared from benzotriazole and (*S*)-phenylglycinol in the presence

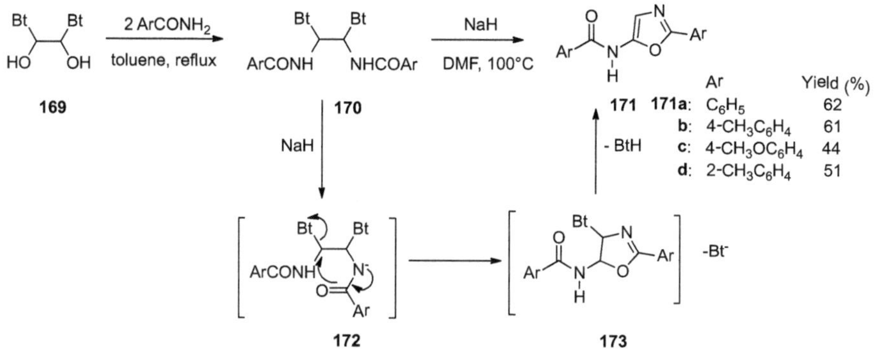

Scheme 36 Synthesis of 2-aryl-5-aroylamino-1,3-oxazoles **171**

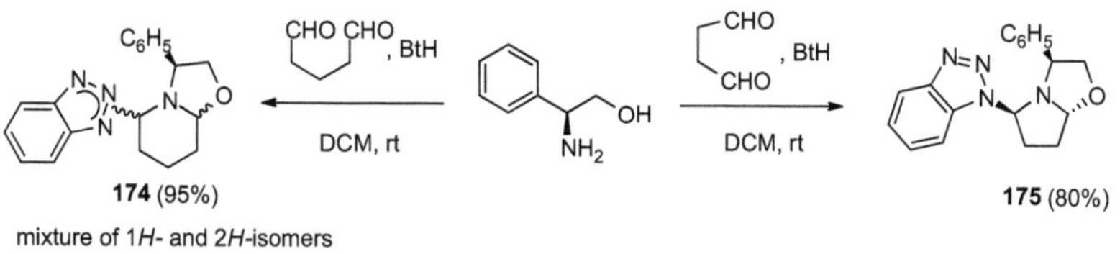

Scheme 37 Synthesis of oxazolidines **173** and **174**

of glutaraldehyde or succinaldehyde, respectively (Scheme 37). Pyridooxazole **174** was obtained as a mixture of several compounds, comprising Bt[1] or Bt[2] isomers and diastereomers, while oxazolo[2,1-*b*]pyrrolidine **175** was isolated as a single diastereomer.

Novel bicycles **177**, rare representatives of the pyrrolo[1,2-*c*]oxazole ring system, were formed in high yields when 1-acylbenzotriazole pyrrole **176** was treated with a strong, non-nucleophilic base (3 equivalents), such as DBU, in the presence of various ketones and aldehydes (Scheme 38) [65]. The scope of this procedure was remarkable since the reaction worked with either enolizable or nonenolizable aldehydes and ketones.

These results enabled the extension of the procedure to the preparation of oxazolo[3,4-*a*]indol-1-ones **179**, which were obtained by reactions of benzotriazol-1-yl(1*H*-indol-2-yl)-methanone **178** with carbonyl compounds. In contrast to **176**, compounds **178**, under the same reaction conditions, reacted sluggishly with ketones and aldehydes to afford oxazolo[3,4-*a*]indol-1-ones **179** in lower yields [65].

The pyrolysis of a number of 1-aroylbenzotriazoles **180** under different conditions gave the corresponding benzoxazole derivatives **181** in 13–18% yields alongside with other products **182–186**, depending on the nature of the substituents on the acyl moiety (Scheme 39) [66].

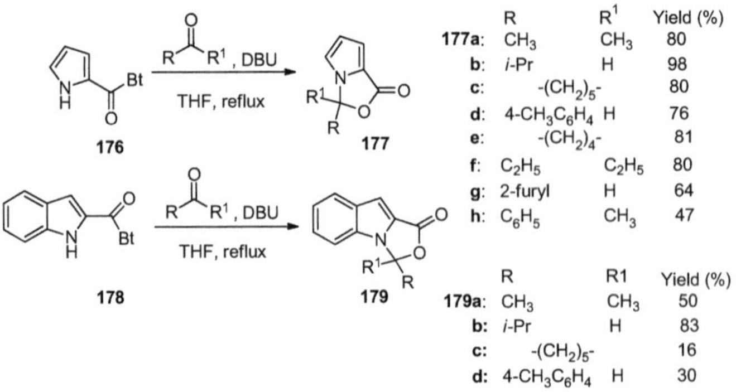

Scheme 38 Synthesis of pyrrolo[1,2-*c*]oxazoles **177** and **179**

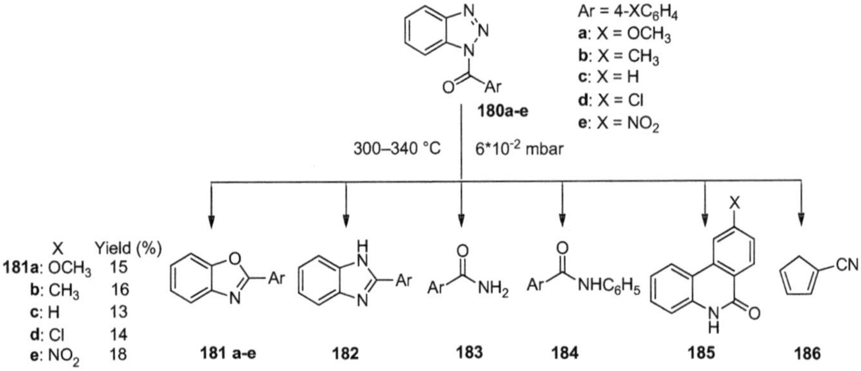

Scheme 39 Thermolysis of *N*-aroylbenzotriazoles **180**

3.1.2 Isoxazoles and Isoxalidines

The labile nature of the benzotriazole-carbon bond in bis(benzotriazol-1-ylmethyl) hydroxylamine **187** makes it useful as a potential synthon for the preparation of 1,3-dipoles such as nitrone **189** (Scheme 2). The +*M* effect of the oxygen to the nitrogen in **188** was expected to enhance the leaving tendency of the benzotriazolate anion, which would result in the formation of the intermediate nitrone **189** after deprotonation (Scheme 40) [67].

The regio- and stereo-specific 1,3-dipolar cycloaddition of the nitrone **189** with several dipolarophiles furnished a library of substituted 2-(benzotriazol-1-ylmethyl) isoxazolidines **190**. Refluxing compound **187** in toluene with methyl acrylate, *N*-methylmaleimide, 2-vinyl- and 4-vinylpyridines, acrylonitrile, or dimethyl fumarate resulted in the regioselective formation of isoxazolidines **190a–f**. The presence of a remaining benzotriazol-1-ylmethyl substituent in products **190** opened up other possibilities for further synthetic manipulation [67].

3,5-Disubstituted isoxazoles **193a–f** were prepared from α-benzotriazolyl-α,-β-unsaturated ketones in a regioselective and efficient fashion [68]. Katritzky

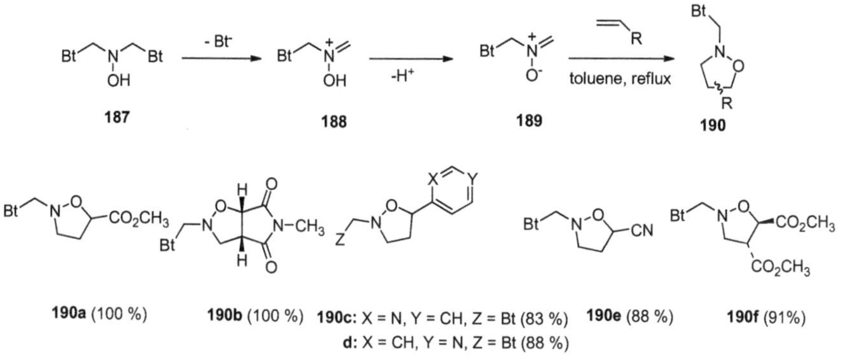

Scheme 40 Synthesis of *N*-(benzotriazol-1-ylmethyl)isoxazolidines **190**

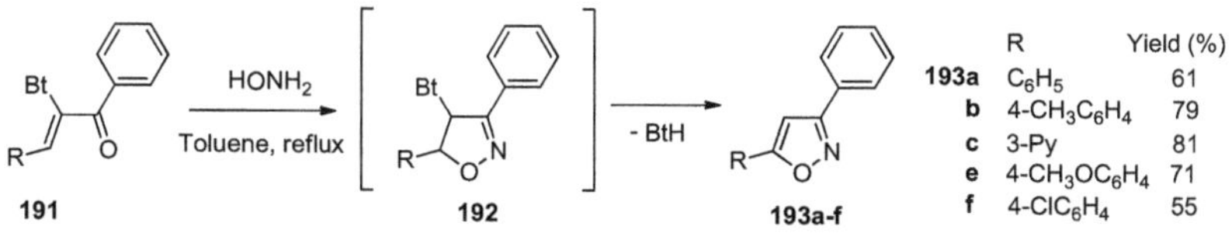

Scheme 41 Synthesis of 3-phenylisoxazoles **193**

reported a new procedure for the preparation of isoxazoles **193** employing α-benzotriazolyl-α,β-unsaturated ketones and hydroxylamine (Scheme 41). The benzotriazolyl moiety behaved as a leaving group and allowed for the direct transformation of the intermediate isoxazolines **192** into the aromatic isoxazoles **193** in 55–81% yields.

A small library of benzotriazole-based electron-rich 1,3-dipolarophiles, namely, *trans*-3-(benzotriazol-l-yl)-l-(*N*-morpholino)prop-l-ene **196**, *trans*-3-(benzotriazol-l-yl)-l-ethoxyprop-l-ene **197**, and *trans*-l,3-bis-(benzotriazol-l-yl)propene **198**, gave the expected 1,3-dipolar cycloaddition products **199–201**, respectively, when reacted with benzonitrile oxide **195**. The latter was formed in situ from benzhydroxamoyl chloride **194** and a triethylamine (Scheme 42) [69]. The treatment of **199–201** with refluxing ethanolic hydrochloric acid each gave the same isoxazole derivative **202** in high yields (90%) [70].

A wide range of allyl or propargyl benzotriazoles **203–207** was used for 1,3-dipolar cycloadditions with nitrile oxides **208** in boiling EtOAc to afford 3,5-substituted isoxazoles **209**, **210** and isoxazolines **212**, **213** in mostly good yields (Scheme 43) [71].

3.1.3 1,2,4- and 1,3,4-Oxadiazoles

A convenient method for the preparation of 1,2,4-oxadiazoles was developed utilizing *N*-acylbenzotriazoles **214** through its reaction with amidoximes **215**. *O*-

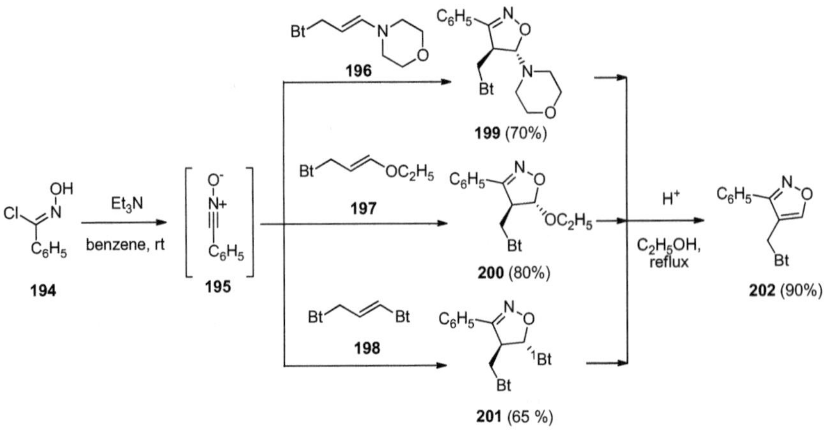

Scheme 42 Synthesis of 3-phenyl-4-(benzotriazole-1-ylmethyl)-isoxazoles **202**

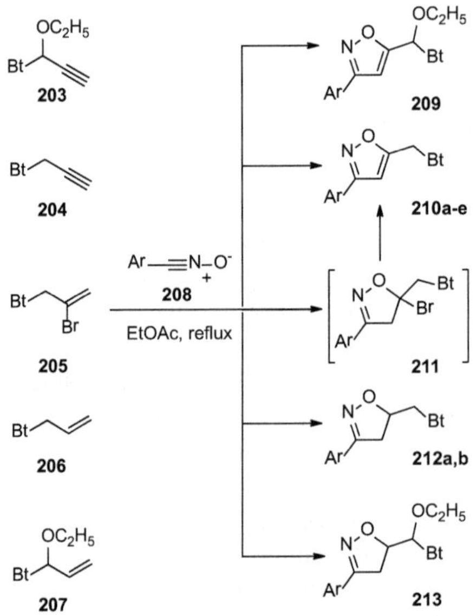

	Ar	Yield (%)
209	4-$CH_3OC_6H_4$	40
210a	C_6H_5	100
210b	4-$CH_3C_6H_4$	94
210c	2-ClC_6H_4	84
210d	3-$NO_2C_6H_4$	52
210e	3-$CH_3OC_6H_4$	67
210f	4-$CH_3OC_6H_4$	95
212a	4-$CH_3C_6H_4$	100
212b	4-$CH_3OC_6H_4$	94
213	4-$CH_3OC_6H_4$	81

Scheme 43 Preparation of 3,5-substituted isoxazoles **209**, **210** and isoxazolines **212**, **213**

acylation of amidoximes **215** with *N*-acylbenzotriazoles **214** proceeded immediately after addition of 1 equiv. of base to the reaction mixture and gave intermediates **216**. Subsequent cyclization occurred upon reflux in ethanol (Scheme 44). These reactions reached completion within 5 min, and the products **217** were precipitated upon addition of water and collected in 70–94% yield.

Aromatic *N*-acylbenzotriazoles **214** were found less reactive for the cyclization reactions than their aliphatic analogs and demanded prolonged reaction times. The

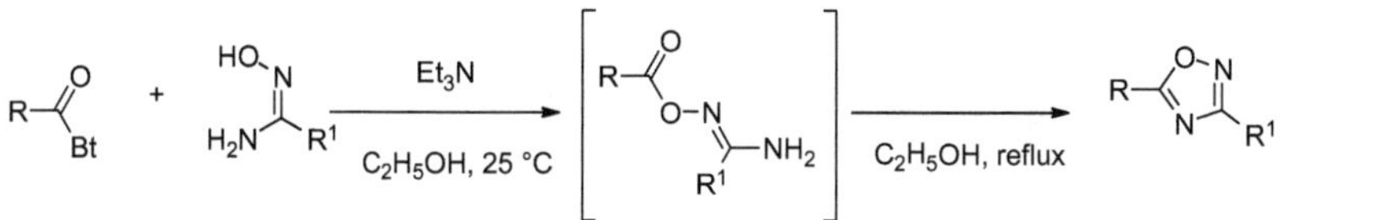

214 215 216 217

	R	R^1	Yield (%)		R	R^1	Yield (%)
217 a:	C_6H_5	4-tolyl	77	**217h**:	Fmoc-*L*-Met	benzyl	85
b:	4-$CH_3OC_6H_4$	4-tolyl	82	**i**:	*Z*-*L*-Phe-*L*-Ala-	4-tolyl	83
c:	2-indolyl	4-tolyl	73	**j**:	*Z*-*D*-Phe-*L*-Ala-	4-tolyl	79
d:	C_6H_5	benzyl	74	**k**:	*Z*-*L*-Ala-*L*-Val-	4-tolyl	87
e:	*Z*-*L*-Ala-	benzyl	87	**l**:	*Z*-*D*-Ala-*L*-Val-	4-tolyl	85
f:	*Z*-*L*-Phe-	benzyl	83	**m**:	*Z*-*L*-Ala-*L*-Phe-	4-tolyl	91
g:	*Z*-*L*-Met-	benzyl	89	**n**:	*Z*-*L*-Ala-*D*-Phe-	4-tolyl	94

Scheme 44 Synthesis of 1,2,4-oxadiazoles **217**

method was successfully employed with *N*-protected (α-aminoacyl)benzotriazoles and dipeptide benzotriazolides and afforded 1,2,4-oxadiazole peptidomimetics with high enantiomeric purity (>97%). This method provided a convenient alternative to other oxadiazole synthetic approaches utilizing acyl halides [72].

The versatile and stable benzotriazole Vilsmeier-type reagent **219** [37] was used in combination with aroyl hydrazides **218** for the high-yielding and convenient preparation of 1,3,4-oxadiazoles **220** (Scheme 45) [73]. Similarly, di(benzotriazol-1-yl)methanimine **221** was successfully utilized for the preparation of 2-amino-5-aryl-1,3,4-oxadiazoles **222** from hydrazides **218** [74].

3.2 Six-Membered Ring with One Oxygen and One Nitrogen

The anion **224** generated from l-[α-(phenylthio)methyl]benzotriazole **223** and LDA reacted with allyl bromide to afford the corresponding allylated intermediates **225**. The *S,N*-acetal **225** was next converted into l-(l,3-butadien-l-yl)benzotriazole **226** in the presence of potassium *tert*-butoxide in 75% yield (Scheme 46). Diene **226** readily reacted with nitrosobenzene under mild conditions to give the *N*-phenyldihydro-l,2-oxazine **227** in high yield (Scheme 46) via a nitroso-Diels–Alder [4+2] cycloaddition [75]. Cycloadduct **227** bearing the benzotriazolyl group possesses a synthetic potential for further transformations involving the nucleophilic displacement of benzotriazole. Stable *N*-(α-thioalkyl)benzotriazoles **223** were conveniently prepared from benzotriazole, thiols, and aldehydes or ketones [76] and enabled the convenient and high-yielding preparation of substituted dihydro-l,2-oxazines.

N-[1-(benzotriazol-1-yl)alkyl]amides **228** were easily prepared from a mixture of amides, aldehydes, and benzotriazole in toluene [60], reacted with both terminal and central acetylenes **230** in refluxing CH_2Cl_2 in the presence of aluminum

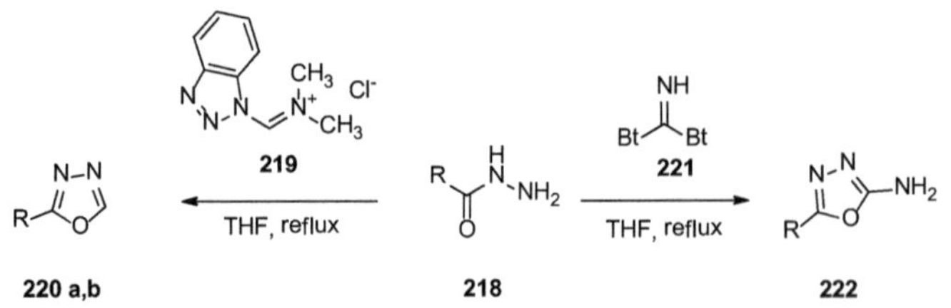

220a: R = C_6H_5 (95%)
b: R = 4-Py (98%)

	R	Yield (%)		R	Yield (%)
222a	C_6H_5	93	g	$3\text{-}NO_2C_6H_4$	96
b	$4\text{-}t\text{-}BuC_6H_4$	92	h	$4\text{-}ClC_6H_4$	96
c	$4\text{-}NH2C_6H_4$	82	i	$2\text{-}ClC_6H_4$	95
d	$4\text{-}NH_2C_6H_4$	92	j	4-Py	97
e	$4\text{-}HOC_6H_4$	93	k	OC_2H_5	99
f	$4\text{-}NO_2C_6H_4$	97			

Scheme 45 Synthesis of 1,3,4-oxadizoles **220** and **222** form acyl hydrazides **218**

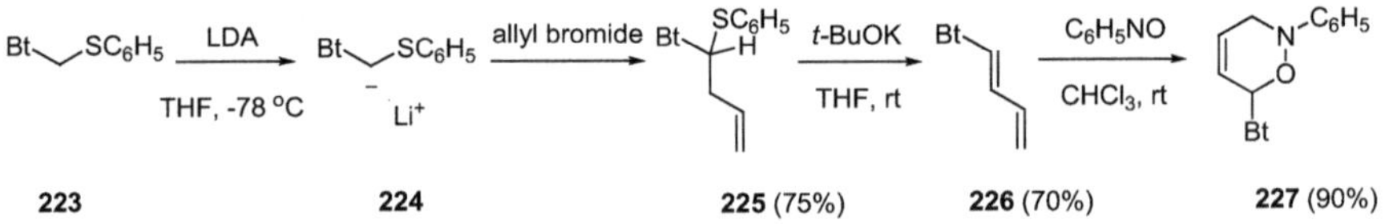

Scheme 46 Synthesis of *N*-phenyl-3,6-dihydro-1,2-oxazine **227**

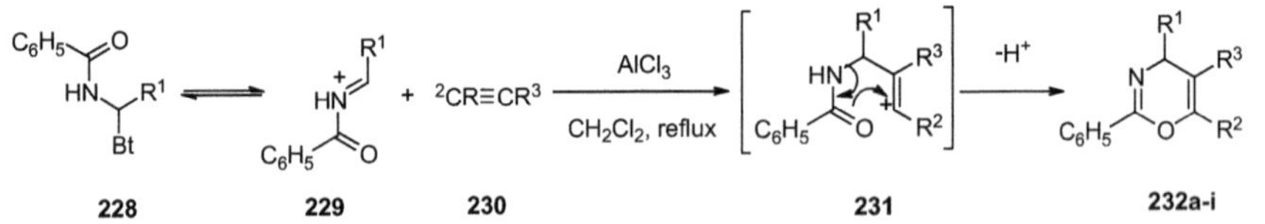

	R^1	R^2	R^3	Yield (%)
232a	C_6H_5	C_6H_5	H	86
b	C_6H_5	C_6H_5	C_2H_5	94
c	$4\text{-}ClC_6H_4$	C_6H_5	H	91
d	$4\text{-}ClC_6H_4$	C_6H_5	C_2H_5	92
e	1-naphthyl	C_6H_5	H	14
f	$4\text{-}CH_3OC_6H_4$	C_6H_5	H	89
g	$4\text{-}CH_3OC_6H_4$	C_6H_5	C_2H_5	94
h	$3\text{-}NO_2OC_6H_4$	C_6H_5	C_2H_5	76
i	C_6H_5	$n\text{-}C_4H_9$	H	83

Scheme 47 Synthesis of 4*H*-1,3-oxazines **232**

chloride (Scheme 47) to give 2,4,5-trisubstituted or 2,4,5,6-tetrasubstituted 4*H*-1,3-oxazines **232** in low to excellent yields (14–94%) [77].

The mechanism of the reaction includes electrophilic addition of the *N*-acyliminium cation intermediate **229** to the triple bond of the acetylene to give an intermediate **231**, followed by the ring closure with formation of the 4*H*-1,3-oxazines (Scheme 47).

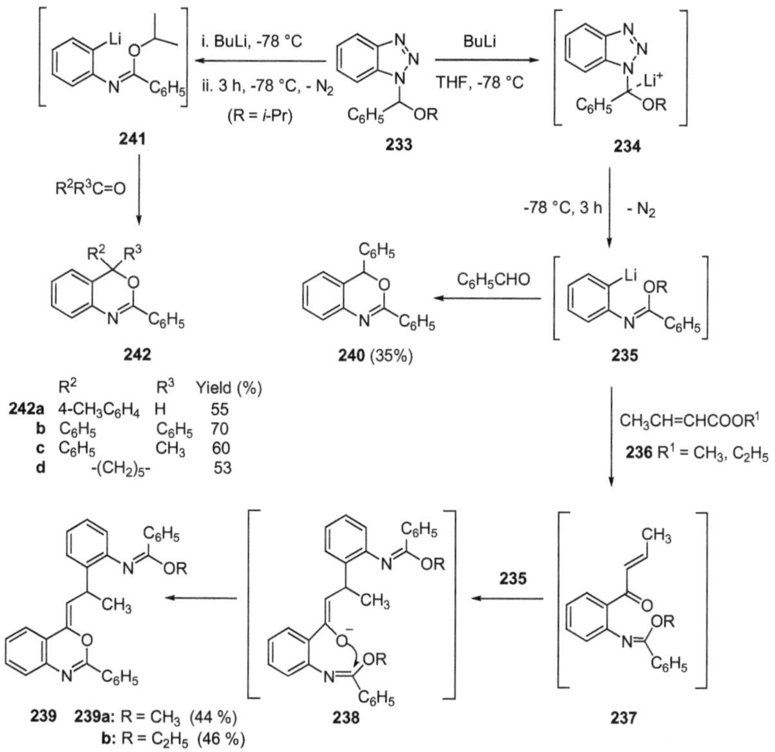

Scheme 48 Synthesis of 4*H*-benzo[*d*][3,1]oxazines **239**, **240,** and **242**

The anions derived from *N*-(α-alkoxyalkyl)benzotriazoles **233** underwent ring opening at low temperatures to give the *ortho*-lithiated anions **235** and **241**, which were successfully trapped with a variety of electrophiles. For instance, the *ortho*-lithiated anion **235** reacted with ethyl crotonate or methyl crotonate to give benzo [*e*]1,3-oxazines **239a,b**, respectively, in 44–46% yield (Scheme 48). Formation of anion **235** was further supported by reaction with benzaldehyde, which gave product **240** (35 %). Treatment of the isopropyl (R^1=*i*-Pr) analogs of **235** with ketones resulted in the formation of 4,4-disubstituted 4*H*-benzo[*d*][3,1]oxazines **242** in 53–70% yield [78].

1,2-Addition of anion **235** to methyl or ethyl crotonate formed an intermediate α,β-unsaturated ketone **237**. The crowded environment of the carbonyl group of **237** directed a second equivalent of **235** to 1,4-addition to give enolate **238**. Following the ring closure of **238** via intramolecular displacement of the ethoxide or methoxide group by the enolate oxygen formed **239**.

The reaction of 2.2 equiv. of isocyanates **245** with *N*-(*o*-hydroxybenzoyl) benzotriazole **243** and *N*-(hydroxynaphthol)benzotriazoles **244** gave benzoxazinones **246** and naphtoxazinones **248**, respectively. The reaction proceeded at room temperature with the formation of a carbamoyl derivative **247** as a side product (Scheme 49) [79].

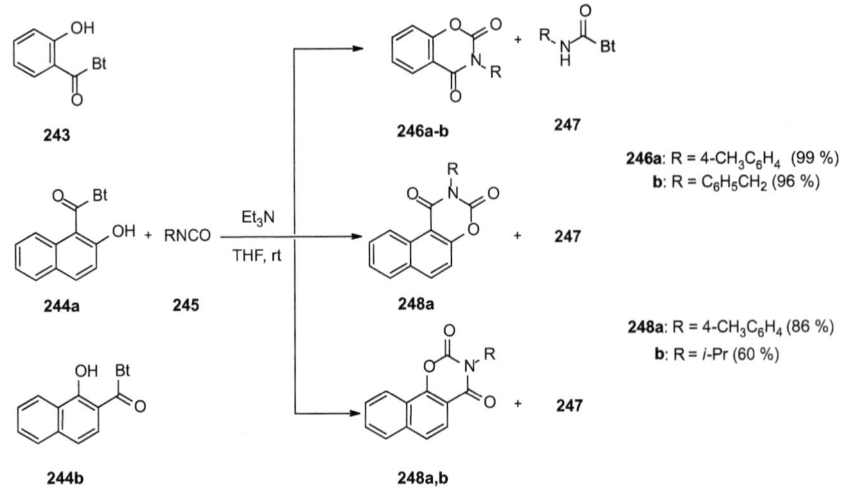

Scheme 49 Synthesis of annulated [*e*]1,3-oxazin-2,4-diones **246** and **248**

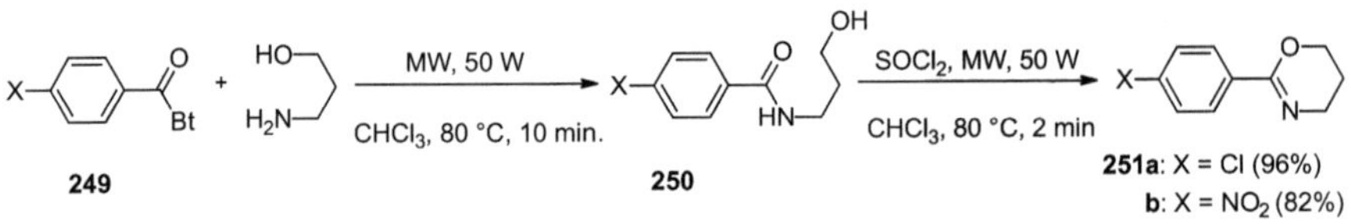

Scheme 50 Synthesis of 2-substituted 5,6-dihydro-4*H*-1,3-oxazines **251**

Another example of successful utilization of 1-aroylbenzotriazols in synthesis of oxazine scaffold is depicted in Scheme 50. Benzoylbenzotriazoles **249** readily reacted with 3-aminopropanol in mild conditions under microwave irradiation to afford 5,6-dihydro-4*H*-1,3-oxazine ring in high (82–96%) yields [59]. Firstly, benzoylbenzotriazoles **249** were converted into the corresponding *N*-(3-hydroxypropyl)benzamides **250**. Following the cyclization into corresponding oxazines **251** required an equivalent amount of thionyl chloride. The procedure was accomplished in a few minutes, providing a facile and convenient synthetic approach towards 5,6-dihydro-4*H*-1,3-oxazines.

3.3 Seven-Membered Ring with One Oxygen and One Nitrogen

1-(2-phenyloxyethyl)-5-benzotriazol-1-yl-2-pyrrolidinone **253** and 3-benzotriazol-1-yl-2-(2-phenyloxyethyl)-1-isoindolinone **254** were readily obtained from the reaction of a mixture of benzotriazole and 2-phenoxyethylamine **252** with either

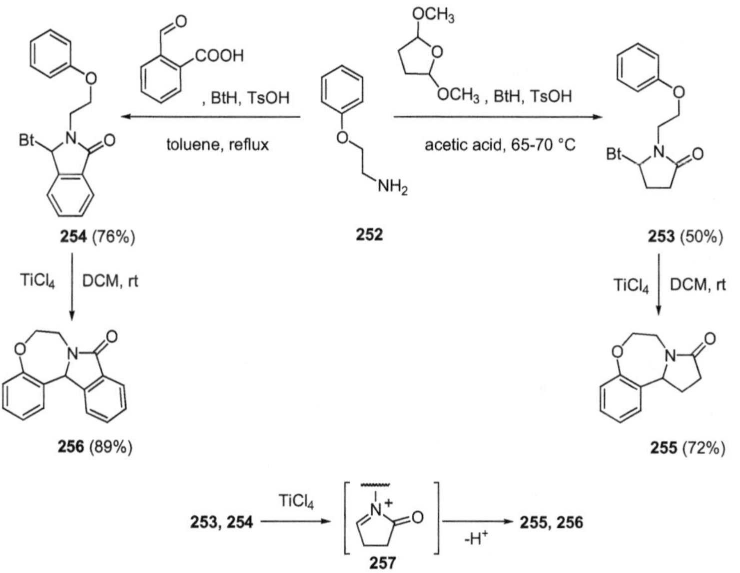

Scheme 51 Lewis acid-catalyzed formation of 2,3,4,5-tetrahydrobenzo[*f*][1,4]oxazepines **255** and **256**

2,5-dimethoxy-2,5-dihydrofuran or 2-formylbenzoic acid, respectively (Scheme 51). The Lewis acid-mediated cyclizations of **253** and **254** produced novel 2,3,4,5-tetrahydrobenzo[*f*][1,4]oxazepine scaffolds **255** and **256** in 72% and 89% yields, respectively. It was proposed that $TiCl_4$ converted the cyclic amide to its corresponding iminium cation **257**, which reacted intramolecularly as a Friedel-Crafts electrophile for the electrophilic aromatic substitution to furnish the oxazepine cycle (Scheme 51) [80].

References

1. Katritzky AR, Rees CW (eds) (1984) Comprehensive Heterocyclic Chemistry, vols 1–8. Pergamon, Oxford
2. Katritzky AR, Rees CW, Scriven EFV (eds) (1996) Comprehensive heterocyclic chemistry II, vols 1–11. Pergamon, Oxford
3. Katritzky AR, Ramsden CA, Scriven EFV, Taylor RJK (eds) (2008) Comprehensive heterocyclic chemistry III (2008) vols 1–15. Elsevier, Oxford
4. Katritzky AR, Wu H, Xie L, Rachwal S, Rachwal B, Jiang J, Zhang G, Lang H (1995) Synthesis 10:1315–1323
5. Katritzky AR, Rachwal S (2010) Chem Rev 110(3):1564–1610
6. Katritzky AR, Rachwal S (2011) Chem Rev 111(11):7063–7120
7. Katritzky AR, Lan X, Fan WQ (1994) Synthesis 1(5):445–456
8. Katritzky AR, Qi M (1998) Tetrahedron 54(12):2647–2668
9. Katritzky AR, Qi M (1998) Coll Czechoslovak Chem Comm 63(5):599–613

10. Katritzky AR, Toader D (2001) Synlett 4:458–466
11. Katritzky AR, Rogovoy BV (2003) Chem Europ J 9(19):4586–4593
12. Katritzky AR, Manju K, Singh SK, Meher NK (2005) Tetrahedron 61(10):2555–2581
13. Katritzky AR, Kirichenko K (2006) Arkivoc 4:119–151
14. Katritzky AR, Angrish P, Todadze E (2009) Synlett 15:2392–2411
15. Avan I, Dennis HC, Katritzky AR (2014) Chem Soc Rev 43(10):3575–3594
16. Katritzky AR, Jiang J (1995) J Org Chem 60:7597–7604
17. Katritzky AR, Xie L, Serdyuk L (1996) J Org Chem 61:7564–7570
18. Katritzky AR, Manju K, Steel PJ (2003) J Org Chem 68:407–411
19. Katritzky AR, Manju K, Gromova AV, Steel PJ (2004) J Org Chem 69:6018–6023
20. Degennaro L, Capriati V, Carlucci C, Florio S, Luisi R, Nuzzo I, Cuocci C (2009) Tetrahedron 65:8745–8755
21. Xiao X, Lin D, Tong S, Mo H (2011) Synlett 19:2823–2826
22. Katritzky AR, Heck KA, Li J, Wells A, Garot C (1996) Synt Commun 26(14):2657–2670
23. Katritzky AR, Maimait R, Denisenko A, Denisenko SN (2001) Arkivoc 5:68–78
24. Wedler C, Kleiner K, Kunath A, Schick H (1996) Liebigs Ann 5:881–885
25. Danheiser RL, Nowick JS (1991) J Org Chem 56:1176–1185
26. Wedler C, Kunath A, Schick H (1995) J Org Chem 60:758–760
27. Zhi J, Melia AT, Guerciolini R, Ching J, Kinberg J, Hauptman JB, Patel IH (1994) Clin Pharm Ther 56:82–85
28. Wedler C, Costisella B, Schick H (1999) J Org Chem 64:5301–5303
29. Pleynet DPM, Dutton JK, Thornton-Pett M, Johnson AP (1995) Tetrahedron Let 36 (35):6321–6324
30. Dutton JK, Pleynet DPM, Johnson P (1999) Tetrahedron 55:11927–11942
31. Katritzky AR, Lan X, Zhang Z (1993) J Heterocycl Chem 30(2):381–387
32. Katritzky AR, Li J, Gordeev MF (1993) J Org Chem 58:3038–3041
33. Katritzky AR, Li J (1995) J Org Chem 60:638–643
34. Katritzky AR, Fali CN, Li J (1997) J Org Chem 62:8205–8209
35. Katritzky AR, Verin SV (1995) J Heterocycl Chem 32:323–328
36. Katritzky AR, Yang B, Jiang J, Steel PJ (1995) Heterocycles 41:765–772
37. Katritzky AR, Cheng D, Leeming P, Ghiviriga I (1996) J Heterocycl Chem 33(6):1935–1941
38. Katritzky AR, Serdyuk L, Xie L (1998) J Chem Soc Perkin Trans 1 6:1059–1064
39. Katritzky AR, Ji Y, Fang Y, Prakash I (2001) J Org Chem 66:5613–5615
40. Katritzky AR, Kirichenko K, Hür D, Zhao X, Ji Y, Steel PJ (2004) Arkivoc 6:27–44
41. Katritzky AR, Jiang J (1995) J Org Chem 60:6–7
42. Katritzky AR, Lan X (1992) Synthesis 8:761–764
43. Katritzky AR, Lan X, Yang JZ, Denisko OV (1998) Chem Rev 98:409–548
44. Katritzky AR, Wang X, Xie L, Toader D (1998) J Org Chem 63:3445–3449
45. Moody CJ, Whitham GH (1992) Reactive intermediates. Oxford University Press, New York
46. Birkett MA, Knight DW, Little PB, Mitchell MB (2000) Tetrahedron 56:1013–1023
47. Knight DW, Little PB (2000) J Chem Soc Perkin Trans 1 15:2343–2355
48. Knight DW, Little PB (1998) Tetrahedron Let 39(28):5105–5108
49. Knight DW, Qing X (2009) Tetrahedron Let 50(26):3534–3537
50. Knight DW, Little PB (1998) Synlett 10:1141–1143
51. Knight DW, Little PB (2001) J Chem Soc Perkin Trans 1 15:1771–1777
52. Katritzky AR, Denisko OV (2002) J Org Chem 67:3104–3108
53. Yu T, Zhao Y, Fan D (2006) J Molec Struc 791:18–22
54. Yu T, Zhang P, Zhao Y, Zhang H, Fan D, Dong W, Ding L (2009) Phosp Sulf Silic 184 (10):2655–2663
55. Katritzky AR, Qi M, Feng D, Nichols DA (1997) J Org Chem 62:4121–4124
56. Katritzky AR, Feng D, Qi M (1997) J Org Chem 62:6222–6225
57. Katritzky AR, Rachwal S, Rachwal BJ (1987) J Soc Perkin Trans 1:799–804
58. Katritzky AR, Feng D, Qi M (1998) Tetrahedron Let 39:6835–6836

59. Katritzky AR, Cai C, Suzuki K, Singh SK (2004) J Org Chem 69:811–814
60. Katritzky AR, Drewniak M (1988) J Chem See Perkin Trans 1:2339–2344
61. Katritzky AR, Chen YX, Yannakopoulou K, Lue P (1989) Tetrahedron Let 30(48):6657–6660
62. Katritzky AR, Wu H, Xie L (1995) J Heterocycl Chem 32:1651–1652
63. Katritzky AR, Qiu G, Yang B, Steel PG (1998) J Org Chem 63:6699–6703
64. Katritzky AR, Cui XL, Yang B, Steel PG (1998) Tetrahedron Let 39:1697–1700
65. Katritzky AR, Singh SK, Bobrov S (2004) J Org Chem 69:9313–9315
66. Al-Awadi NA, George BJ, Dib HH, Ibrahim MR, Ibrahim YA (2005) Tetrahedron 61:8257–8263
67. Katritzky AR, Hitchings GJ, Zhao X (1990) J Chem Soc Perkin Trans 1:2371–2377
68. Katritzky AR, Wang M, Zhang S, Voronkov MV (2001) J Org Chem 66:6787–6791
69. Kuehne ME, Weaver SJ, Franz P (1964) J Org Chem 29:1582–1586
70. Katritzky AR, Musgrave RP, Breytenbach JC (1996) J Heterocycl Chem 33:1637–1646
71. Katritzky AR, Button MAC, Denisenko SN (2000) J Heterocycl Chem 37:1505–1510
72. Katritzky AR, Shestopalov AA, Suzuki K (2005) Arkivoc 7:36–55
73. Katritzky AR, Huang TB, Voronkov MV (2000) J Org Chem 65:2246–2248
74. Katritzky AR, Vvedensky V, Cai X, Rogovoy B, Steel PJ (2002) Arkivoc 6:82–90
75. Katritzky AR, Gupta V, Gordeev M (1993) J Heterocycl Chem 30(4):1073–1077
76. Katritzky AR, Offerman RJ, Cabildo P, Soleimen M (1988) Recl Trav Chim PayBass 107:641–645
77. Katritzky AR, Pernak J (1992) Fan WQ. J pract Chem 334:114–118
78. Katritzky AR, Zhang G, Jiang J (1995) J Org Chem 60:7625–7630
79. Katritzky AR, Singh SK, Akhmedova R, Cai C, Bodroy S (2007) Arkivoc 6:6–13
80. Katritzky AR, Xu YJ, He HY, Mehta S (2001) J Org Chem 66:5590–5594

Top Heterocycl Chem (2016) 43: 177–234
DOI: 10.1007/7081_2015_178

Published online: 3 October 2015

Benzotriazole-Mediated Synthesis of Nitrogen-Containing Heterocycles

Judit K. Beagle

Abstract Application of benzotriazole methodology for the preparation of nitrogen-containing heterocyclic compounds is summarized in this review. The characteristic advantages of benzotriazole are first briefly considered followed by an overview of its use in the synthesis of heterocycles organized by the ring size. Finally, the use of benzotriazole in the synthesis of bicyclic systems is showcased in only selected examples.

Keywords Benzotriazole · Benzotriazole-mediated synthesis · Heterocycles

Contents

J.K. Beagle (✉)
University of Dayton, 300 Collage Park, Dayton, OH 45459-2357, USA
e-mail: jbeagle1@udayton.edu

1 Introduction

Heterocyclic compounds are a major class of organic molecules characterized by a cyclic structure that includes at least one atom of an element other than carbon. Heterocycles are of great importance biologically and industrially. The majority of pharmaceuticals and biologically active agrochemicals are heterocycles [1, 2], and also many additives and modifiers used in industrial applications ranging from cosmetics, information storage, and plastics are heterocycles [3, 4]. Therefore, the interest in the synthesis of heterocycles has greatly attracted attention for decades.

Alan R. Katritzky and his group dedicated their research to the study of chemical properties of benzotriazole over the last three decades. Benzotriazole intermediates are now commonly used for introduction of a variety of functional groups into molecules. The five major roles of benzotriazole in organic transformations are illustrated on Fig. 1 [5–10].

This review only gives a short insight into the various known methods in which benzotriazole-mediated syntheses can lead to nitrogen-containing heterocycles, organized by increasing ring size and increasing number of heteroatoms in the ring.

2 The Construction of Three-Membered Ring

2.1 Aziridines

Aziridines are of great importance in natural products [11, 12] and versatile synthetic intermediates, used as chiral auxiliaries, reagents, and ligands [13, 14] and as monomers for polymer synthesis [15, 16]. It has been demonstrated that benzotriazole-mediated synthesis of aziridines is possible from 1-(triphenylphosphoranylideneaminomethyl)benzotriazole **1**, a synthon equivalent

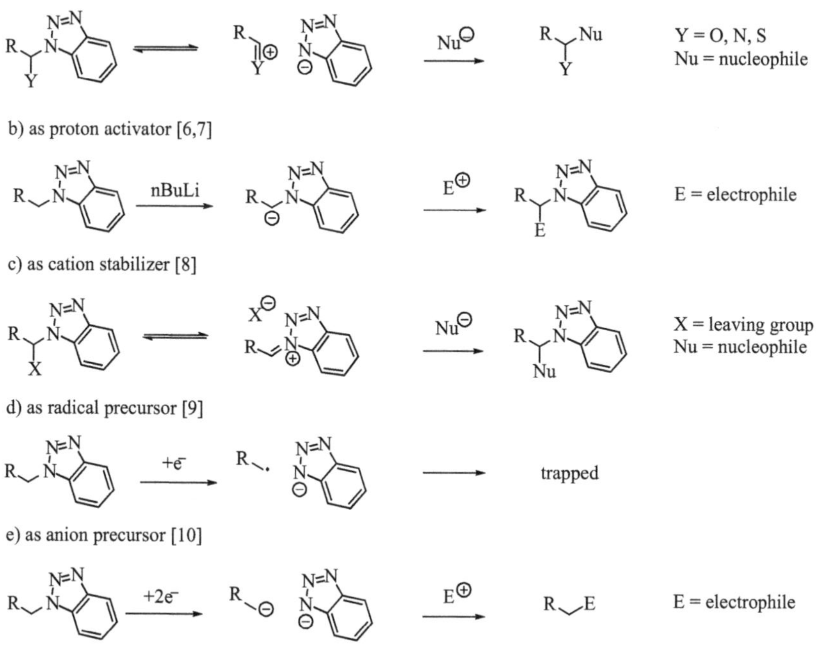

Fig. 1 Reactivity profile of benzotriazole

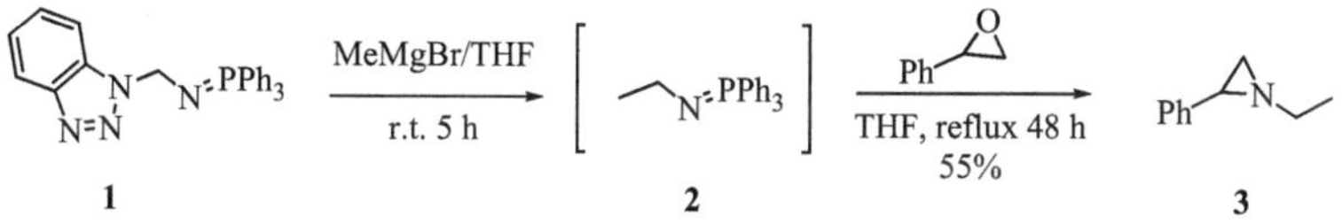

Scheme 1 Preparation of aziridines **3** from phosphazene **1**

to $NH_2CH_2^+$ [17]. Preparation of **1** was accomplished in four steps starting from benzotriazole in high yield [18–20]. Treatment of phosphazene **1** with Grignard reagent displaced the benzotriazole moiety to give an iminophosphorane **2** intermediate. Further reaction with epoxides under reflux condition formed aziridines **3** in 55% yield (Scheme 1).

The anion-stabilizing property of benzotriazole makes it possible to easily access benzotriazole-substituted aziridines which are valuable starting materials for the synthesis of pyrroles [21]. Two methods were developed, one involving a carbenoid intermediate **5** and the 1,2-dihalide route. 1-Chloromethylbenzotriazole [18] **4** can be lithiated in the presence of diarylimine, and the very unstable carbenoid intermediate **5** can be captured to form 1,3-diaryl-substituted aziridines **6** as mixtures of *cis*- and *trans*-isomers (Scheme 2).

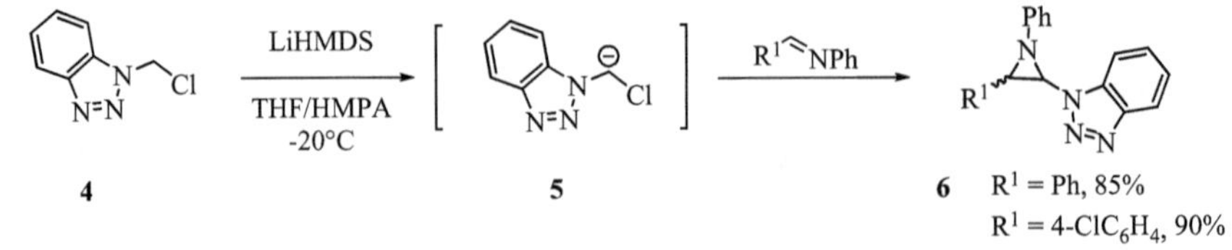

Scheme 2 Preparation of aziridines **6** from 1-chloromethylbenzotriazole **4**

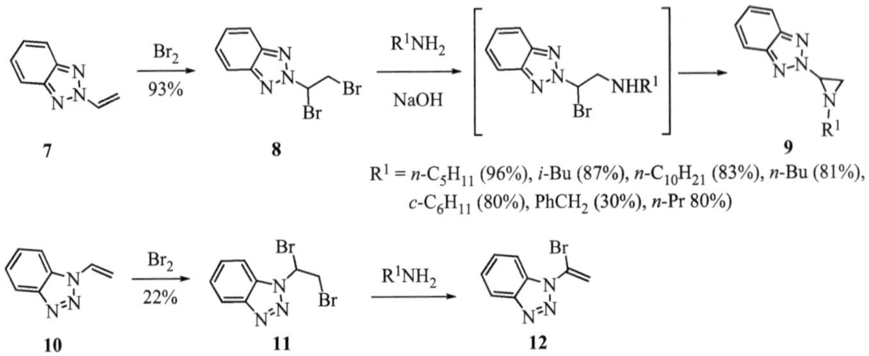

Scheme 3 Preparation of aziridines **9** from 2-vinylbenzotriazole **7**

The above-described method is limited to diarylimines; to extend the scope of possible products, a different approach was developed. 2-Vinylbenzotriazole [22] **7** was easily brominated and then reacted with alkyl amines in the presence of sodium hydroxide to afford 2-(benzotriazol-2-yl)aziridines **9** in good yields (Scheme 3). 1-Vinylbenzotriazole **10**, however, did not give aziridines but 1-(1-bromovinyl)benzotriazole **12** (Scheme 3). This behavior can be attributed to the higher acidity of the α-proton of the benzotriazol-1-yl isomer compared to the benzotriazol-2-yl isomer [6, 23, 24], which suppressed the cyclization [21].

2.2 *Azirines*

Azirines are important for their versatile chemical and biological behavior [25, 26]. Azirines have been shown to naturally occur in antibiotics [27–30], precursors for the synthesis of aziridines [14, 31–33], and, when derivatized with phosphorus substituents, shown to regulate many important biological functions [34]. *N*-Acylmethylbenzotriazoles [35, 36] **13** with hydroxylamine hydrochloride provided oximes **14** as single isomers. They can be converted into their tosylates **15**; however, a large excess of potassium hydroxide directly gave 2*H*-azirines **16** in moderate yields (Scheme 4) [37].

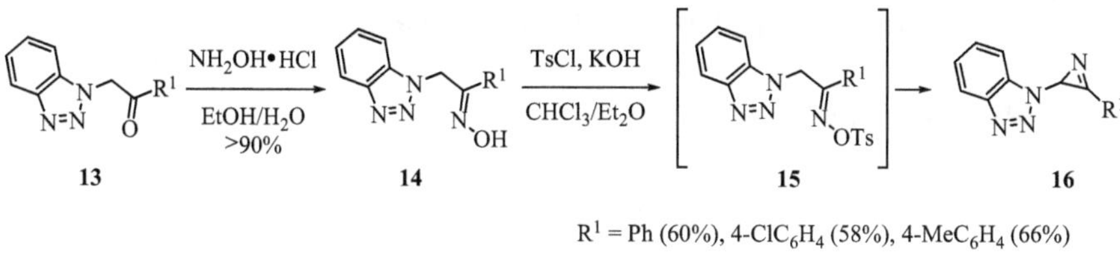

Scheme 4 Preparation of azirines **16**

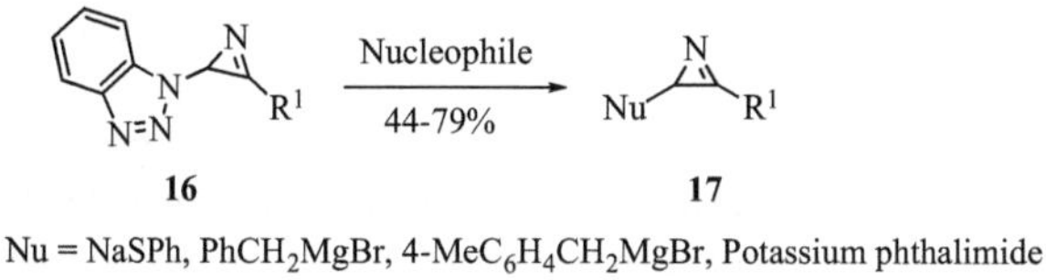

Scheme 5 Preparation of disubstituted azirines **17**

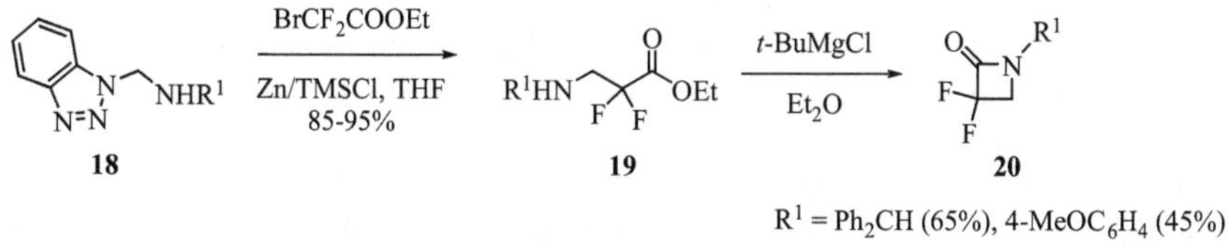

Scheme 6 Preparation of 3,3-difluoroazetidin-2-one derivatives **20**

The benzotriazole moiety was then readily substituted by nucleophiles to give disubstituted azirines **17** (Scheme 5) [37].

3 The Construction of Four-Membered Ring

Benzotriazole methodology has been used for the synthesis of 3,3-difluoroazetidin-2-one derivatives. Several other attempts at the synthesis of azetidines have failed. 3,3-Difluoroazetidin-2-one derivatives are of interest either as enzyme inhibitors [38] or as intermediates for modified peptides [39]. *N*-Substituted 1-(aminomethyl) benzotriazoles [40, 41] **18** were treated with ethyl bromodifluoroacetate under Reformatsky-type conditions to give *N*-substituted ethyl 3-amino-2,2-difluoropropionates **19**. Cyclization of **19** promoted by *tert*-butylmagnesium chloride furnished *N*-protected 3,3-difluoroazetidin-2-ones **20** (Scheme 6) [42].

4 The Construction of Five-Membered Ring

4.1 Pyrrolidines

Pyrrolidines represent an important class of heterocyclic compounds as advanced intermediates in the synthesis of natural products and structural components in a number of biologically active pharmaceuticals [43–48]. Pyrrolidines monosubstituted at C-2 were easily generated from *N*-Boc-*N*-(benzotriazol-1-ylmethyl-3-chloropropylamine [49] **21** which in the presence of *n*-BuLi generated unstable anion **22** and then underwent intramolecular cyclization to form pyrrolidine **23** (Scheme 7) [50].

The benzotriazole moiety is readily substituted by Grignard reagents in the presence of zinc chloride to furnish 2-substituted pyrrolidines **24** (Scheme 8) [50].

In another approach, protected pyrrolidinones **25** were reduced by DIBAL-H to protected 2-hydroxypyrrolidines **26**, which upon condensation with benzotriazole was converted to pyrrolidine **23**, in case the protecting group is Boc, and enantioselectively, using a chiral auxiliary, instead to **27**. Treatment of either protected pyrrolidines with (tributylstannyl)lithium afforded protected 2-stannylpyrrolidines **28**. The (*S*) configuration is predominant (80:20) when the chiral auxiliary was used, which can be easily removed by DIBAL-H at elevated temperature (Scheme 9) [51].

Another asymmetric synthetic route to pyrrolidines can be accomplished by the double Robinson–Schopf condensation of benzotriazole **29**, (*S*)-phenylglycinol **30**, and hydrolyzed 2,5-dimethoxytetrahydrofuran **31**, the synthetic equivalent of succindialdehyde (Scheme 10) [52].

Oxazolopyrrolidine **32** is a convenient synthon for asymmetric syntheses of 2-substituted and 2,5-disubstituted pyrrolidines. Thus, treatment with allyltrimethylsilane in the presence of a Lewis acid gave the separable mixture of the

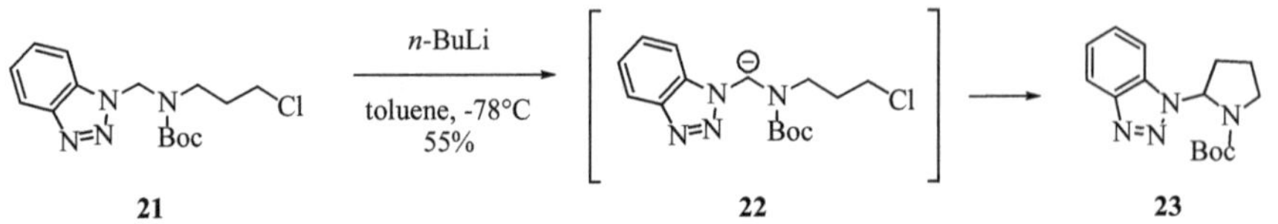

21 **22** **23**

Scheme 7 Preparation of pyrrolidine **23**

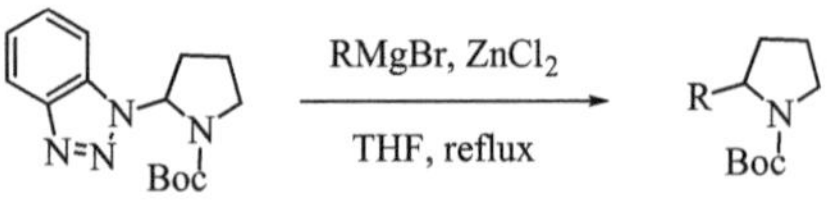

23 **24**

R = Ph (86%), 4-MeC_6H_4 (66%), 4-$MeOC_6H_4$ (70%), 4-ClC_6H_4 (62%), PhC≡C (77%)

Scheme 8 Substitution of benzotriazole to form 2-substituted pyrrolidines **24**

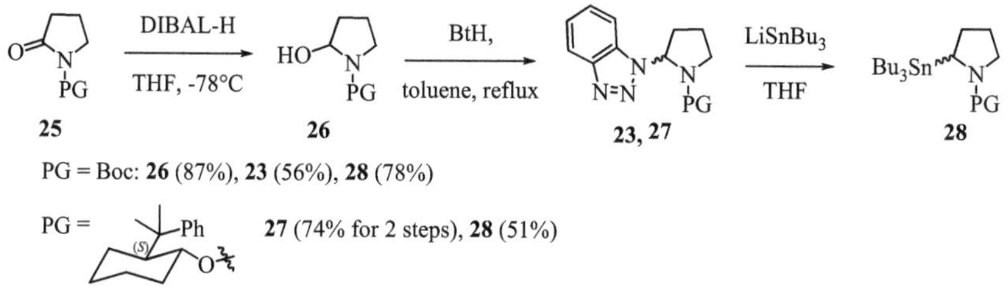

Scheme 9 Preparation of 2-(trimethylstannyl)pyrrolidines **28**

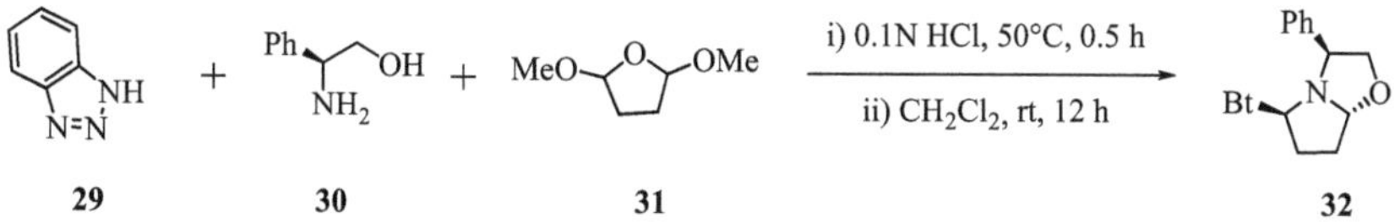

Scheme 10 Preparation of chiral pyrrolidine, (3*S*, 5*R*, 7a*R*)-5-(benzotriazol-1-yl)-3-phenyl[2,1-b] oxazolopyrrolidine **32**

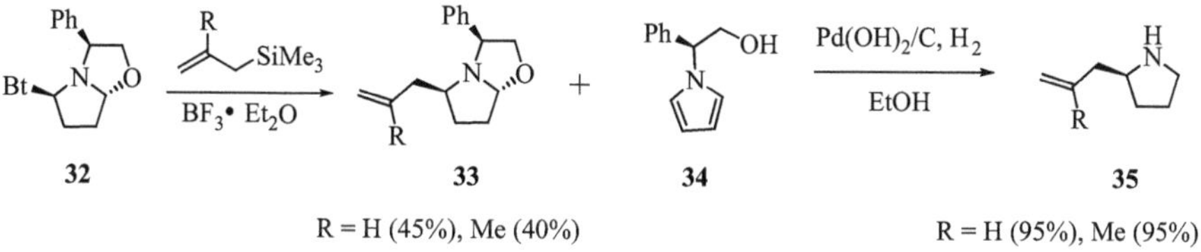

Scheme 11 Preparation of chiral 2-alkylpyrrolidines **35**

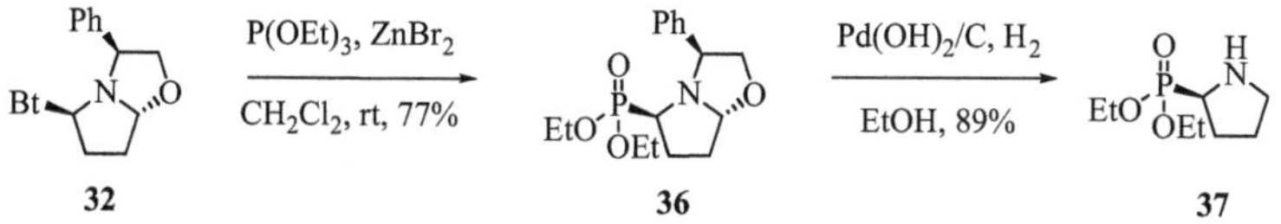

Scheme 12 Preparation of chiral pyrrolidine phosphonate **37**

substitution **33** and the elimination products **34**. Hydrogenation of **33** cleaved the chiral auxiliary and reduced the double bond to form 2-alkylpyrrolidines **35** (Scheme 11) [52].

The Arbuzov reaction converted **32** into the 2-substituted pyrrolidine phosphonate **36** in the presence of zinc bromide as Lewis acid. Hydrogenation removed the chiral auxiliary to furnish diethyl (*R*)-tetrahydropyrro-2-ylphosphonate **37** (Scheme 12) [52].

Excess Grignard reagent was reacted with **32** to give mixtures of the *cis*-**38** and *trans*-**39** products in good total yields. The *cis*-diastereomer was isolated as the

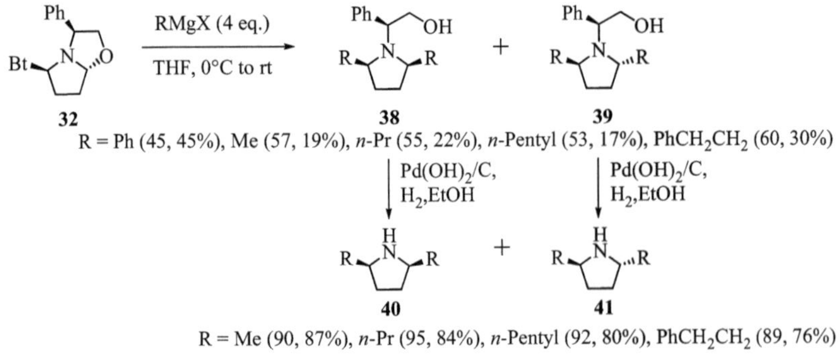

Scheme 13 Preparation of 2,5-disubstituted pyrrolidines

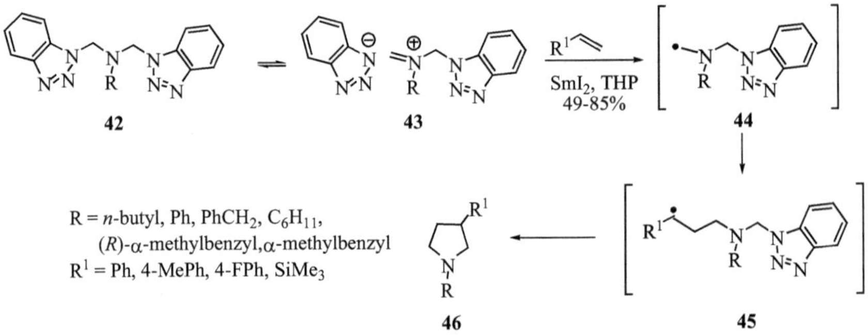

Scheme 14 Preparation of 1,3-disubstituted pyrrolidines **46**

major product. Hydrogenation gave the corresponding *cis*- and *trans*-2,5-disubstituted pyrrolidines **40** and **41** (Scheme 13) [52].

N,*N*-Bis(benzotriazol-1-ylmethyl)amines **42** [18, 53] exist in equilibria with their ionic form **43** in solution. Such solutions were treated with samarium diiodide to generate radicals **44** which were rapidly reacted with various alkenes to provide more stable radicals **45**. Consecutive ionization and reduction of the second benzotriazolylmethyl group provided a diradical that coupled intramolecularly to form 1,3-disubstituted pyrrolidines **46** in 49–85% yields (Scheme 14) [54].

α-Aminoalkenyl benzotriazoles **47** [55–58] reacted readily with samarium diiodide to generate radicals **48**. Electron-deficient alkenes were able to trap the radical intramolecularly to form the cyclized radical **49**. Final reduction by a second molecule of SmI_2 provided 1,2,3-trisubstituted pyrrolidines **50** (Scheme 15). For pyridyl and aliphatic substituents, the *cis*-stereoisomer predominates, but for most aromatic substituents, the ratio is reversed [57].

The same method can be used to form 1,2,4-trisubstituted pyrrolidines. Thus, condensation of amines **51** with *N*-(hydroxymethyl)benzotriazole gave tertiary amines **52**, which upon treatment with samarium diiodide furnished pyrrolidines **53** (Scheme 16) [59].

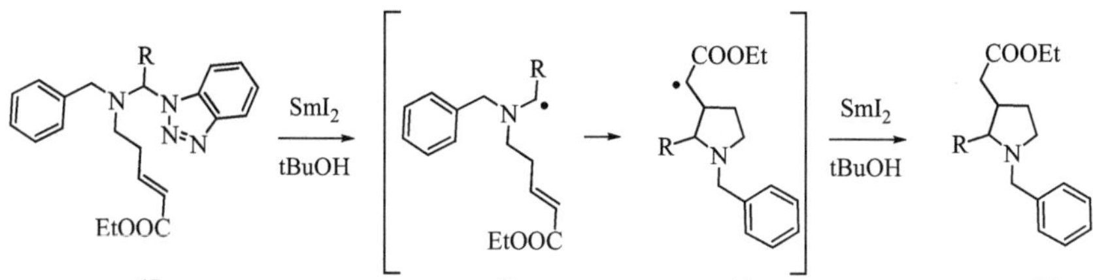

47 48 49 50

R = H (70%), *n*-Pr (67%), *i*-Pr (74%), Ph$(CH_2)_2$ (49%), 3,4-$(MeO)_2C_6H_3(CH_2)_2$ (72%), *trans*-non-3-enyl (44%), pent-4-ynyl (68%), $(CH_2)_4$COOEt (65%), CH_2OCH_2Ph (29%), Ph (56%), 3-ClC_6H_5 (60%), 2-naphthyl (53%), 3-pyridyl (65%), 2-pyridyl (33%)

Scheme 15 Preparation of 1,2,3-trisubstituted pyrrolidines **50**

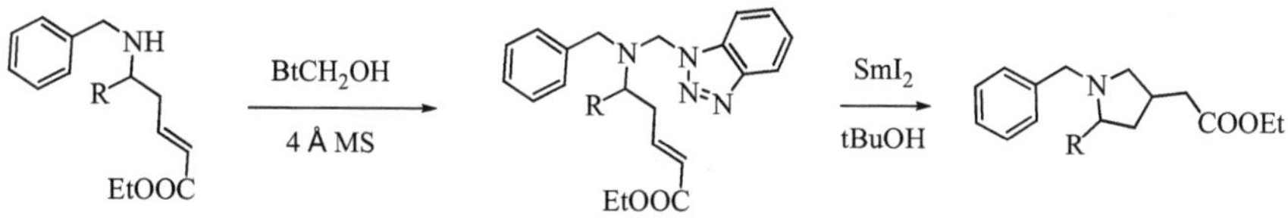

51 52 53

R = nPr (75%), Ph (79%), 3-pyridyl (66%)

Scheme 16 Preparation of 1,2,4-trisubstituted pyrrolidines **53**

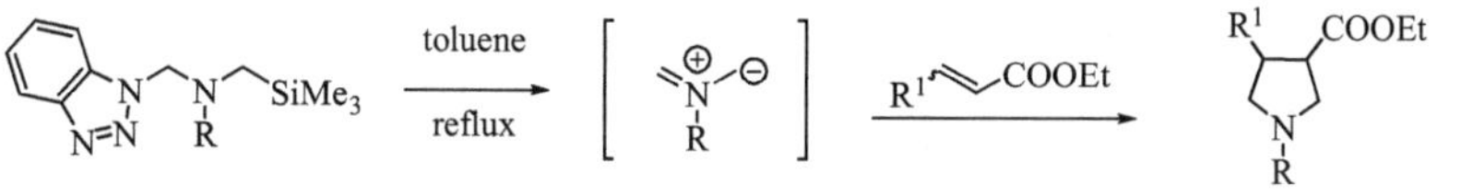

54 55 56

R = cyclohexyl, R^1 = COOEt (Z) (84%), Ph (Z) (71%)
R = n-hexyl, R^1 = COOEt (Z) (89%), COOEt (E) (87%)
R = allyl, R^1 = Ph (Z) (90%); R = sBu, R^1 = COOEt (Z) (88%)

Scheme 17 Preparation of 1,3,4-trisubstituted pyrrolidines **56**

Thermal desilylation of benzotriazol-1-ylmethylaminosilanes **54** [41, 60–63] generated azomethine ylides **55**, which readily reacted with electron-deficient alkenes to produce pyrrolidines **56** (Scheme 17). The cycloaddition proceeded with retention of the olefinic dipolarophile configuration [64].

Lithiation of **57** [22], followed by alkylation with 1-bromo-3-chloropropane derivatives, gave the corresponding chlorides **58** in good yields. Reaction between chlorides **58** and primary amines was carried out in DMF at 80°C to give intermediate **59** in situ, followed by the addition of methanol, $Pd(OAc)_2/Ph_3P$, and K_2CO_3, which resulted in the cyclization to furnish 2-vinylpyrrolidines **60** (Scheme 18) [65].

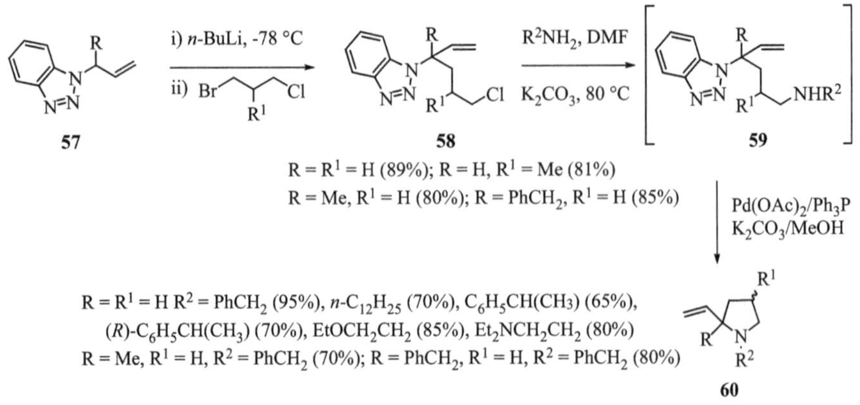

Scheme 18 Preparation of 2-vinylpyrrolidines **60**

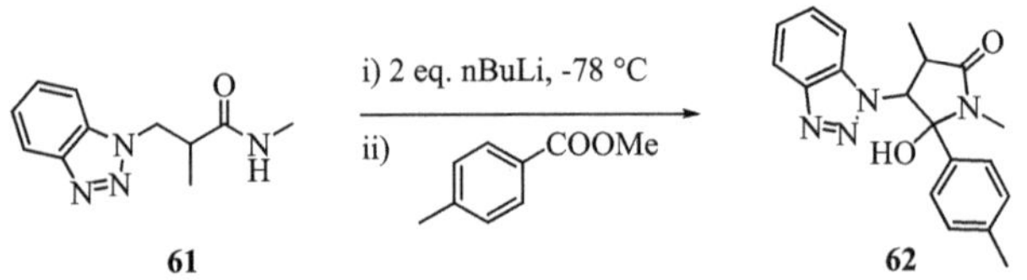

Scheme 19 Preparation of benzotriazole-substituted pyrrolidinone **62**

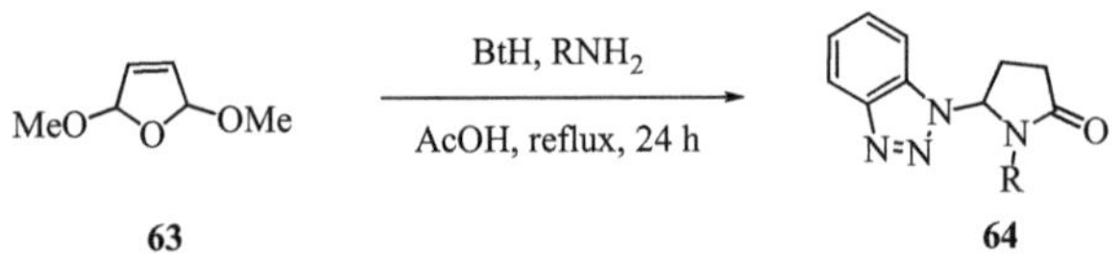

R = Ph (70%), $PhCH_2$ (69%), $PhCH_2CH_2$ (66%), 4-$MeOC_6H_4CH_2$ (64%), 4-$MeOC_6H_4CH_2CH_2$ (63%), $CH_3COOCH_2CH_2$ (65%), 3,4-$(MeO)_2C_6H_3CH_2$ (74%), (S)-$PhCH(CH_3)$ (67%), (S)-$PhCH_2CH(COOMe)$ (66%), (*S*)-$PhCH(AcOCH_2)$ (60%), 2-pyridylCH_2CH_2 (66%)

Scheme 20 Preparation of 5-(benzotriazol-1-yl)-1-substituted-pyrrolidin-2-ones **64**

4.2 *Pyrrolidinones*

Pyrrolidinones possess varied biological activities and have been used as pharmaceuticals [66–69]. They can also be intermediates in the synthesis of pyrrolidine alkaloids and γ-amino acids [70–74]. Benzotriazole can undergo Michael addition with *N*-methylmethacrylamide at 150°C to afford β-benzotriazol-1-ylpropionamide **61** in 45% yield. The dianion generated from **61** was trapped by methyl 4-methylbenzoate to give 2-pyrrolidinone **62** in 63% yield (Scheme 19) [75].

The reaction of 2,5-dimethoxy-2,5-dihydrofuran **63**, benzotriazole, and primary amines in refluxing AcOH gave 5-benzotriazolyl-1-substituted-pyrrolidin-2-ones **64** in good yields, with strong preference for the benzotriazol-1-yl isomer (Scheme 20) [76].

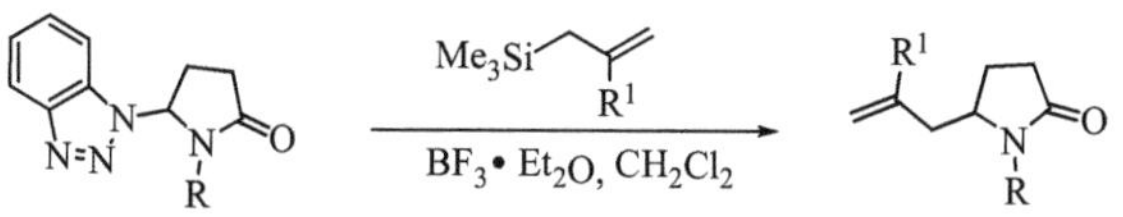

64 65

R^1 = H, R = $PhCH_2$ (90%), $PhCH_2CH_2$ (90%), 4-$MeOC_6H_4CH_2CH_2$ (85%), (S)-$PhCH_2CH(COOMe)$ (49%)
R^1 = Me, R = $PhCH_2CH_2$ (84%), 4-$MeOC_6H_4CH_2$ (80%), 4-$MeOC_6H_4CH_2CH_2$ (80%)

Scheme 21 Preparation of 1-substituted-5-allylpyrrolidinones **65**

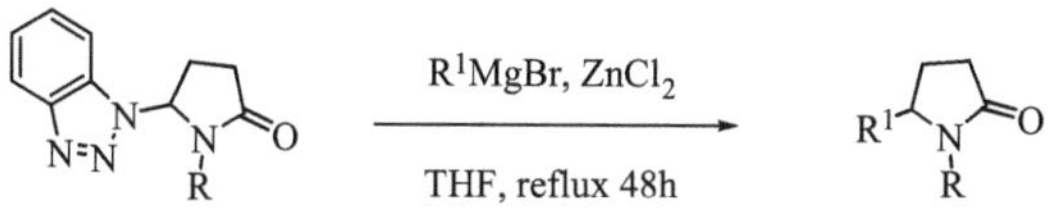

64 66

R = $PhCH_2$, R^1 = $CH_3C{\equiv}C$ (49%), R = 4-$MeOC_6H_4CH_2CH_2$, R^1 = $CH_3C{\equiv}C$ (71%), CH_2=CH (80%), CH_3CH=CH (52%), $PhCH_2$ (50%), cyclo-C_5H_9 (57%), n-C_5H_{11} (73%), $CH(COOEt)_2$ (67%)

Scheme 22 Preparation of 1,5-disubstituted pyrrolidinones **66**

Substitution of the benzotriazolyl moiety in **64** with nucleophiles gave 5-substituted 2-pyrrolidinones **65–67**. Since both benzotriazolyl isomers have similar reactivity, the isomeric mixture can be used in this step. Treatment of intermediates **64** with 4 equiv. of trimethylallyl- or trimethyl(2-methylallyl)silane in the presence of a Lewis acid furnished 1-substituted-5-allylpyrrolidinones **65** (Scheme 21) [76].

Grignard reagents cannot be used for the substitution of the benzotriazole moiety in **64**, since they react with carbonyl groups. Treatment with organozinc reagents afforded pyrrolidinones **66** in good yields (Scheme 22) [76].

The treatment of **64** in dry THF with triethylphosphite in the presence of zinc bromide produced diethyl 1-substituted-5-oxo-2-pyrrolidinylphosphonates **67** in moderate to good yields (Scheme 23) [76].

N-Protected glutamic acid **68** can be converted to its dibenzotriazolide derivative **69** using thionyl chloride and benzotriazole. Condensation with L-amino acids in aqueous acetonitrile in the presence of triethylamine gave 2-pyrrolidinones **70** (Scheme 24) [77].

4.3 *Pyrrolines*

The condensation reaction of benzotriazole, aryl aldehydes, and ammonia results in the formation of *N*-arylmethylene[(benzotriazol-1-yl)arylmethyl]amines **71** [78] in quantitative yields. Anion **72** derived from imine **71** adds readily to electron-deficient double bonds. In the case of methyl acrylate, pyrrolidine anion **73**

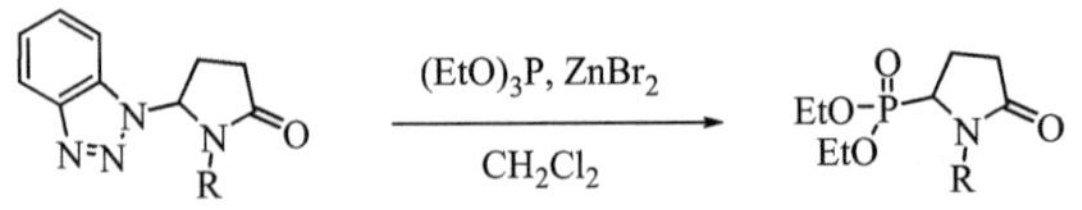

64 67

R = $HOCH_2CH_2$ (49%), 4-$MeOC_6H_4CH_2$ (79%), 4-$MeOC_6H_4CH_2CH_2$ (76%), 3,4-$(MeO)_2C_6H_3CH_2$ (78%), (*S*)-$PhCH_2CH(COOMe)$ (67%), (*S*)-$PhCH(CH_2OH)$ (85%)

Scheme 23 Preparation of diethyl 1-substituted-5-oxo-2-pyrrolidinylphophonates **67**

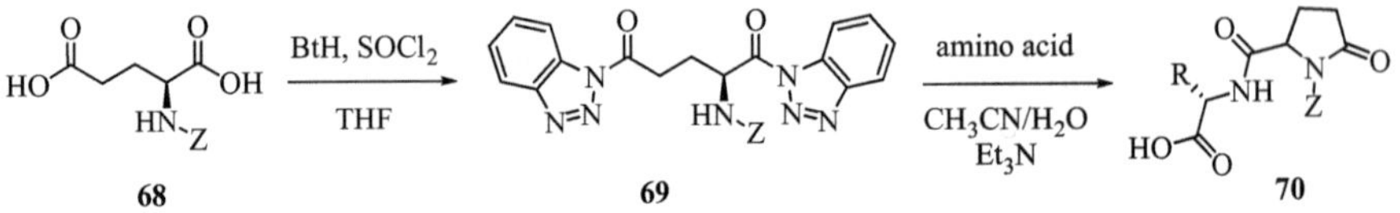

68 69 70

R = Me (58%), $PhCH_2$ (88%), *i*-Pr (71%)

Scheme 24 Preparation of Z-pyroglutamyl pseudopeptides **70**

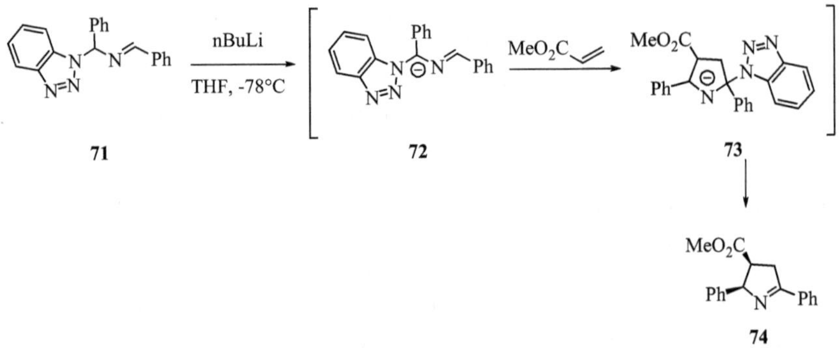

Scheme 25 Preparation of 1,2-pyrroline **74**

eliminates a benzotriazolide anion spontaneously to give 1,2-pyrroline **74** with a *cis*-relative configuration (Scheme 25) [79].

Pyrroline **75** with a *trans*-relative configuration was prepared by the reaction of anion **72** with acrylonitrile. The reaction with dimethyl fumarate leads to pyrroline **76** with *trans,trans*-relative configuration (Scheme 26) [79].

Alkylation of benzotriazole with 2,3-dichloro-1-propene provided 1-(2-chloropropen-3-yl)benzotriazole **77** in 56% yield and its isomer in 30% yield. The solution of **77** in DMSO in the presence of sodium hydroxide was boiled to afford 1-allenylbenzotriazole **78** in 90% yield. Treatment of **77** with LDA resulted in the in situ formation of allene **78**, in which the second equivalent of LDA removes the allenyl proton to form the corresponding anion. This anion can be successfully trapped by various electrophiles. When *N*-(4-methoxybenzylidene) aniline was used as the electrophile, allenyl product **79** rearranges partially to its

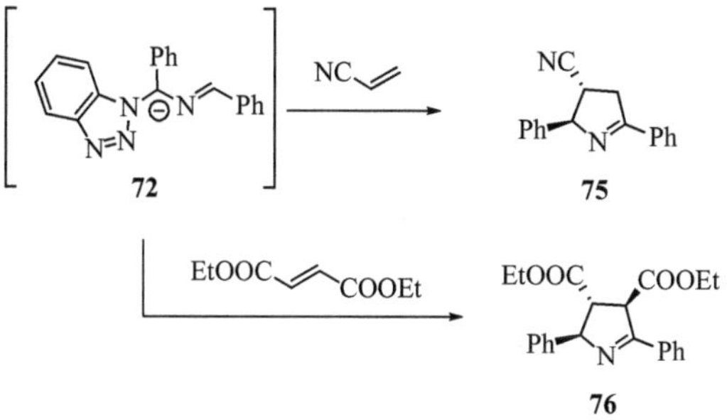

Scheme 26 Preparation of pyrrolines **75**, **76**

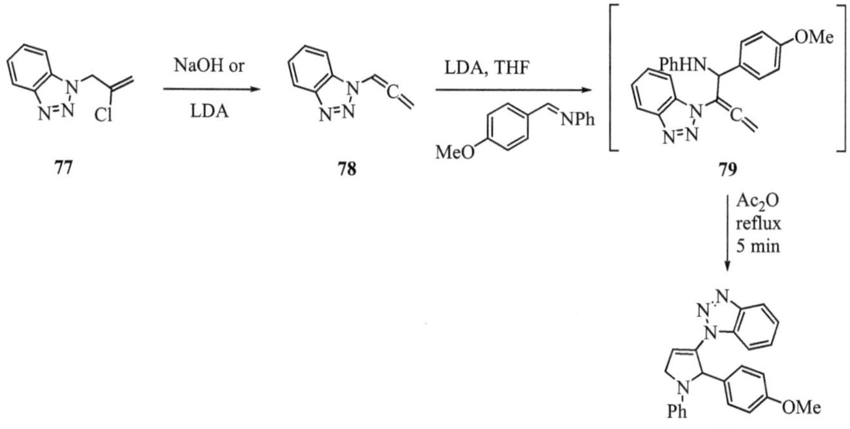

Scheme 27 Preparation of 3-pyrroline **80**

propargyl isomer. Heating the allenyl product in acetic anhydride gave 3,4-pyrroline **80** in 30% overall yield (Scheme 27) [80].

4.4 Pyrroles

Pyrrole and its derivatives are of great importance in organic and biochemistry and have found many applications in medicine and technology [81, 82]. Many pyrrole derivatives have shown interesting biological properties [83–87], pharmaceutical importance, [88, 89], and use as monomers for conducting polymers and nonlinear optics material [90].

The anion generated in situ from 1-propargylbenzotriazole **81** [22] reacted readily with *N*-tosylarylimines to give corresponding sulfonamides **82** in high yields. Upon heating compounds **82** in ethanolic sodium hydroxide, an acetylene–

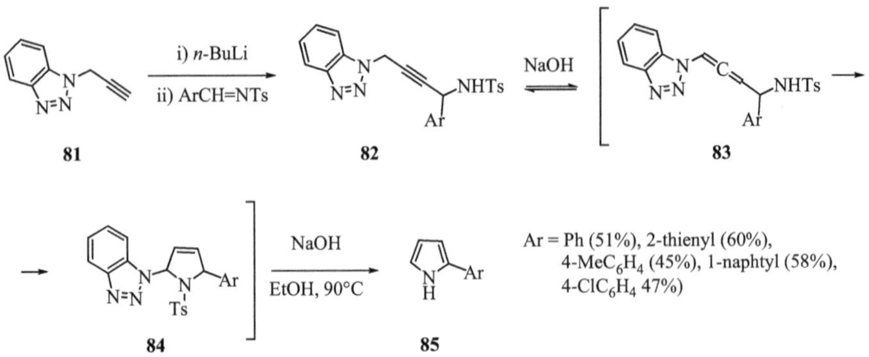

Scheme 28 Preparation of 2-aryl- or 2-heteroarylpyrroles **85**

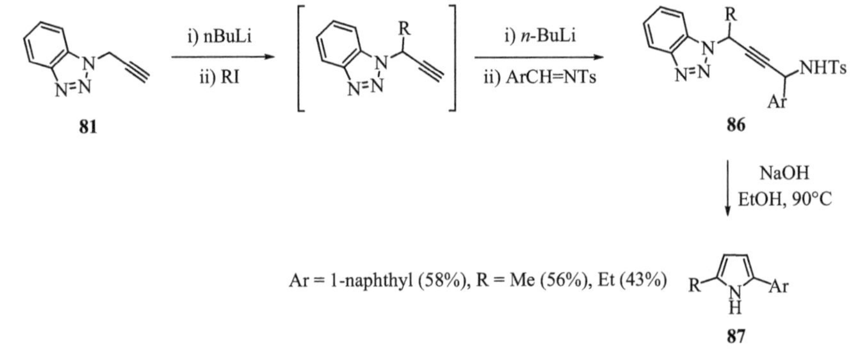

Scheme 29 Preparation of 5-alkyl-2-(1-naphthalyl)pyrroles **87**

allene isomerization took place to form intermediate **83**, which further cyclized to 2,5-dihydropyrroles **84**. This was then followed by heteroaromatization and *N*-deprotection to afford 2-aryl- or 2-heteroarylpyrroles **85** (Scheme 28) [91].

When 1-propargylbenzotriazole **81** was treated with 2 equiv. of *n*-butyllithium, followed by the addition of alkyl iodide, the alkylation takes place selectively in 1-position forming monoanions [22], which can be further reacted with *N*-tosylarylimines to give the corresponding sulfonamides **86**. Compound **86** then can be cyclized in a similar manner as shown above to give 5-alkyl-2-(1-naphthalyl) pyrroles **87** (Scheme 29) [91].

Treatment of 1-(3-morpholinoprop-2-enyl)benzotriazole **88** [92] with *n*-butyllithium, followed by the addition of diarylimines, yielded diamines **89**, which underwent cyclization in situ to give 1,2-diarylpyrroles **90** in 60–68% total yields (Scheme 30) [93].

Alkoxy analogues of enamine **88** can be applied in such reactions as well. Thus, the condensation of benzotriazole with acrolein diethyl acetal gave α-ethoxy derivative **91**, which underwent a zinc bromide-promoted rearrangement to afford

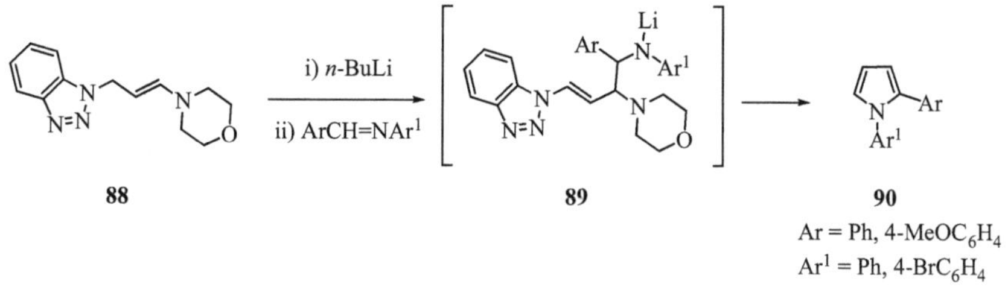

Scheme 30 Preparation of 1,2-diarylpyrroles **90**

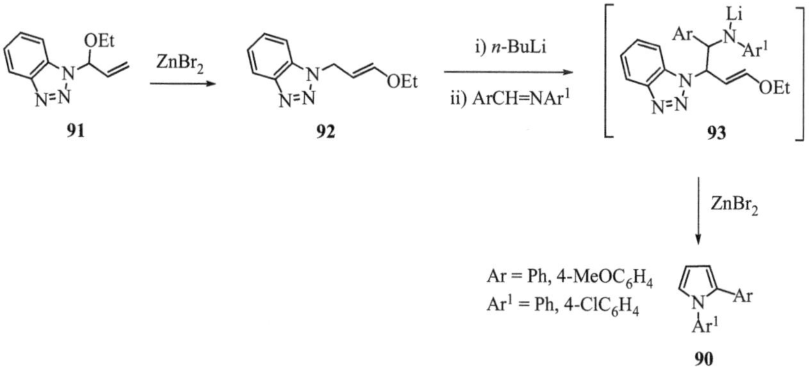

Scheme 31 Preparation of 1,2-diarylpyrroles **90** from **91**

(γ-ethoxyallyl)benzotriazole **92**. Treatment of **92** with *n*-butyllithium, followed by the addition of diarylimines and heating in the presence of zinc bromide, provided 1,2-diarylpyrroles **90**, through intermediate **93**, which underwent the equivalent of intramolecular S_N2' with the elimination of benzotriazole and ethanol (Scheme 31) [94].

A similar cyclization is also possible in the absence of the second leaving group (OEt or morpholinyl). Thus, lithiated *N*-allylbenzotriazole **94** [22] was reacted with diarylimines to give intermediates **95** in moderate yields. Cyclization of **95** occurred in the presence of Pd(II) catalyst and a weak base with the simultaneous oxidation by copper chloride to furnish 1,2-diarylpyrroles **90** (Scheme 32) [95].

The alkylation of **94** with methyl iodide gave 2-(buten-3-yl)benzotriazole **96** [22]. Lithiation of **96** followed by the addition of diarylimines furnished intermediate **97**, which were difficult to purify but could be used crude in the subsequent cyclization reaction, thus affording 1,2,3-trisubstituted pyrroles **98** in low yields (Scheme 33) [95].

Treatment of *N*-(benzotriazol-1ylmethyl)thiobenzamide **99** [96, 97] with *n*-butyllithium, followed by the addition of methyl iodide, gave *S*-methyl thioimidate **100**. Reaction of thioimidate **100** with a Michael acceptor in the presence of sodium hydride gave 2,3,4-trisubstituted pyrroles **102**, through the expected intermediate **101** by elimination of benzotriazole and methanethiol (Scheme 34) [98].

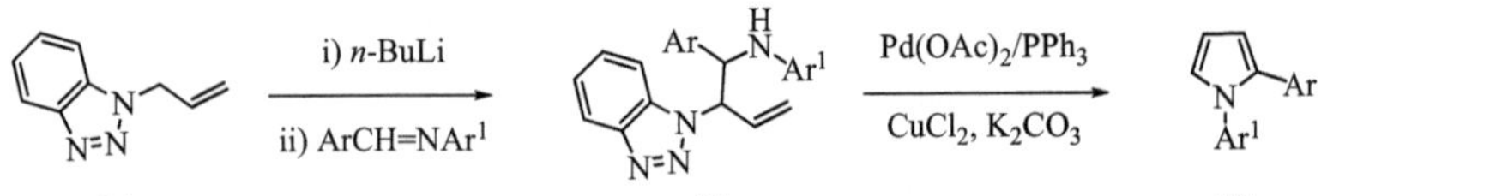

94 95 90

Ar = Ph, 4-FC$_6$H$_4$, 4-ClC$_6$H$_4$, 4-MeC$_6$H$_4$, 4-MeOC$_6$H$_4$, 2-furyl, 3-thiophenyl, 3-pyridyl
Ar^1 = Ph, 2-ClC$_6$H$_4$, 3-ClC$_6$H$_4$, 4-ClC$_6$H$_4$

Scheme 32 Preparation of 1,2-diarylpyrroles **90** from **94**

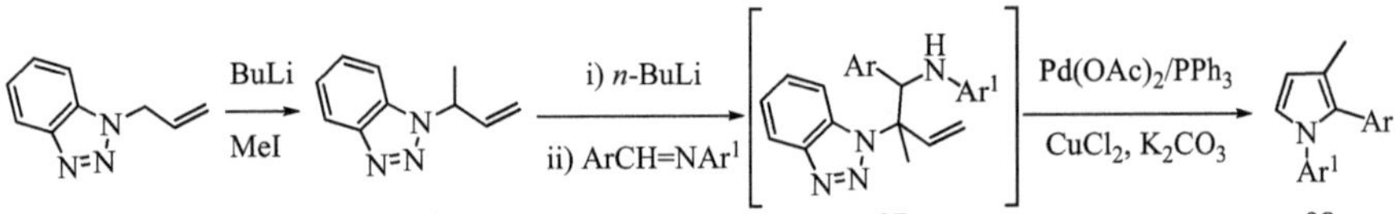

94 96 97 98

Ar = Ar^1 = Ph (26%); Ar = 3-thiophenyl, Ar^1 = Ph (24%)

Scheme 33 Preparation of 1,2,3-trisubstituted pyrroles **98**

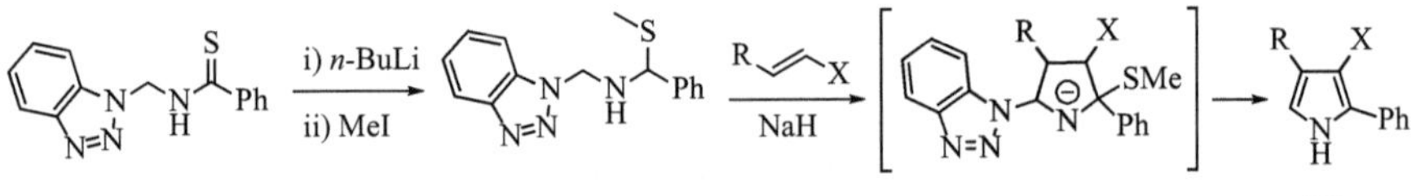

99 100 101 102

R = H, X = CN (88%), 4-pyridyl (64%), 2-pyridyl (36%)
R = Ph, X = COOMe (83%), COOEt (93%), COPh (75%)
R = Me, X = CN (85%), R = X = COOMe (79%)

Scheme 34 Preparation of 2,3,4-trisubstituted pyrroles **102**

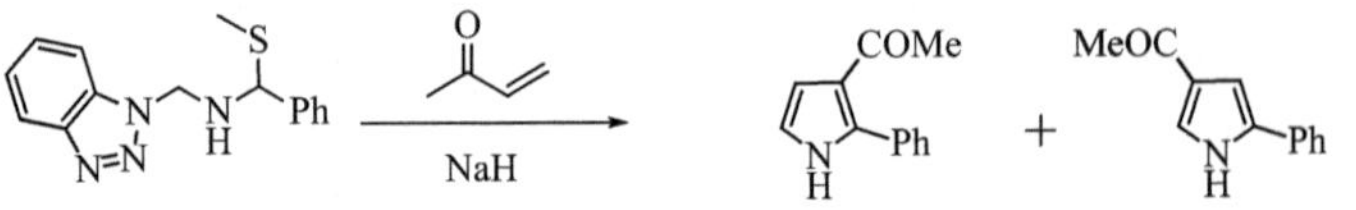

100 **103** (34%) **104** (18%)

Scheme 35 Reaction of **100** with methyl vinyl ketone

When methyl vinyl ketone was used as Michael acceptor with thioimidate **100** under similar conditions, two regioisomers were generated (Scheme 35) [98].

The use of potassium *tert*-butoxide as a base in similar reactions allowed the reactions to take place in a "one-pot" procedure. Thus, thioamides **105** with a variety of R^1 and R^2 substituents were prepared according to literature procedure [98–101]. Treatment of **105** with potassium *tert*-butoxide, followed by the addition of methyl iodide, furnished the corresponding *S*-methylthioamidates **106**.

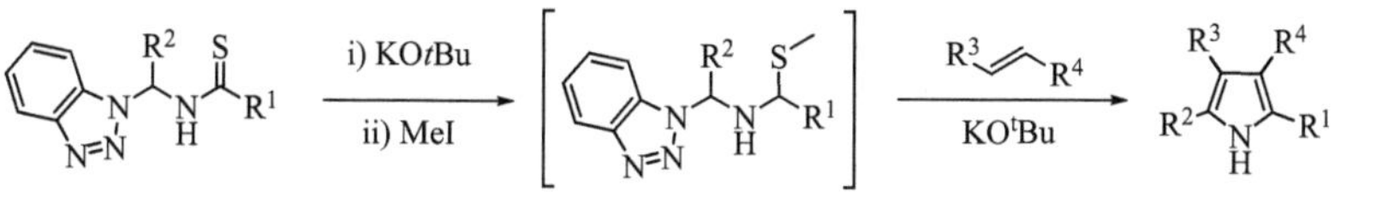

105 106 107

$R^2 = H, R^1 = Ph, R^3 = Ph, R^4 = CN$ (99%), $SO_2C_6H_5$ (90%), COOEt (99%), $R^3 = 2\text{-}C_4H_3S$, $R^4 = COPh$ (96%), $R^3 = 2\text{-}C_4H_3O$, $R^4 = COOH$ (63%), $R^3 = CF_3$, $R^4 = COOEt$ (20%), R^1 = 3-pyridyl, $R^3 = R^4 =$ COOMe (60%), $R^1 = 2\text{-}C_4H_3O$, $R^3 = Ph$, $R^4 = COOEt$ (65%), $R^1 = 4\text{-}MeOC_6H_4$, $R^3 = Ph$, $R^4 = COOEt$ (87%)

$R^2 = 2,4\text{-}Cl_2C_6H_3$, $R^1 = R^3 = Ph$, $R^4 = CN$ (63%)

Scheme 36 Preparation of tri- and tetrasubstituted pyrroles **107**

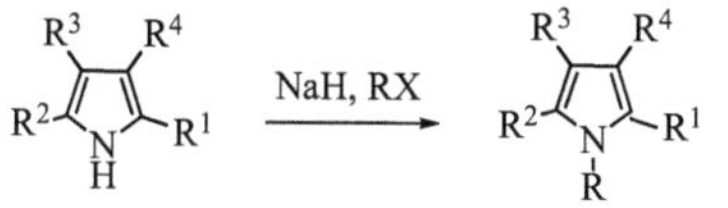

107 108

$R^1 = Ph, R^2 = H, R^3 = Ph, R^4 = SO_2C_6H_5$, R = Ac (82%),
$R^1 = Ph, R^2 = H, R^3 = Ph, R^4 = COOEt$, R = Me (89%),
$R^1 = Ph, R^2 = H, R^3 = 2\text{-}C_4H_3O, R^4 = COOH$, R = Me (80%),
$R^1 = Ph, R^2 = H, R^3 = CF_3, R^4 = COOEt$, R = Me (67%),
$R^1 = 2\text{-}C_4H_3O, R^2 = H, R^3 = Ph, R^4 = COOEt$, R = Me (85%)

Scheme 37 Preparation of *N*-acyl- and *N*-alkylpyrroles **108**

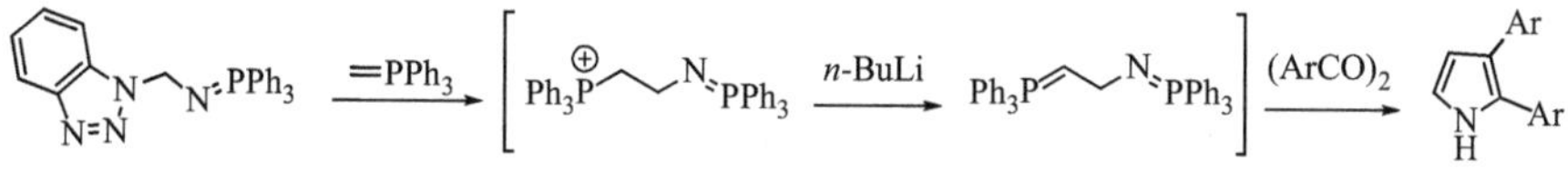

1 109 110

Ar = Ph (67%), 4-MeC_6H_4 (75%), 2-pyridyl (58%)

Scheme 38 Preparation of 2,3-diarylpyrroles **110**

Conversion of **106** into the desired pyrroles **107** can be achieved by the addition of another 3 equiv. of KO^{t}Bu and an activated olefin to the reaction mixture. This procedure worked well for the introduction of aromatic or heteroaromatic groups as R or R^1, but only a trace amount of the desired product was isolated with aliphatic substituents (Scheme 36) [101].

N-acylation and *N*-alkylation of isolated pyrroles **107** can be achieved in the presence of sodium hydride and an acylation or alkylation agent to give *N*-acyl- and *N*-alkylpyrroles **108** (Scheme 37) [101].

1-(Triphenylphosphoranylideneaminomethyl)benzotriazole **1** [17, 20, 102, 103] and methylenetriphenylphosphorane gave intermediate **109**, which after deprotonation with *n*-butyllithium and stirring with diaryl 1,2-diones furnished 2,3-diarylpyrroles **110** (Scheme 38) [104].

In the presence of base, benzotriazol-1-ylmethyl isocyanide **111** [105] was added readily to electron-deficient double bonds, and the resultant pyrroline intermediate **112** spontaneously eliminated benzotriazole to give 3,4-disubstituted pyrroles **113** (Scheme 39) [106].

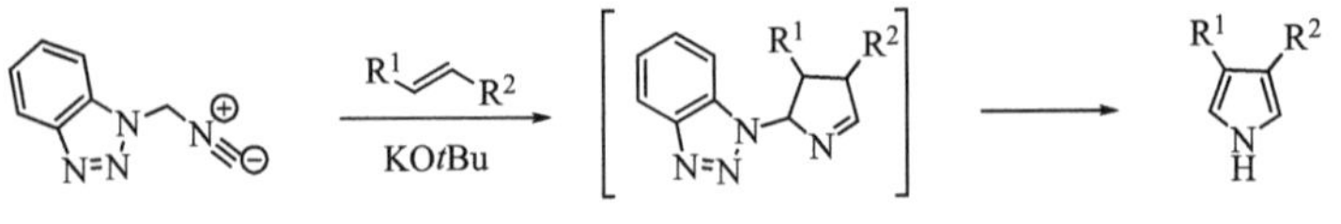

111 **112** **113**

R^1 = Me, R^2 = CN (92%), R^1 = Ph, R^2 = CN (81%), R^1 = Ph, R^2 = COOMe (40%)

Scheme 39 Preparation of 3,4-disubstituted pyrroles **113**

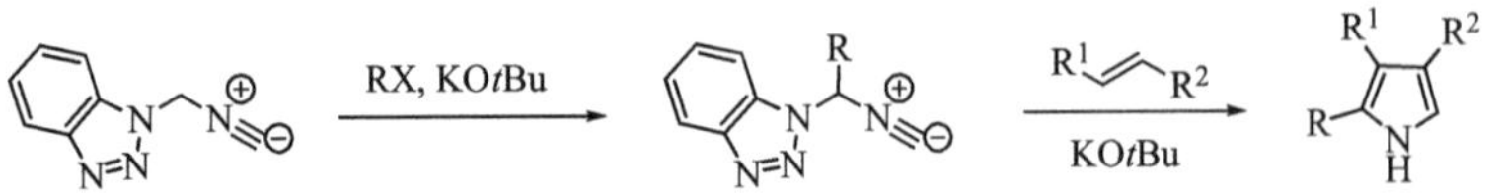

111 **114** **115**

R = Me, R^1 = H, R^2 = COOMe (30%)

Scheme 40 Preparation of 2,3,4-trisubstituted pyrroles **115**

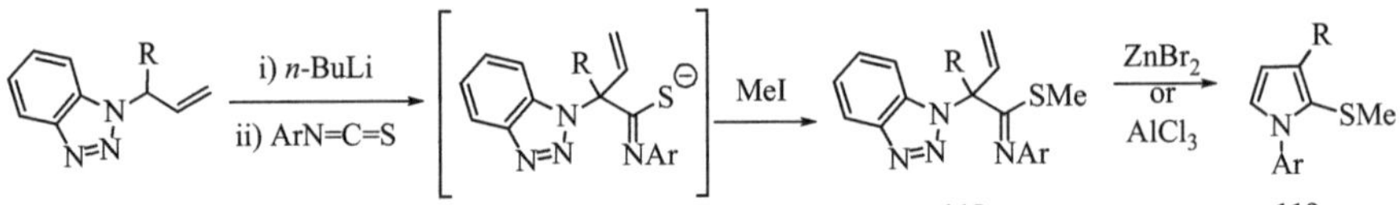

116 **117** **118** **119**

R =OEt, Ar = Ph (65%), 3-FC_6H_4 (59%), 4-ClC_6H_4 (33%), 4-FC_6H_4 (38%), 2-ClC_6H_4 (21%), R = CH_2CHCH_2, Ar = Ph (33%), R = nPr, Ar = Ph (39%)

Scheme 41 Preparation of 2-(methylthio)pyrroles **119**

Isocyanide **111** underwent alkylation on the methylene group to give **114**, which upon deprotonation by additional KO^{t}Bu was added to electron-deficient double bonds to produce 2,3,4-trisubstituted pyrroles **115** (Scheme 40) [106].

1-(1-Alkylprop-2-enyl)benzotriazoles **116** [22] can be lithiated and then added to the C=N bond of isothiocyanates to produce anions **117**, which were methylated in situ to form methyl 2,2-disubstituted 3-buteniminothiates **118**. Treatment of **118** with a Lewis acid resulted in cyclization to 2-(methylthio)pyrroles **119**. The low stability contributed to the low yields of the electron-rich pyrroles during purification (Scheme 41) [107].

The reaction of 1-benzylbenzotriazoles **120** [36, 108] with *n*-butyllithium generated α-lithiobenzylbenzotriazoles, which were treated with 2-bromoacetaldehyde diethyl acetal to give 3-(benzotriazol-1-yl)-3-arylpropanal diethyl acetals **121**. Intermediates **121** were treated without isolation with another equivalent of *n*-butyllithium followed by the addition of *N*-benzylideneaniline to afford adduct **122**. An intramolecular cyclization was promoted by formic acid in ethanol, followed by the spontaneous elimination of benzotriazole and ethanol to furnish 1,2,3-triarylpyrroles **124** (Scheme 42) [109].

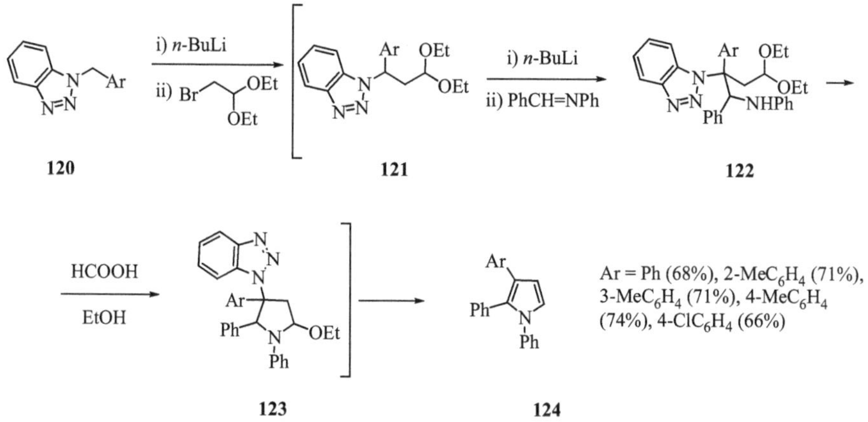

Scheme 42 Preparation of 1,2,3-triarylpyrroles **124**

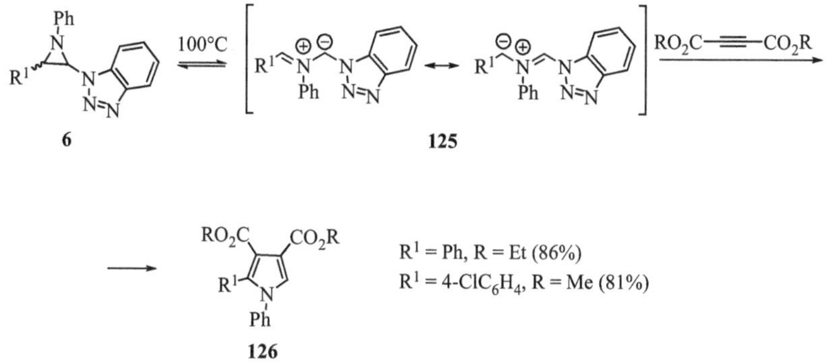

Scheme 43 Preparation of tetrasubstituted pyrroles **126**

Reaction between **6** and dialkyl acetylenedicarboxylates at 100°C gave pyrroles **126**, presumably via the formation of an azomethine ylide **125** followed by a polar [2+3] cyclization and aromatization (Scheme 43) [21].

1-Alkyl-2-(benzotriazol-2-yl)aziridines **9** required higher temperature and longer reaction times to complete the sequence with diethyl acetylenedicarboxylate to yield pyrrole-dicarboxylic esters **129** with methoxycarbonyl groups at the 2- and 3-positions. At a higher temperature, N–C bond underwent heterolytic cleavage to form **127**, which reacted with acetylene to give pyrrolines **128**. In the final step, intermediate **128** was aromatized by the loss of benzotriazole to form pyrrole **129** (Scheme 44) [21].

Treatment of *N*-(benzotriazol-1-ylmethyl)amides **130** [110] with phosphorus pentachloride provided the corresponding imidoyl chlorides **131**, which were sensitive to moisture and were used immediately. Solutions of imidoyl chlorides **131** were treated with KO^{t}Bu followed by the reaction with a dipolarophile to give benzyl pyrrole esters **132** (Scheme 45) [111].

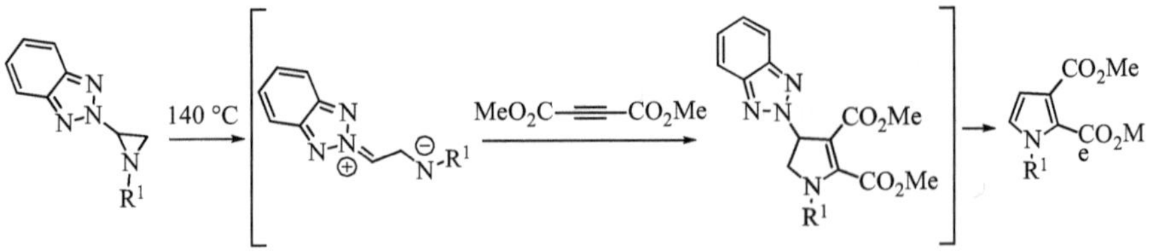

9 127 128 129

$R^1 = n\text{-}C_5H_{11}$ (80%), *i*-Bu (53%), $n\text{-}C_{10}H_{21}$ (25%), *n*-Bu (53%)

Scheme 44 Preparation of pyrrole-dicarboxylic esters **129**

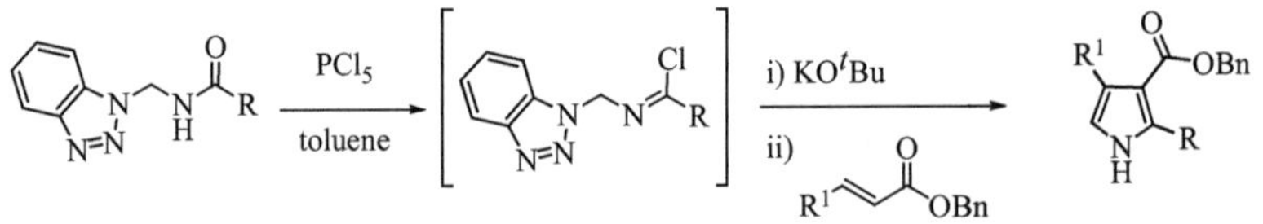

130 131 132

R^1 = R = Ph (90%), R = *p*-tolyl (81%), *p*-anisyl (85%), 2-furyl (76%), R^1 = *p*-tolyl, R = propyl (81%), 2-thienyl (86%)

Scheme 45 Preparation of benzyl pyrrole esters **132**

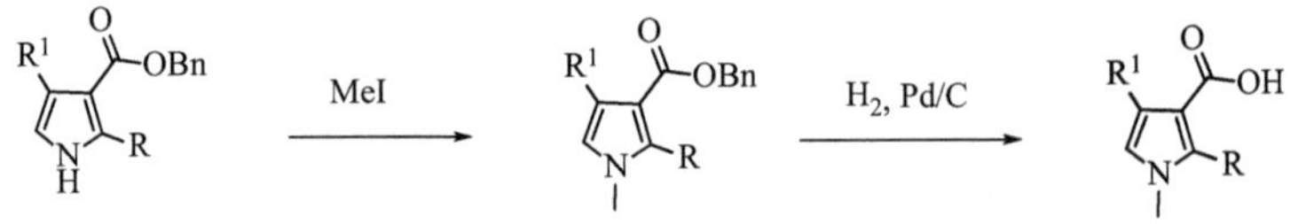

132 133 134

R^1 = R = Ph (90%), R = *p*-tolyl (93%), *p*-anisyl (95%), 2-furyl (85%), R^1 = *p*-tolyl, R = propyl (91%), 2-thienyl (96%)

R^1 = R = Ph (96%), R = *p*-tolyl (99%), *p*-anisyl (90%), 2-furyl (93%), R^1 = *p*-tolyl, R = propyl (97%)

Scheme 46 Preparation of pyrrole carboxylic acids **134**

Pyrroles **132** can be alkylated in situ or after isolation to provide *N*-alkylpyrroles **133**. The hydrogenation of **133** readily gave pyrrole carboxylic acids **134** (Scheme 46) [111].

The anion generated from **71** added readily to dimethyl acetylenedicarboxylate to give tetrasubstituted pyrrole **137**. The reaction presumably proceeded through a pyrrolinyl anion **135**, which spontaneously expelled a benzotriazolide anion to generate **136**, which eventually rearranged to the more stable aromatic tautomer **137** (Scheme 47) [79].

Under special circumstances, the benzotriazolyl substituent may be retained. Thus, 1-(benzotriazol-1-yl)acetic acid **138** was reacted with phosphorus oxychloride in DMF to produce a vinamidinium salt. Vinamidinium salts can be precipitated as perchlorides [112] or hexafluorophosphates [113]. The condensation

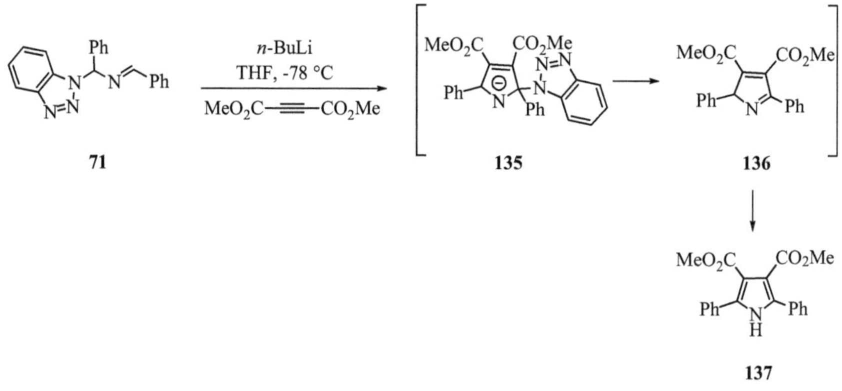

Scheme 47 Preparation of tetrasubstituted pyrrole **137**

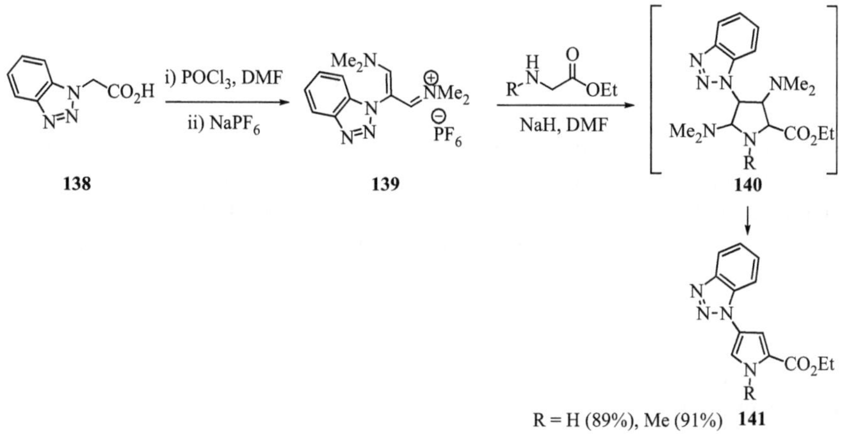

Scheme 48 Preparation of 4-benzotriazol-1-ylpyrrole derivatives **141**

of 2-(1-benzotriazolyl)vinamidinium hexafluorophosphate salt **139** with glycine ethyl ester and sarcosine ethyl ester yielded the corresponding 4-benzotriazol-1-ylpyrrole derivatives **141**. The reaction possibly proceeded through pyrrolidine **140** intermediate, which eliminated two molecules of dimethylamine, leaving the benzotriazolyl moiety unaffected. The role of benzotriazole in this reaction is the activation of the α-proton (Scheme 48) [114].

4.5 Pyrazolidine

Pyrazolidines have been widely studied as fungicides, herbicides, anti-inflammatories, antibacterials, anesthetics, and anticonvulsants [115, 116]. *N*-(1-Benzotriazolylalkyl)-*N*,*N*-disubstituted hydrazines **142** were prepared by the

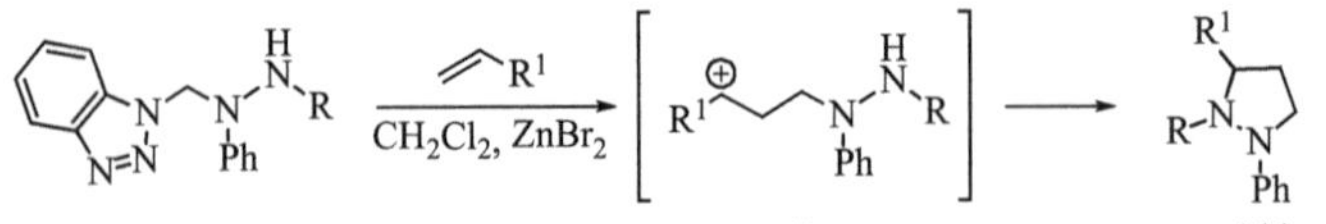

R = Ph, R^1 = 2-pyrrolidonyl (53%), *N*-methylacetamidyl (43%),
R = PhNHCO, R^1 = OEt (60%)

Scheme 49 Preparation of pyrazolidines **144**

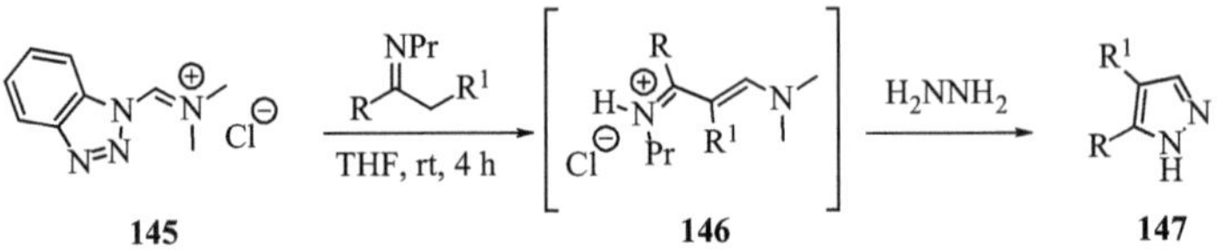

R = Et, R^1 =Me (31%), R = Ph, R^1 = Et (38%), R = *i*-Pr, R^1 =Me (34%), R = *i*-Bu, R^1 = Me (83%)

Scheme 50 Preparation of 3,4-disubstituted pyrazoles **147**

condensation of the appropriate hydrazines, aldehydes, and benzotriazole in good yields. Treatment of **142** with electron-rich alkenes in the presence of Lewis acids gave pyrazolidines **144**. The nitrogen atom in hydrazines **142** readily attacked the cationic intermediates **143** formed from starting material **142** and the alkene (Scheme 49) [117].

4.6 Pyrazole

The wide range of biological activities of pyrazoles has made them popular synthetic targets [118–120]. The most popular method for the preparation of pyrazoles unsubstituted at the nitrogen atoms involves condensation of hydrazine with β-dicarbonyl compounds. Easily accessible synthons for β-dicarbonyl compounds are enamine–iminium salts **146**, prepared from a Vilsmeier-type salt **145** [121] and *N*-propylimines. Treatment of **146** in situ with hydrazine gave 3,4-disubstituted pyrazoles **147** (Scheme 50) [122].

2-(1-Benzotriazolyl)vinamidinium hexafluorophosphate salts **139** reacted readily with aryl hydrazines to give 4-benzotriazol-1-ylpyrazoles **148** (Scheme 51) [114].

In another approach, 4-benzotriazol-1-ylpyrazoles **148** can be prepared starting from benzotriazolylacetone **149** [6, 123]. Condensation of benzotriazolylacetone **149** with dimethylformamide dimethyl acetal (DMFDMA) in boiling xylene yielded enaminone **150**, which further reacted with phenylhydrazine to give 1,4,5-trisubstituted pyrazole **151** (Scheme 52) [124].

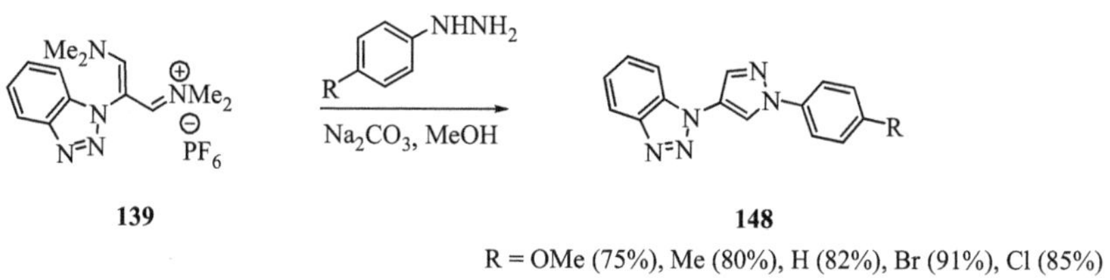

Scheme 51 Preparation of 4-benzotriazolylpyrazoles **148**

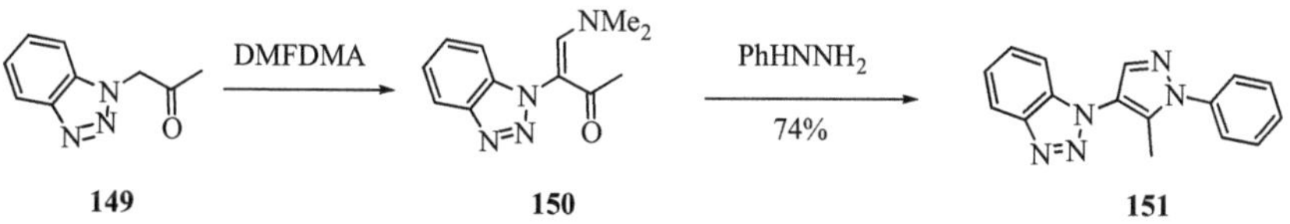

Scheme 52 Preparation of 1,4,5-trisubstituted pyrazole **151**

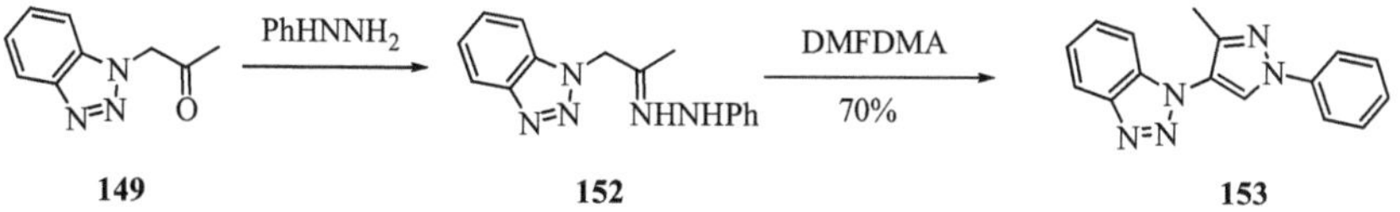

Scheme 53 Preparation of 1,3,4-trisubstituted pyrazole **153**

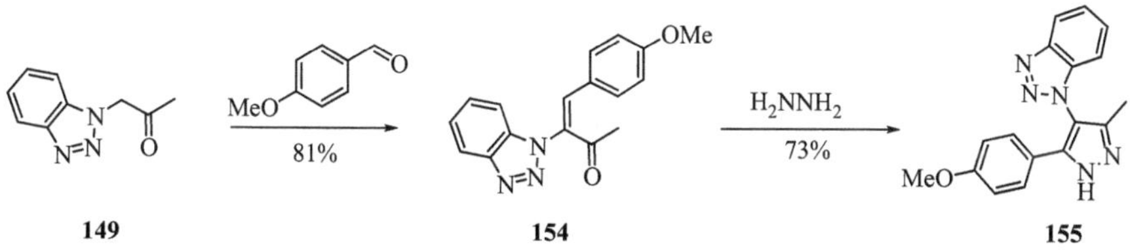

Scheme 54 Preparation of 3,4,5-trisubstituted pyrazole **155**

Carrying out the same reactions but in reverse order resulted in the formation of a different regioisomer, namely, 1,3,4-trisubstituted pyrazole **153** (Scheme 53) [124].

The condensation of equimolar amounts of **149** and with *p*-anisaldehyde in the presence of triethylamine gave benzotriazolylchalcone **154**, which reacted readily with hydrazine to furnish 3,4,5-trisubstituted pyrazole **155**. The formation of **155** is assumed to proceed via the condensation of hydrazine with the carbonyl group in chalcone **154** and subsequent cyclization to pyrazole **155** (Scheme 54) [125].

Treatment of **149** with an excess of *p*-nitrobenzaldehyde under the same reaction conditions afforded dienone **156**, which was further reacted with hydrazine to give pyrazolylbenzotriazole **157**. The formation of **157** is assumed to proceed via initial addition of hydrazine to the carbonyl group and the resulting product then cyclized to pyrazole (Scheme 55) [125].

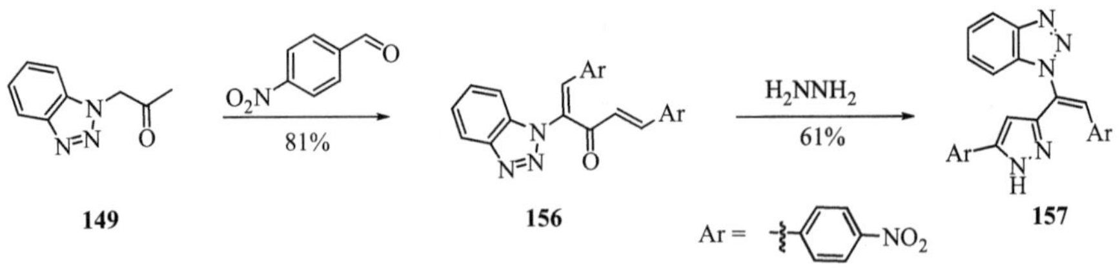

Scheme 55 Preparation of pyrazolylbenzotriazole **157**

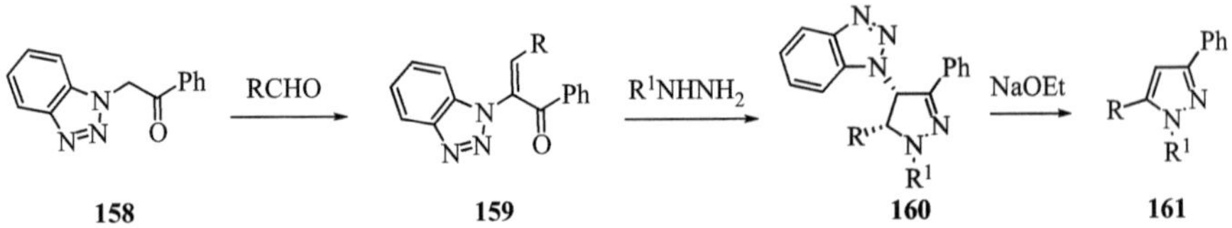

R^1 = Me, R = Ph (90%), *p*-tolyl (86%), *i*-Pr (50%)
R^1 = Ph, R = Ph (84%), *p*-tolyl (81%), *i*-Pr (60%), 3-pyridyl (94%)

Scheme 56 Preparation of 1,3,5-trisubstituted pyrazoles **161**

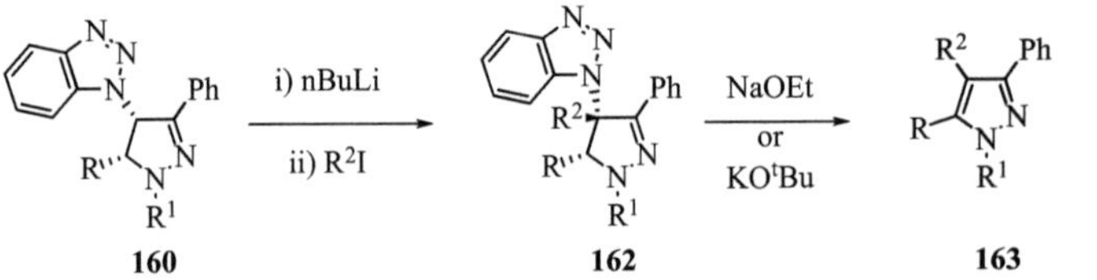

R^1 = Me, R = p-tolyl, R^2 = Me (85%)
R^1 = Ph, R = Ph (84%), R^2 = Me (91%)
R^1 = Ph, R = p-tolyl, R^2 = Me (85%), nBu (95%), allyl (93%), benzyl (95%)

Scheme 57 Preparation of 1,3,4,5-tetrasubstituted pyrazoles **163**

The α-benzotriazolyl-α,β-unsaturated ketones **159** were generated stereoselectively from benzotriazolylacetophenone **158** [123] and the corresponding aldehyde using piperidines as a base. The *Z*-configuration of the double bond was demonstrated in all cases. Compounds **159** react regio- and stereoselectively with *N*-phenyl- or *N*-methylhydrazine in the presence of catalytic amount of NaOEt to form stable intermediates **160**. Elimination of benzotriazole from pyrazolines **160** in the presence of base furnishes 1,3,5-trisubstituted pyrazoles **161** (Scheme 56) [126, 127].

Pyrazolines **160** can be further functionalized by alkylation at the 4-position with alkyl iodides in the presence of *n*-butyllithium to afford compounds **162**, which can eliminate benzotriazole in the presence of NaOEt or KO^tBu to give 1,3,4,5-tetrasubstituted pyrazoles **163** (Scheme 57) [128].

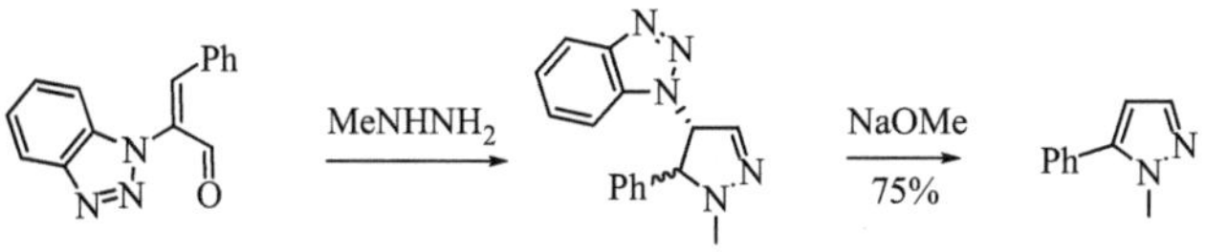

164 165 166

Scheme 58 Preparation of 1,5-disubstituted pyrazoles **166**

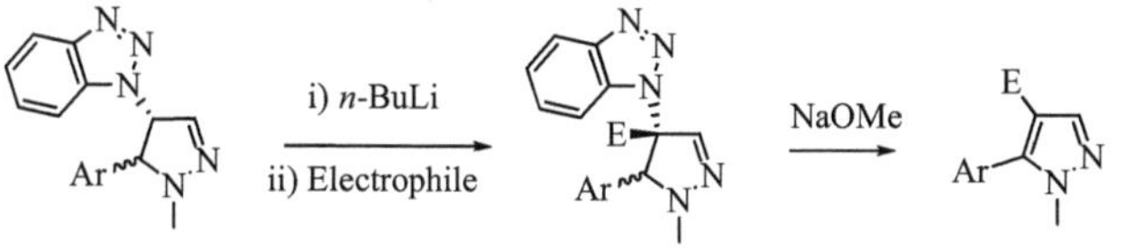

165 167 168

Ar = Ph, E = Me (63%)
Ar = 4-MeC$_6$H$_4$, E = Me (79%), Bu (52%)
Ar = C$_6$H$_5$CH$_2$CH$_2$, E = Me (60%), (CH$_3$)$_2$CH(CH$_2$)$_2$ (53%), n-hexyl (57%), Et (58%), benzyl (55%)

Scheme 59 Preparation of 1,4,5-trisubstituted pyrazoles **168**

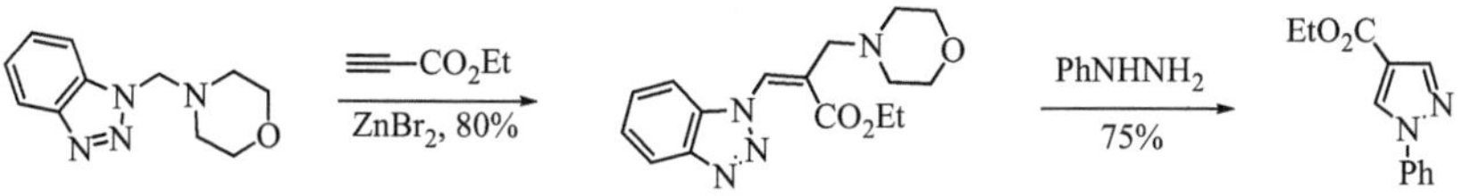

169 170 171

Scheme 60 Preparation of 1,4-disubstituted pyrazole **171**

α-Benzotriazol-1-yl-α,β-unsaturated aldehydes **164** [126] reacted with *N*-methylhydrazine to form stable intermediates **165** as a single 1,4,5-regioisomer. Pyrazolidines **165** can be converted further to 1,5-disubstituted pyrazoles **166** with sodium ethoxide (Scheme 58) [127].

4,5-Dihydro-1*H*-pyrazoles **165** were functionalized further by alkylation or acylation at 4-position with alkyl iodides, bromides, or acyl chlorides in the presence of *n*-butyllithium to afford intermediates **167** as mixtures of diastereomers, which after elimination of benzotriazole gave 1,4,5-trisubstituted pyrazoles **168** (Scheme 59) [127].

Addition of *N*-(benzotriazol-1-ylmethyl)morpholine **169** [129, 130] to ethyl propionate in the presence of zinc bromide catalyst provided adduct **170**, as mainly *E* isomer. Heating **170** with phenylhydrazine gave 1,4-disubstituted pyrazole **171**. The reaction presumably proceeded through a cycloaddition between **170** and hydrazine followed by the elimination of benzotriazole and morpholine to furnish pyrazole **171** (Scheme 60) [131].

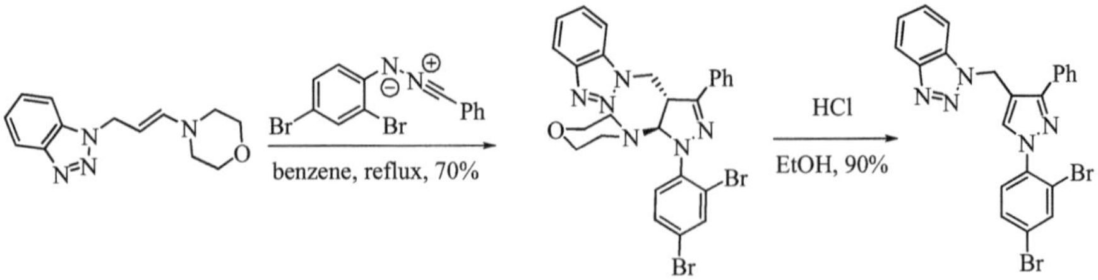

88 **172** **173**

Scheme 61 Preparation of 1,3,4-trisubstituted pyrazole **173**

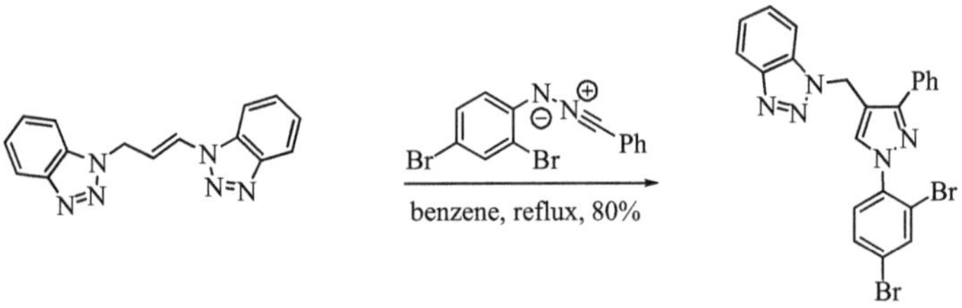

174 **173**

Scheme 62 Direct preparation of 1,3,4-trisubstituted pyrazole **173**

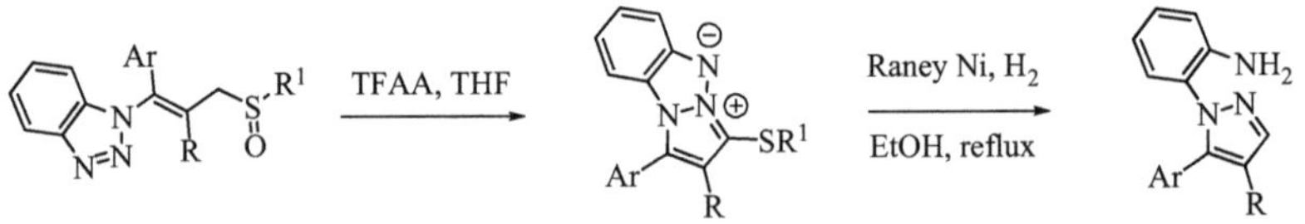

175 **176** **177**

Ar = Ph, R = Ph (89%), *t*-Bu (86%),
Ar = 4-FC_6H_4, R = Me (89%),
Ar = 4-$MeOC_6H_4$, R = 4-MeC_6H_4 (92%)

Scheme 63 Preparation of 1-(2-aminophenyl)pyrazoles **177**

1,3-Dipolar cycloaddition of 1-(3-morpholin-4-ylallyl)benzotriazole **88** [92] to nitrilimine [132] gave pyrazoline **172**, which eliminated morpholine to provide 1,3,4-trisubstituted pyrazole **173** (Scheme 61) [133].

1,3-Bisbenzotriazol-1-ylpropene **174** reacted with nitrilimine to give pyrazole **173** directly (Scheme 62) [133].

In a special case, two benzotriazolyl nitrogens were used to generate the pyrazole ring. Thus, sulfoxides **175** underwent cyclization to form triazapentalenes **176** in the presence of trifluoroacetic anhydride. The reaction is believed to proceed through the acylation of the sulfoxide oxygen atom and generation of a carbocation (Pummerer reaction), which attacks the *N*-2 atom of benzotriazole. Hydrogenation of **176** over Raney nickel generated 1-(2-aminophenyl)pyrazoles **177** (Scheme 63) [134].

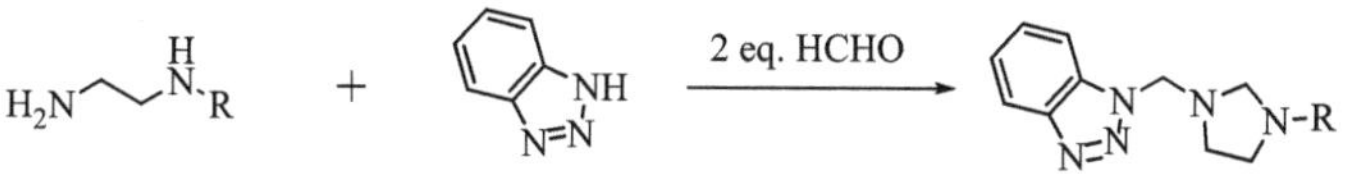

178 179

R = Ph (96%), Et (85%), $PhCH_2$ (92%)

Scheme 64 Preparation of 1-substituted 3-(benzotriazol-1-ylmethyl)imidazolidines **179**

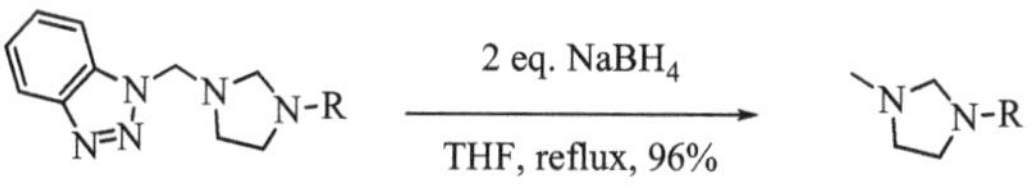

179 180 R = Ph

Scheme 65 Preparation of 1-phenyl-3-methylimidazolidine **180**

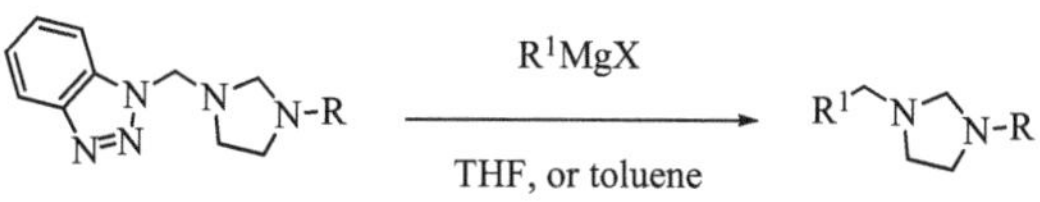

179 181

R = Ph, R^1 = *n*-Bu (80%), $PhCH_2CH_2$ (96%), $PhCH_2$ (96%), 4-$MeOC_6H_4$ (81%), PhCC (80%), CH_2CH (75%)
R = Et, R^1 = $PhCH_2$ (75%), 4-MeC_6H_4 (71%)
R = $PhCH_2$, R^1 = $PhCH_2$ (79%), CH_2CH (63%), PhCC (65%), *n*-C_5H_{11} (80%)

Scheme 66 Preparation of unsymmetrical 1,3-disubstituted imidazolidines **181**

4.7 *Imidazolidines*

Imidazolidines are important building blocks in biologically active compounds [135–138] and carriers of pharmacologically active carbonyl compounds [139, 140]. Mannich condensation of *N*-substituted 1,2-ethylenediamines **178** with 1 equiv. of benzotriazole and 2 equiv. of formaldehyde gave *N*-substituted 3-(benzotriazol-1-ylmethyl)imidazolidines **179** (Scheme 64) [141].

Nucleophilic substitution of benzotriazole in **179** afforded variety of 1,3-disubstituted imidazolidines. Thus, treatment of 1-(benzotriazolylmethyl)-3-phenylimidazolidine **179** (R=Ph) with 2 equiv. of sodium borohydride replaced the benzotriazolyl group with hydrogen to provide 1-phenyl-3-methylimidazolidine **180** (Scheme 65) [141].

Nucleophilic substitutions of **179** with Grignard reagents led to the replacement of the benzotriazole moiety with alkyl, vinyl, aryl, phenylethynyl, and benzyl group and furnished unsymmetrical 1,3-disubstituted imidazolidines **181** (Scheme 66) [141].

Additional reactions of **179** with sodium cyanide, thiophenol and sodium hydride, or triethyl phosphate and zinc bromide demonstrate easy substitution of

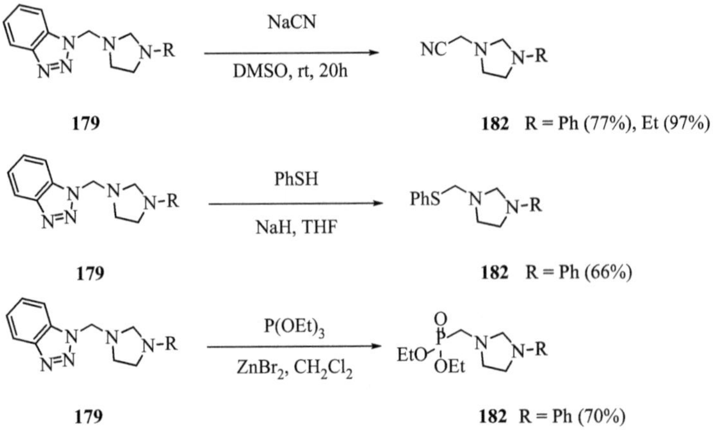

Scheme 67 Preparation of 1,3-disubstituted imidazolidines **182**

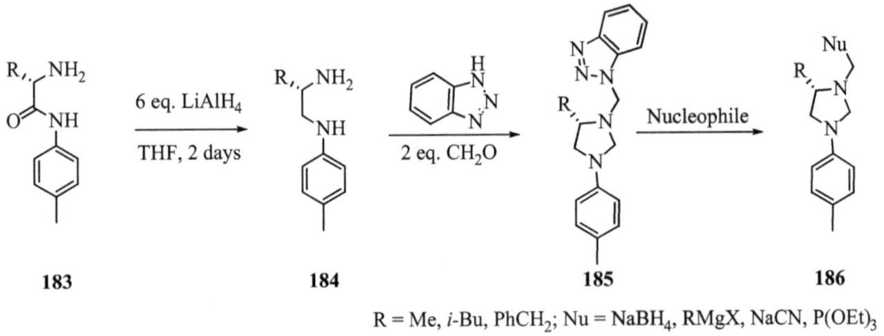

Scheme 68 Preparation of optically active imidazolidines **186**

the benzotriazole moiety to furnish 1,3-disubstituted imidazolidines **182** (Scheme 67) [141].

α-Amino amides **183** [142] can be reduced with $LiAlH_4$ to afford chiral diamines **184**, which reacted readily with benzotriazole and formaldehyde to generate benzotriazol-1-yl intermediates **185**. Nucleophilic substitution of **185** with Grignard reagents, triethylphosphite, or sodium cyanide gave optically active imidazolidines **186** (Scheme 68) [141].

Condensation of ethylenediamine with 3 equiv. of formaldehyde and 2 equiv. of benzotriazole provided 1,3-bis(benzotriazol-1-ylmethyl)imidazolidines **187**. The reaction of crude **187** with KCN in acetonitrile gave dinitrile **188** and with Grignard reagents in THF afforded 1,3-dialkylimidazolidines **189** (Scheme 69) [53].

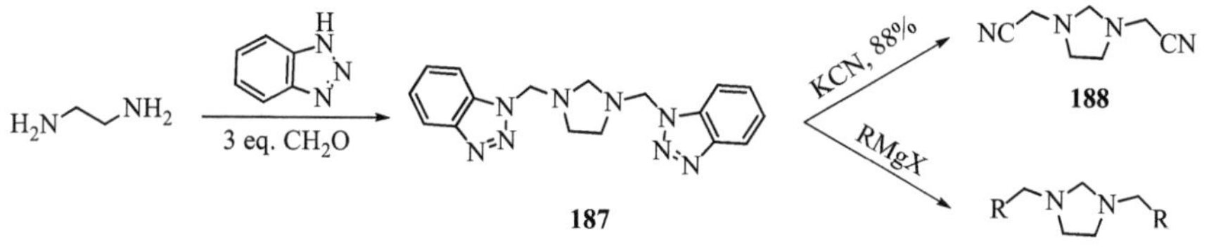

189

R = Et (72%), Bu (68%), *n*-octyl (75%), vinyl (74%)

Scheme 69 Preparation of dinitrile **188** and 1,3-dialkylimidazolidines **189**

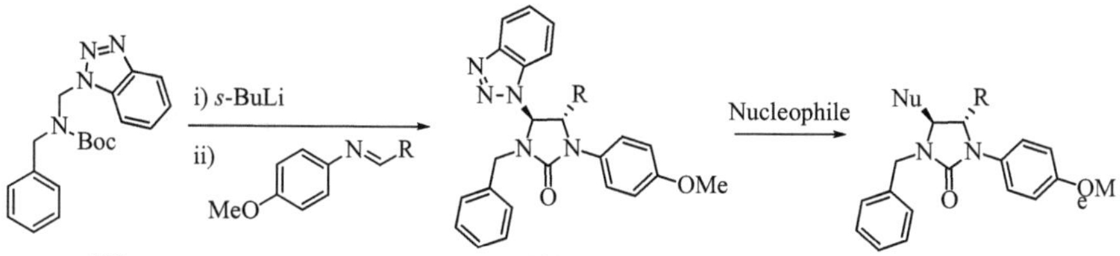

190 **191** **192**

R = Ph, 4-MeC_6H_4, 4-$MeOC_6H_4$, 4-FC_6H_4, 2-furyl; Nu = RMgX, allyltrimethylsilane, vinyloxysilanes

Scheme 70 Preparation of imidazolidinone derivatives **192**

4.8 Imidazolidinones

Hydantoin (2,4-imidazolidinone) derivatives display diverse pharmacological properties, such as anticonvulsive, antimicrobial, and skeletal muscle relaxant among many others [143–145]. *N*-Boc-benzylamine [146] reacted with formaldehyde and benzotriazole to furnish *N*-Boc-*N*-(benzotriazolylmethyl)benzylamine **190**, which upon treatment with *sec*-butyllithium followed by the addition of imines gave benzotriazolylimidazolidinone derivatives **191**. Subsequent treatment of **191** with nucleophiles provided products **192** stereoselectively keeping the *trans*-relative stereochemistry of the substituents on C-4 and C-5 of the starting material (Scheme 70) [147].

When the Vilsmeier-type salt **145** [121] is treated with triethylamine, it generated an aminocarbene, which underwent [1+2+2] cycloaddition with phenylisocyanate to give imidazolidine-2,4-diones **193**. Quenching the reaction with tetraethylammonium sulfide resulted in reduction of the functionality on C-5 to give 1,3-diphenylhydantoin **194**. On treatment of **193** with sodium borohydride, only the benzotriazole moiety was reduced to afford 5-(dimethylamino)hydantoin **195**. Similarly, the replacement of benzotriazole can be done with Grignard reagents to furnish hydantoin **196**. Analogous quenching of the reaction with methanol or morpholine replaced the benzotriazole moiety without affecting the dimethylamino group. However, when propylamine is used as a nucleophile, both the benzotriazolyl and the dimethylamino groups were eliminated to yield 5-(*N*-

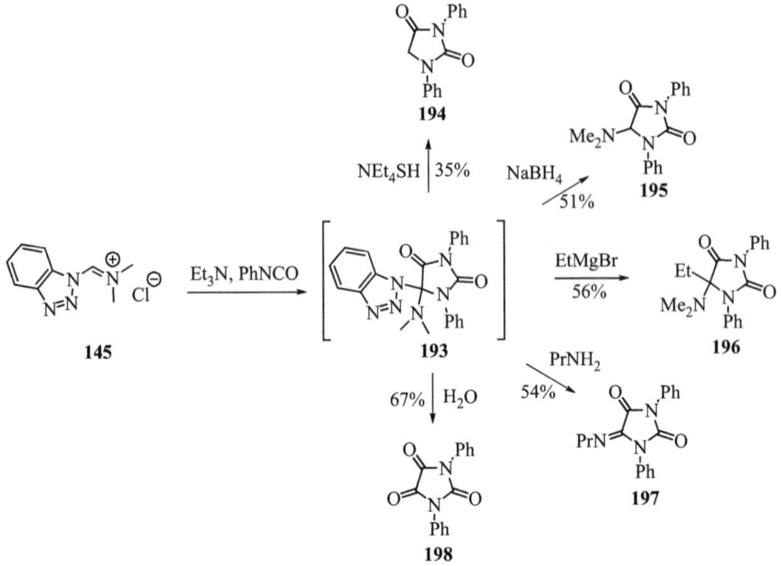

Scheme 71 Preparation of imidazolidinone derivatives **193–198**

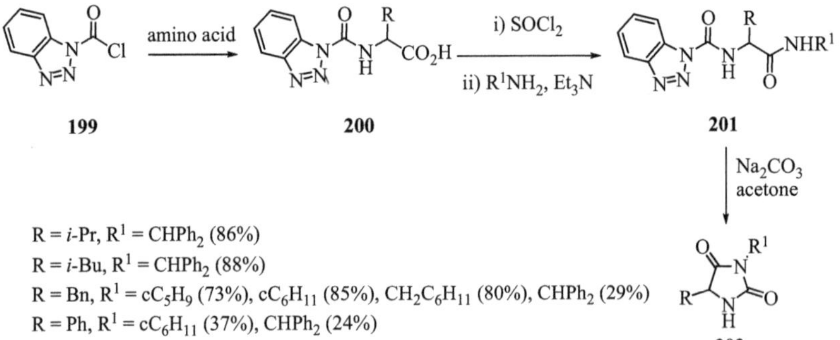

R = *i*-Pr, R^1 = $CHPh_2$ (86%)
R = *i*-Bu, R^1 = $CHPh_2$ (88%)
R = Bn, R^1 = cC_5H_9 (73%), cC_6H_{11} (85%), $CH_2C_6H_{11}$ (80%), $CHPh_2$ (29%)
R = Ph, R^1 = cC_6H_{11} (37%), $CHPh_2$ (24%)

Scheme 72 Preparation of disubstituted imidazolidinones **202**

propylimino) derivative **197**. A simple workup of the reaction mixture containing **193** with water resulted in imidazolidine-2,4,5-trione **198** (Scheme 71) [121].

A different approach toward the synthesis of hydantoins started with the reaction of 1-(chloroformyl)benzotriazole **199** [148–150] with amino acids to give derivatives **200**, which further reacted with amines in the presence of thionyl chloride to furnish *N*-(1-benzotriazolecarbonyl)-amino acid amides **201**. Intramolecular cyclization of **201** in the presence of sodium carbonate gave disubstituted hydantoins **202** (Scheme 72) [151, 152].

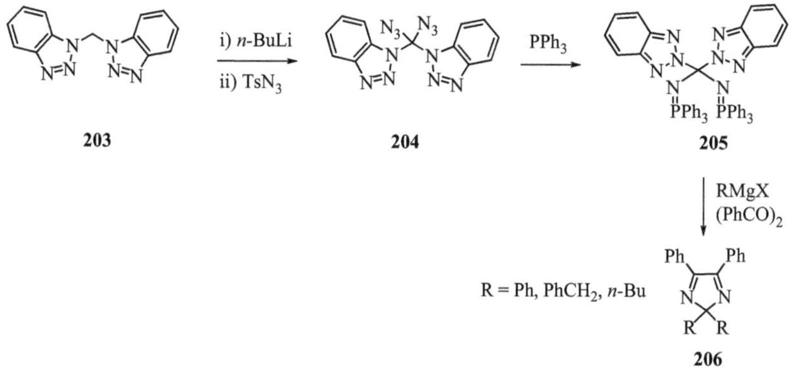

Scheme 73 Preparation of imidazoles **206**

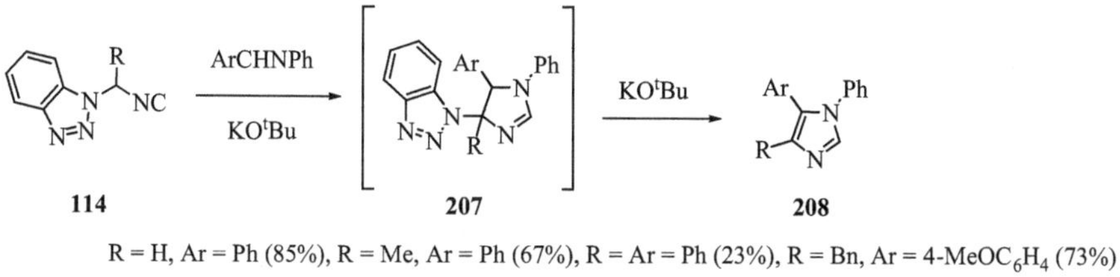

Scheme 74 Preparation of 1,5-diarylimidazoles **208**

4.9 Imidazole

Imidazoles have important features in a variety of medicinal agents, such as antifungal, antibacterial, anti-inflammatory, and analgesic among many others [153–157]. Lithiation of bisbenzotriazolylmethane **203** followed by the addition of tosyl azide gave diazide **204**, which further reacted with triphenylphosphine to yield bis(triphenylphosphoranylidene) derivative **205**. Due to steric hindrance, the benzotriazole fragment rearranged to its 2-yl form as revealed by NMR studies. Consecutive treatments of crude **205** with Grignard reagents and benzil gave 2*H*-imidazoles **206** in 15–21% yield (Scheme 73) [158].

The anion derived from isocyanide **114** on treatment with KO^{t}Bu was added readily to Schiff bases to form imidazolines **207**, which under basic conditions eliminated benzotriazole to produce 1,5-diarylimidazoles **208** (Scheme 74) [106].

Displacement of benzotriazole in **1** by primary amines provided compounds **209**, which were treated in situ with benzil to give imidazoles **210** (Scheme 75) [102].

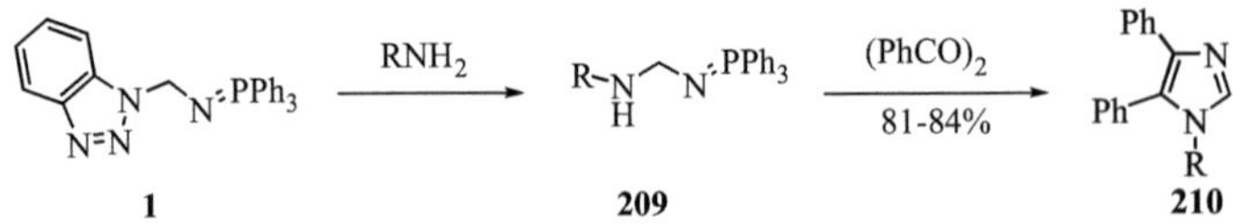

1 209 210

R = n-$C_{12}H_{25}$, c-C_7H_{13}, $PhCH_2$, 4-(Me_2N)C_6H_4

Scheme 75 Preparation of imidazoles **210**

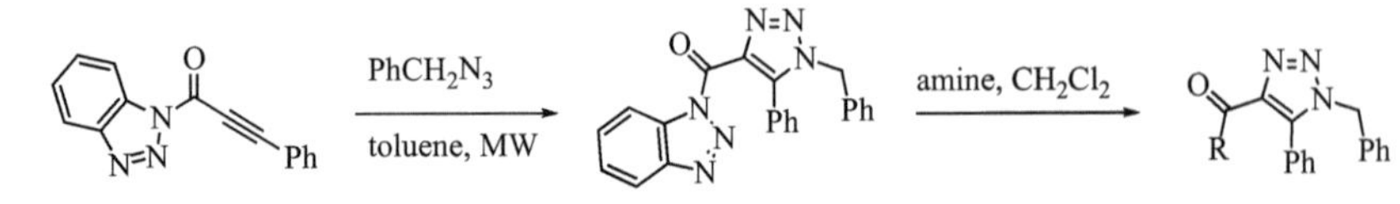

211 212 213

R = morpholinyl (71%), NHp-C_6H_4Cl (62%), NHCH(CH_3)Ph (54%), $NHCH_2Ph$ (91%)

Scheme 76 Preparation of *C*-carbamoyl 1,2,3-triazoles **213**

4.10 Triazoles

Triazoles possess therapeutic value, are synthetic intermediates in the preparation of medicinal compounds, and have found applications in the chemical industry [159–161]. The thermal [3+2] cycloaddition of benzyl azide with *N*-propioloylbenzotriazole **211** gave the benzotriazolylcarbonyl-substituted 1,2,3-triazole **212**, which can be further reacted with amines to provide the corresponding *C*-carbamoyl 1,2,3-triazoles **213** (Scheme 76) [162].

Benzotriazole-1-carboximidamides **214** [163, 164] can be easily converted into acyl derivatives **215**, which served as precursors in the synthesis of triazoles. Thus, cyclization of acyl derivatives **215** with hydrazines led to 3-amino-1,2,4-triazoles **216** as single regioisomers, when unsubstituted or methyl-substituted hydrazine is used. In contrast, phenyl or benzyl hydrazine reacted with acyl derivatives **215** with the competitive formation of **216** and 3-benzotriazole-substituted triazole **217**. The lower the electron-donating character of the substituent of the hydrazine, the more of the side product **217** is formed (Scheme 77) [165].

When the previously described reactions were carried out on a solid support, no chromatographic purification of intermediates or final products is needed. Thus, amines **218** bound to StratoSpheres PL-FMP resin [166] were treated with bisbenzotriazolylmethanimine **219** to give carboxyimidamides **220**, which were subsequently treated with acyl chlorides to provide acylated products **221**. The resins containing **221** were reacted with hydrazines in the presence of DBU to furnish 3-amino-1,2,4-triazole derivatives **222**. The cleavage of the resin was carried out with TFA to yield final products **223** (Scheme 78) [167].

In the reaction of imine **219** with methylhydrazine, the formation of triazole **225** was observed, probably because intermediate **224** spontaneously eliminated one

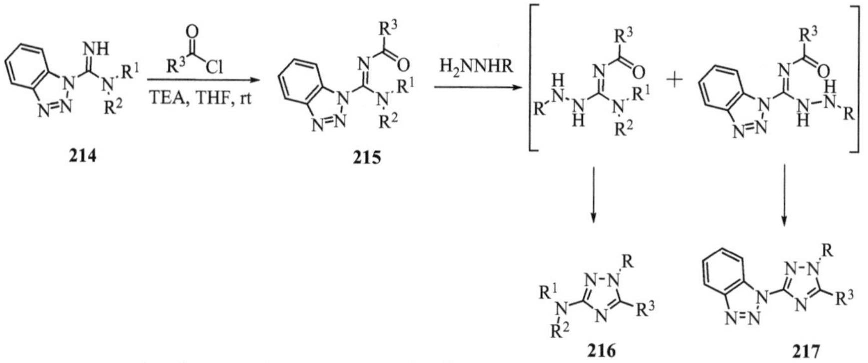

R = H, $R^1 = R^2$ = allyl, R^3 = Ph (72, 0%), $R^1 = R^2$ = morpholinyl, R^3 = Ph (84, 0%)
R = Me, $R^1 = R^2$ = Et, R^3 = *t*-Bu (75, 0%), $R^1 = R^2$ = morpholinyl, R^3 = 1-naphtalenyl (90, 0%), R^3 = Et (82, 0%), $R^1 = R^2$ = allyl, R^3 = Ph (84, 0%), $R^1 = R^2$ = *i*-Pr, R^3 = Ph (95, 0%)
R = CH_2Ph, $R^1 = R^2$ = Et, R^3 = *t*-Bu (36, 30%), $R^1 = R^2$ = morpholinyl, R^3 = 1-naphtalenyl (45, 17%), $R^1 = R^2$ = morpholinyl, R^3 = 1-thiophenyl (42, 29%)
R = Ph, $R^1 = R^2$ = *i*-Pr, R^3 = Ph (25, 13%)

Scheme 77 Preparation of 3-amino-1,2,4-triazoles **216**

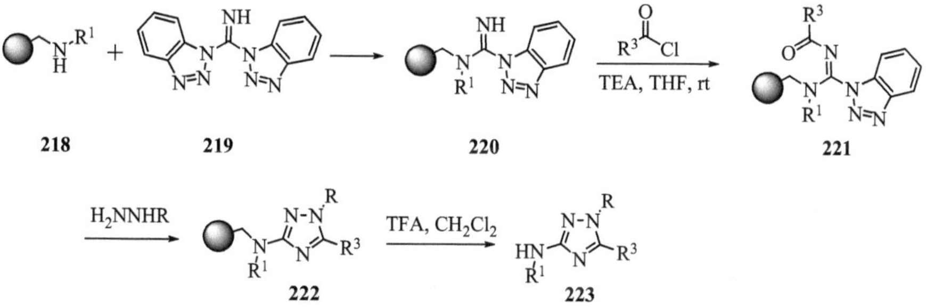

R^1 = 4-MeBn, R^3 = 4-MePh, R = Me (65%), Bn (45%), 4-MeOPh (45%)
R^1 = *s*-butyl, R^3 = *c*-hexyl, R = Me (65%), Bn (45%), 4-MeOPh (45%)

Scheme 78 Preparation of 3-amino-1,2,4-triazole derivatives **223**

benzotriazole fragment to form the aromatic benzotriazole-substituted triazole **225** (Scheme 79) [168].

When aryl hydrazines were used under similar conditions, the initial formation of intermediate **224** took place, which then underwent cyclization to form 5-hydrazino-1,2,4-tiazoles **226**. Spontaneous oxidation of **226** led to 5-azo-1,2,4-triazole derivatives **227** (Scheme 80) [168].

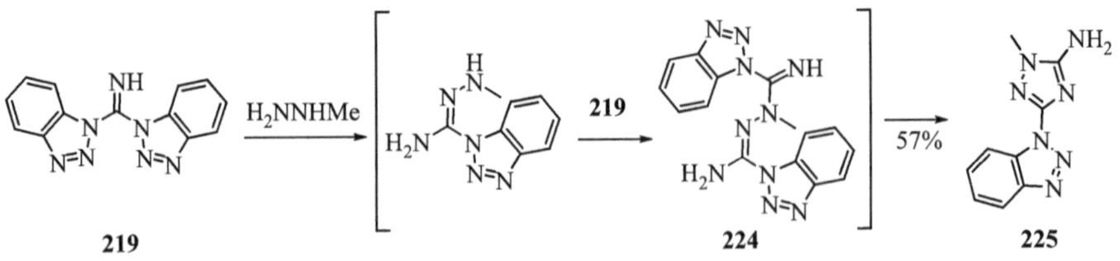

Scheme 79 Preparation of benzotriazole-substituted triazole 225

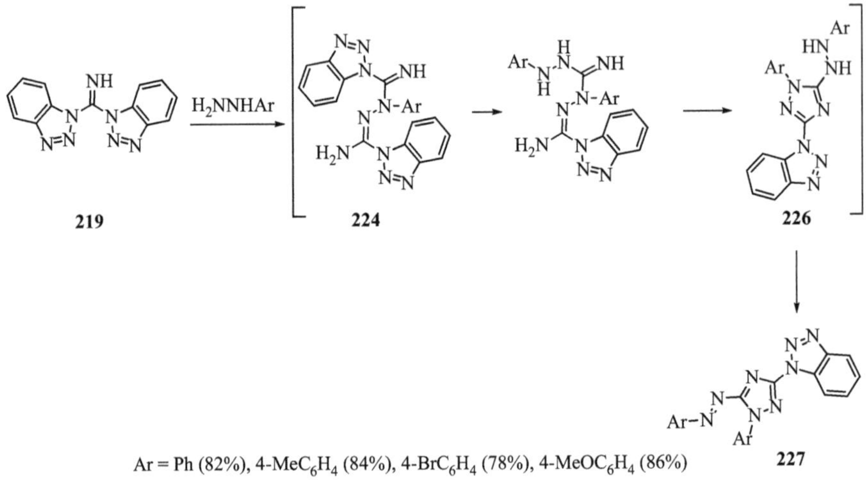

Ar = Ph (82%), 4-MeC_6H_4 (84%), 4-BrC_6H_4 (78%), 4-MeOC_6H_4 (86%)

Scheme 80 Preparation of 5-azo-1,2,4-triazole derivatives **227**

4.11 *Tetrazole*

Tetrazoles are of general interest since they are often used in medicinal chemistry as (bio)isosteres of the carboxylic group. Moreover, chiral α-aminomethyl tetrazoles are useful as catalyst for enantioselective aldol-type condensations [169–172]. In a Ugi-type reaction, **111** reacted with an enamine and trimethylsilyl azide in methanol to form tetrazole derivative **228**, which further can be treated with diluted acid to remove the benzotrazol-1-ylmethyl group and furnish α-aminomethyl tetrazoles **229** (Scheme 81) [173].

1-Imidoylbenzotriazoles **230** only reacted with sodium azide in the presence of TFA to give 1,5-disubstituted tetrazoles **231** after the elimination of benzotriazole (Scheme 82) [174].

1-Chlorobenzotriazole **232**, a mild source of the electrophilic chloronium cation, reacted with 1,3,5-triarylformazans **233** to give 2,3,5-triaryl-2*H*-tetrazolium chlorides **235** through *N*-chloroformazan intermediate **234**. When R is a strongly electron-withdrawing substituent, it facilitates the formation of compound **235** (Scheme 83) [175].

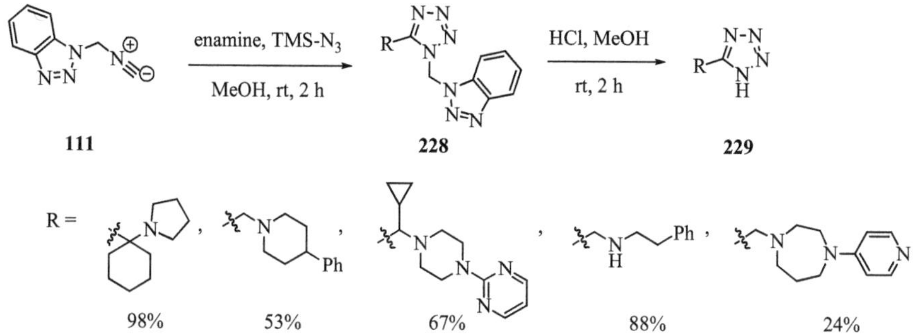

Scheme 81 Preparation of α-aminomethyl tetrazoles **229**

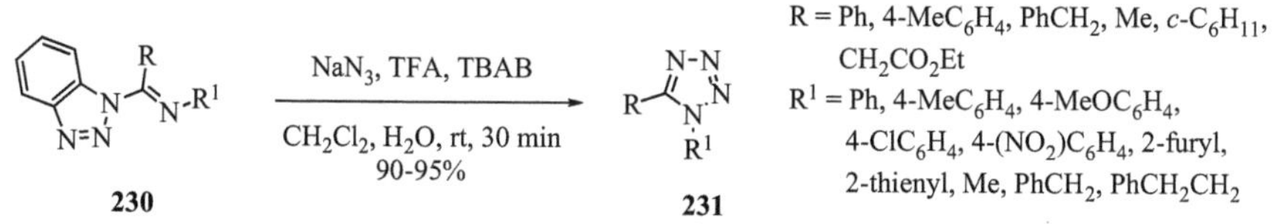

Scheme 82 Preparation of 1,5-disubstituted tetrazoles **231**

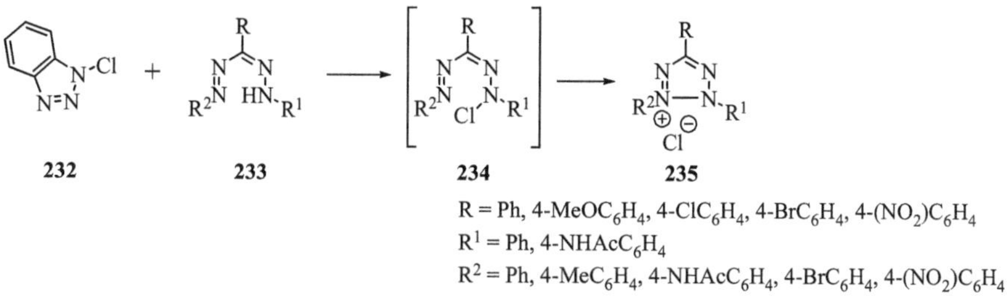

Scheme 83 Preparation of 2,3,5-triaryl-2H-tetrazolium chlorides **235**

5 The Construction of Six-Membered Ring

5.1 *Piperidines*

Functionalized piperidines are frequently encountered structural motifs in bioactive compounds and natural products. The piperidine ring is often present in drug-like molecules as substructures [176–179]. Condensation of benzotriazole, glutaraldehyde, and amines in water gave 2,6-bisbenzotriazolylpiperidines **236** as a mixture of six stereoisomers (*cis-/trans*-benzotriazol-1-yl/2-yl). However, after reduction with sodium borohydride, *N*-substituted piperidines **237** were produced as single compounds (Scheme 84) [180].

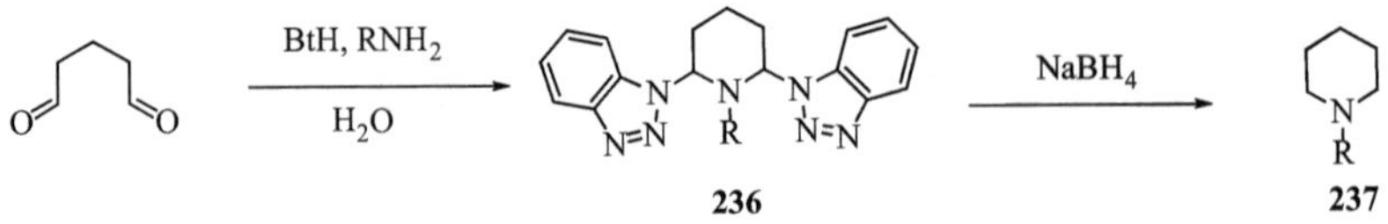

R = Ph (80%), $PhCH_2$ (78%), 3-MeC_6H_4 (68%), *n*-Bu (59%), *i*-Pr (55%), 2-pyridyl (64%)

Scheme 84 Preparation of *N*-substituted piperidines **237**

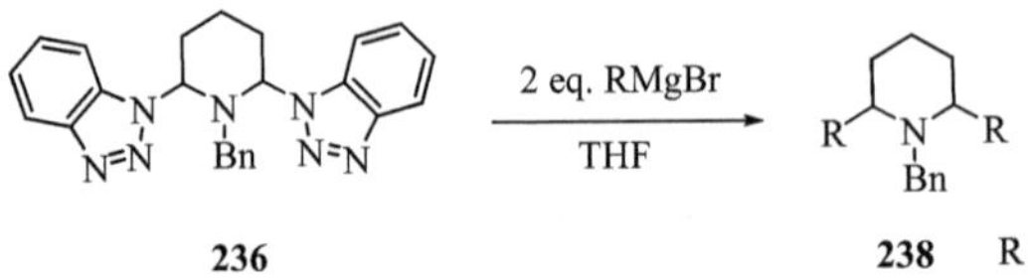

238 R = Me (48%), Bu (45%)

Scheme 85 Preparation of 1-benzyl-2,6-dialkylpiperidines **238**

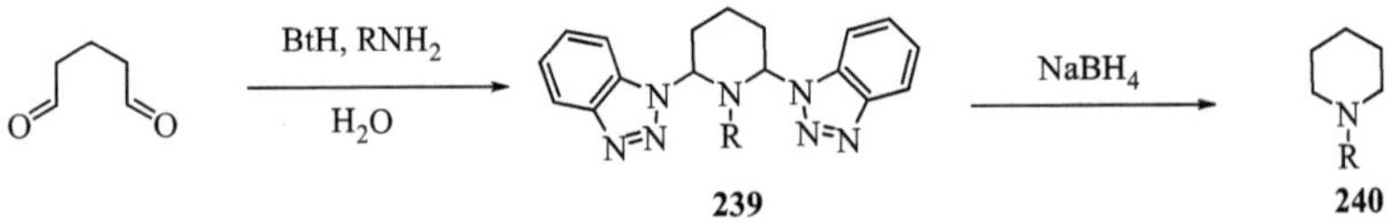

R = PhNH (87%), Me_2N (79%), PhCONH (82%), MeCONH (84%), EtO_2CNH (88%), Me_3CO_2CNH (85%), $PhCH_2O_2CNH$ (90%)

Scheme 86 Preparation of 1-amino- or 1-amidopiperidines **240**

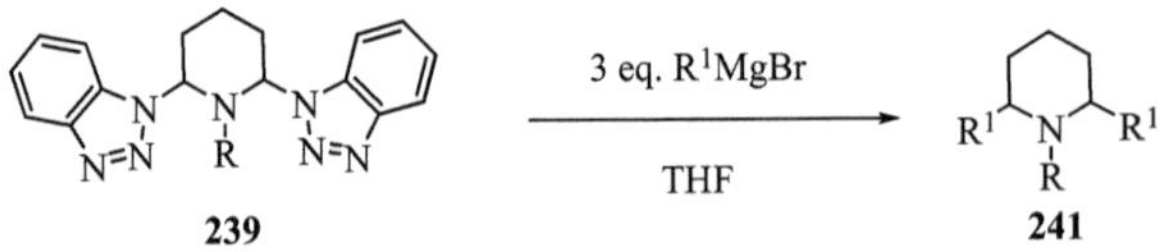

R = PhCONH, R^1 = Me (70%), Et (72%), Ph (49%), Me_3CO_2CNH, R^1 = $PhCH_2$ (52%), Ph (75%)

Scheme 87 Preparation of 1-amido-2,6-dialkylpiperidines **241**

Treatment of **236** with Grignard reagents allowed the formation of 1-benzyl-2,6-dialkylpiperidines **238** having *cis*-relative stereochemistry of the substituents (Scheme 85) [180].

Similarly to amines, hydrazines and hydrazides also reacted with benzotriazole and glutaraldehyde to give bisbenzotriazole derivative **239** as a mixture of isomers, which were substituted by sodium borohydride reduction of the benzotriazolyl groups to form 1-amino- or 1-amidopiperidines **240** (Scheme 86) [180].

Treatment of **239** with Grignard reagents allowed the formation of 1-amido-2,6-dialkylpiperidines **241** as single isomers (Scheme 87) [180].

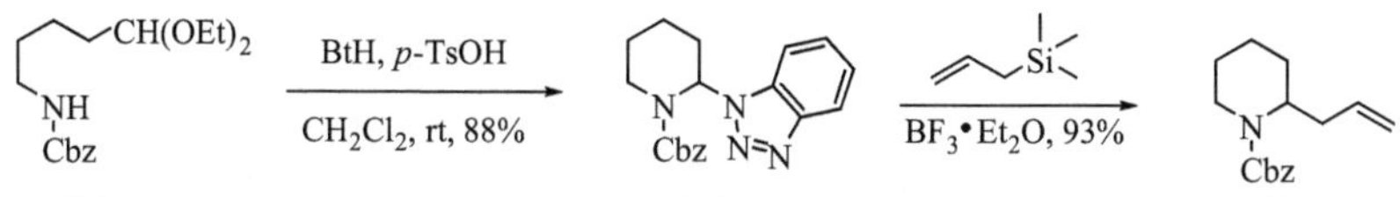

242 243 244

Scheme 88 Preparation of *N*-Cbz-protected 2-allylpiperidine **244**

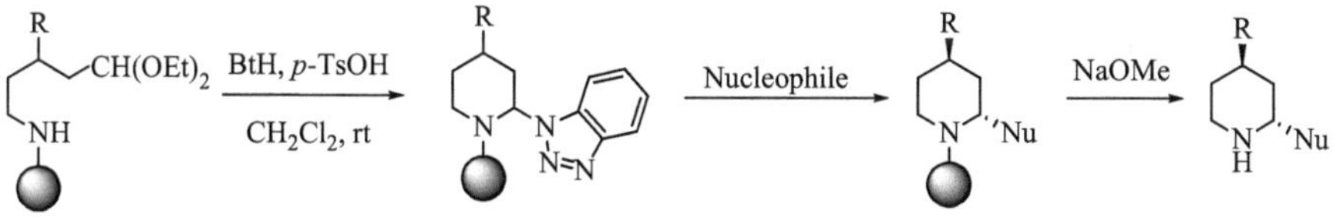

245 246 247 248

R = H, Nu = allyl (78%)
R = Ph, Nu = allyl (80%), methallyl (76%), 2-furanyl (54%), Et (14%)
R = 4-pyridyl, Nu = allyl (71%), R = 4-BrC_6H_4, Nu = allyl (85%), R = *i*-Pr, Nu = allyl (86%)

Scheme 89 Preparation of 2,4-disubstituted piperidines **248**

The reaction of *N*-Cbz-protected acetal **242** with benzotriazole in the presence of *p*-TsOH gave piperidine derivatives **243**. Treatment with allyltrimethylsilane in the presence of boron trifluoride etherate allowed substitution of the benzotriazole moiety to furnish *N*-Cbz-protected 2-allylpiperidine **244** (Scheme 88) [181].

The reaction shown above can be facilitated by using solid support in place of the protecting group. Thus, aminoacetals were attached to sulfonylethoxycarbonyl-modified polystyrene. The obtained resin-bound acetals **245** reacted similarly with benzotriazole in the presence of *p*-TsOH to give derivatives **246**, which were subsequently subjected to the reaction with nucleophile to furnish the resin-bound products **247**. Finally, the products were cleaved from the solid support using sodium methoxide to furnish 2,4-disubstituted piperidines **248** exclusively as the *trans*-stereoisomer (Scheme 89) [181].

Using nonracemic Betti base **249** as the amine component in the condensation with benzotriazole and glutaraldehyde allowed the formation of chiral 2-substituted piperidines, through the initial formation of the 2-benzotriazolylpiperidine **250** with *R,R* configuration of its stereocenters. Treatment with sodium cyanoborohydride cleaved selectively the benzotriazole moiety to obtain derivative **251**, which subsequently reacted with Grignard reagents to furnish pure diastereomers of piperidines **252**. Final deprotection by hydrogenation released chiral piperidines **253** (Scheme 90) [182].

In another approach to chiral 2-substituted piperidines, condensation of (*S*)-2-phenylglycinol with glutaraldehyde and benzotriazole was carried out in methanol to provide intermediate **254**, which was further reacted with nucleophiles to furnish protected products **255**. Final deprotection by hydrogenation gave chiral 2-substituted piperidines **256** (Scheme 91) [183]. A similar synthesis starting from (*R*)-2-phenylglycinol led to the opposite enantiomer of **256** [184].

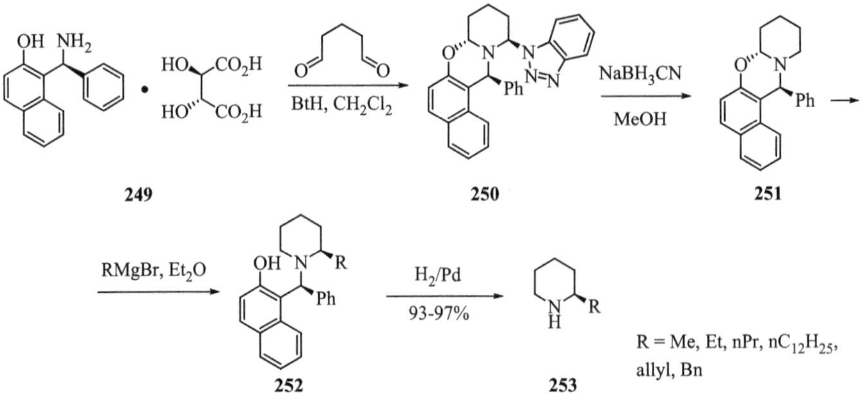

R = Me, Et, nPr, $nC_{12}H_{25}$, allyl, Bn

Scheme 90 Preparation of chiral 2-alkylpiperidines **253**

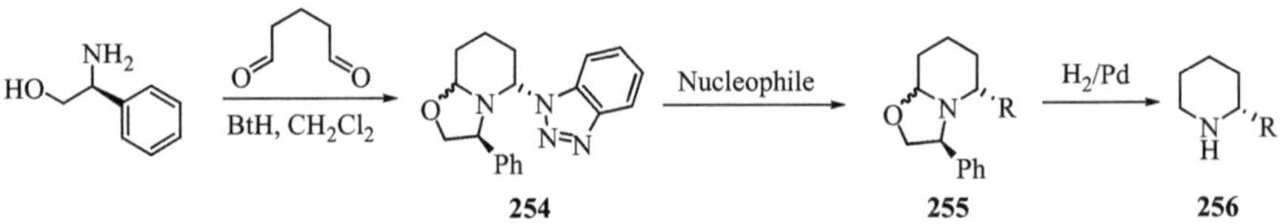

R = nPr (95%), Ph (92%), $PO(OEt)_2$ (68%)

Scheme 91 Preparation of chiral 2-substituted piperidines **256**

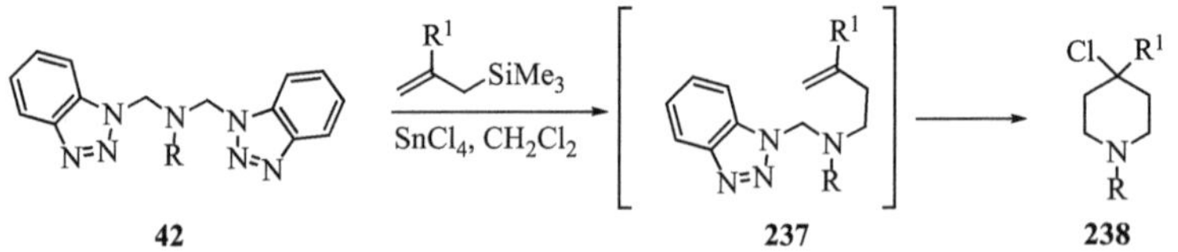

R = $PhCH_2$, R^1 = H (62%), Me (68%)
R = PhCHMe, R^1 = H (58%), Me (68%)
R = $PhCH_2CH_2$, R^1 = H (64%)

Scheme 92 Preparation of 4-chloropiperidines **238**

Piperidine rings also can be assembled through carbon–carbon bond formation reactions. Thus, *N,N*-bis(benzotriazolylmethyl)amines **42** underwent cyclization with allylsilanes in the presence of a catalytic amount of $SnCl_4$ to give 4-chloropiperidines **238**. This [3+3] cyclization is believed to proceed through intermediate **237** (Scheme 92) [185].

The reduction of benzotriazolyl derivatives of 4-penten-1-ylamine **239** with samarium diiodide generated radical **240** which were trapped by the alkenyl group to produce radical **241**. Subsequent reaction with an electrophile in the

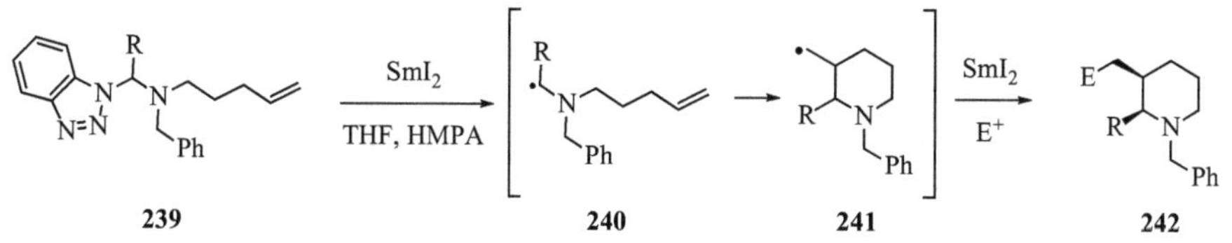

R = *i*-Pr, E = H (52%), R = Pr, E = H (56%), EtC(OH)Et (51%), MeC(OH)Et (43%), MeC(OH)Pr (41%), PhCHOH (17%), MeCHOH (21%), *i*-PrNHCO (25%)

Scheme 93 Preparation of 1,2,3-trisubstituted piperidines **242**

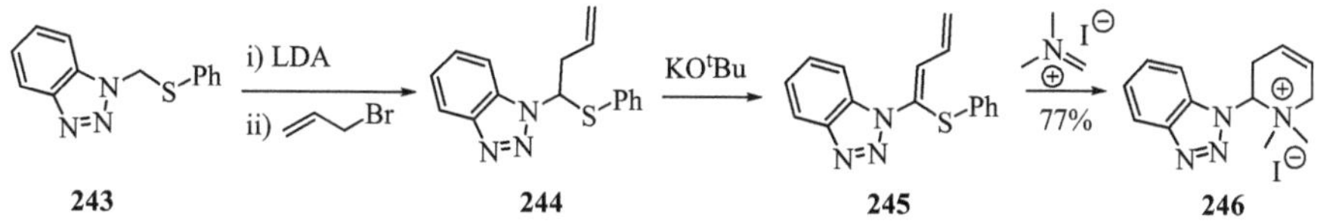

Scheme 94 Preparation of tetrahydropyridinium salt **246**

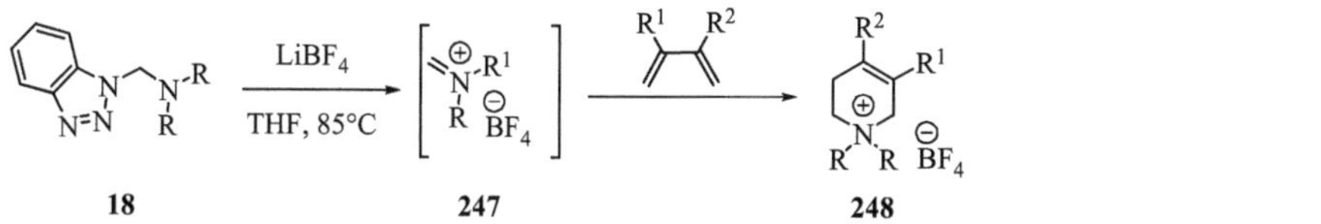

R = morpholinyl, R^1 = Me, R^2 = H (79%), $R^1 = R^2$ = Me (90%)
R = Me, R^1 = Me, R^2 = H (70%), $R^1 = R^2$ = Me (87%)

Scheme 95 Preparation of 1,2,5,6-tetrahydropyridinium salts **248**

presence of excess SmI_2 furnished 1,2,3-trisubstituted piperidines **242**. The *cis*-stereoisomer is predominant (Scheme 93) [186].

5.2 *Tetrahydropyridines*

The alkylation of lithiated benzotriazol-1-ylmethyl phenyl sulfide **243** with allyl bromide provided compound **244**, which was treated with potassium *tert*-butoxide to give 1-butadien-1-ylbenzotriazole **245**. Hetero-Diels–Alder reaction of butadiene **245** with Eschenmoser's salt furnished tetrahydropyridinium salt **246** (Scheme 94) [187].

N-Substituted 1-(aminomethyl)benzotriazoles [41] **18** can be converted to Eschenmoser's salts **247** using lithium tetrafluoroborate, which can further be reacted with dienes in hetero-Diels–Alder reactions to give 1,2,5,6-tetrahydropyridinium salts **248** (Scheme 95) [188].

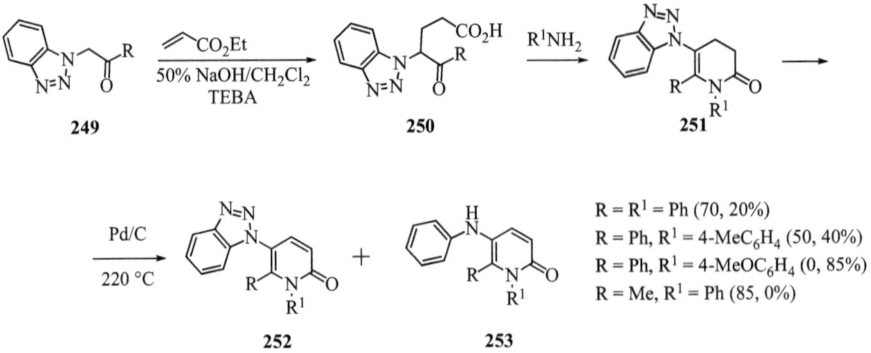

Scheme 96 Preparation of 2-pyridones **252** and **253**

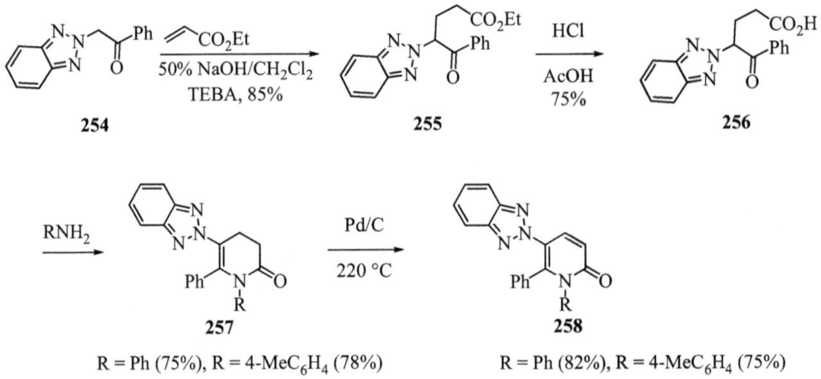

Scheme 97 Preparation of benzotriazole-substituted 2-pyridone derivatives **258**

5.3 Pyridones

Benzotriazol-1ylmethyl ketones **249** underwent Michael addition with ethyl acrylate in the presence of phase transfer catalyst to give the corresponding esters, which were hydrolyzed in situ to δ-ketoacids **250**. The conversion of ketones **250** to 3,4-dihydro-2-pyridones **251** was accomplished by the reaction with aniline or benzylamine in toluene. Dehydrogenation of **251** was carried out at 220°C in the presence of 10% Pd/C to result in partial nitrogen loss from the benzotriazole ring, giving rise to the mixture of 2-pyridones **252** and **253** (Scheme 96) [189].

The same reaction sequence can be carried out starting from the benzotriazol-2-yl analogue of ketone **249**. Condensation with ethyl acrylate gave ester **255**, which can be hydrolyzed under acidic conditions to yield δ-ketoacid **256**. Reaction of acid **256** with aniline or *p*-toluidine provided 3,4-dihydro-2-pyridones **257**, which can be dehydrogenated to give exclusively the benzotriazole-substituted 2-pyridone derivatives **258** (Scheme 97) [189].

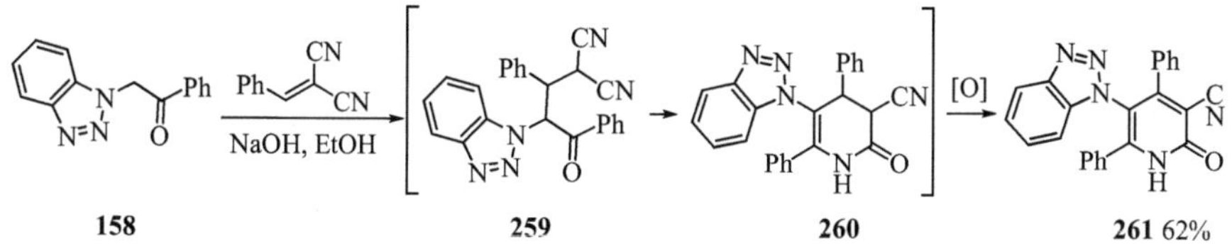

Scheme 98 Preparation of tetrasubstituted 2-pyridinone **261**

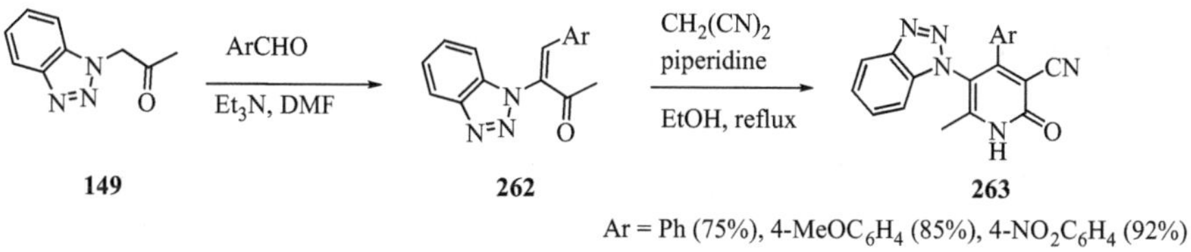

Scheme 99 Preparation of 2-pyridinones **263**

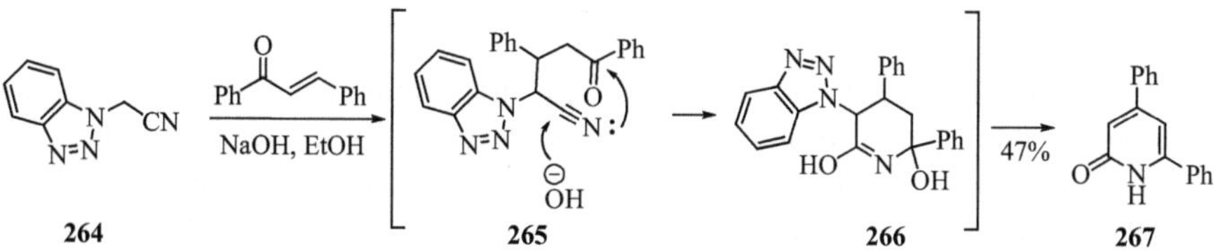

Scheme 100 Preparation of 4,6-diphenyl-2-pyridone **267**

Benzotriazolylacetophenone **158** [123] reacted with benzylidenemalononitrile to give intermediate **259**, which subsequently underwent cyclization to give 3,4-dihydro-2-pyridinone **260**, in which benzotriazole was located in a position where elimination is not possible. However, under the reaction conditions, **260** was rapidly oxidized to furnish tetrasubstituted 2-pyridinone **261** (Scheme 98) [190, 191].

1-(1-Benzotriazolyl)acetone **149** readily reacted with aromatic aldehydes to form 3-(1-benzotriazolyl)chalcones **262**. Malononitrile in the presence of piperidine converted the chalcones **262** into the corresponding 2-pyridinones **263** through similar intermediates depicted above (Scheme 99) [125].

The Michael addition of benzotriazol-1-ylacetonitrile **264** [19] to chalcone gave compound **265**, which in the presence of a catalytic amount of sodium hydroxide cyclized to tetrahydropyridine **266**. Spontaneous elimination of benzotriazole and water furnished 4,6-diphenyl-2-pyridone **267**. The method did not seem general as other α,β-unsaturated ketones failed to furnish the corresponding 2-pyrridinones (Scheme 100) [192].

A more general synthesis of 4,6-diaryl-2-pyridinones involved the Michael addition of benzotriazol-1-ylacetamide **268** to chalcones. The immediately formed tetrahydropyridines **269** eliminated benzotriazole and water to produce 4,6-diaryl-2-pyridinones **270** (Scheme 101) [192].

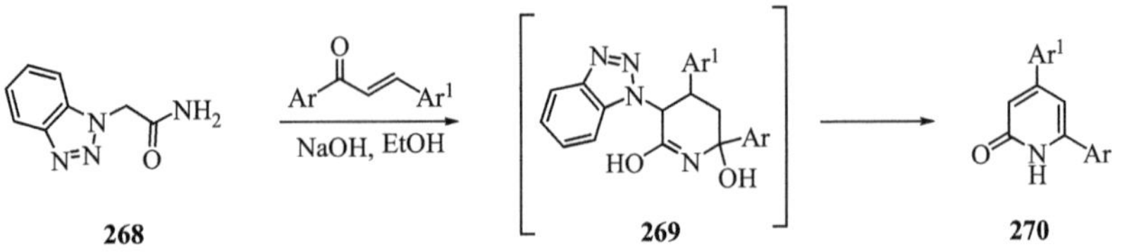

Ar = Ph, Ar[1] = Ph (92%), 4-$NO_2C_6H_4$ (57%), 4-ClC_6H_4 (53%)
Ar = 4-MeC_6H_4, Ar[1] = Ph (85%), Ar = 4-$MeOC_6H_4$, Ar[1] = 4-MeC_6H_4 (54%)
Ar = Ar[1] = 4-ClC_6H_4 (75%), Ar = 1-naphthyl, Ar[1] = 3-$NO_2C_6H_4$ (31%)

Scheme 101 Preparation of 4,6-diaryl-2-pyridinones **270**

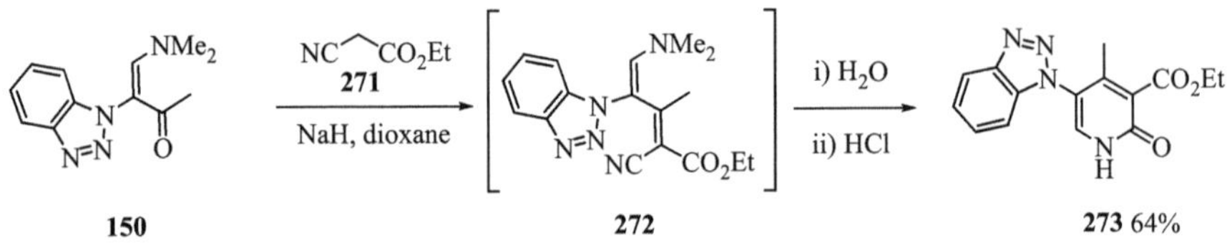

Scheme 102 Preparation of 5-(benzotriazol-1-yl)-2-pyridinone **273**

Enaminone **150** [125] reacted readily with ethyl cyanoacetate **271**; however, the outcome of the reaction strongly depended on the conditions. Thus, in the presence of sodium hydride, the carbonyl group reacted first to give intermediate **272**, which underwent cyclization after the hydrolysis of the cyano group to provide 2-pyridinone **273** by elimination of the dimethylamino group (Scheme 102) [190].

5.4 *Pyridines*

Substituted pyridines are important because of their biological activity and optical properties. Their quaternary salts have diverse synthetic applications [193, 194]. If the reaction shown on Scheme 102 was carried out in the presence of acetic acid and ammonium acetate, the process started from the Michael addition of ethyl cyanoacetate **271** to the carbon–carbon double bond of **150**, followed by the condensation of the carbonyl group with ammonia, generated imine **274**. Finally, the addition of the imine to the cyano group and the elimination of dimethylamine resulted in the corresponding 2-iminopyridine, which tautomerized to the more stable 2-aminopyridine **275** (Scheme 103) [190].

The substitution of the chlorine in 1-(α-chloroalkyl)benzotriazoles **276** [195] with sodium azide proceeded smoothly to yield stable azides **277**, which were subsequently treated with triphenylphosphine in a Staudinger phosphenimide-forming reaction to provide 1-[α-(phosphoranylideneamino)alkyl]-benzotriazoles **278**. Compound **278** upon treatment with sodium hydride eliminated benzotriazole to give (*N*-vinylimino)phosphoranes **279** as a mixture of *E* and *Z* isomers. The

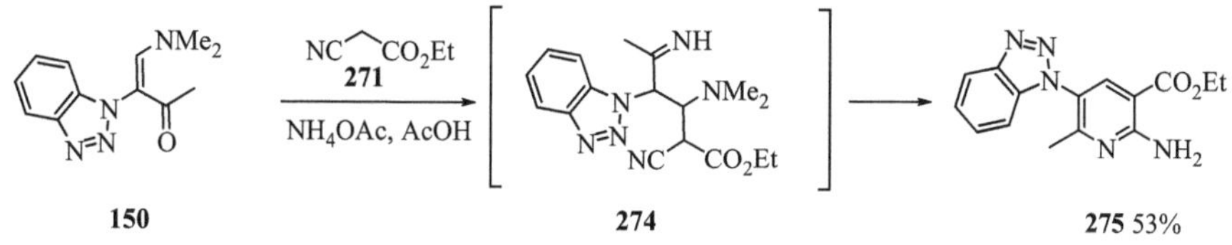

Scheme 103 Preparation of 2-aminopyridine 275

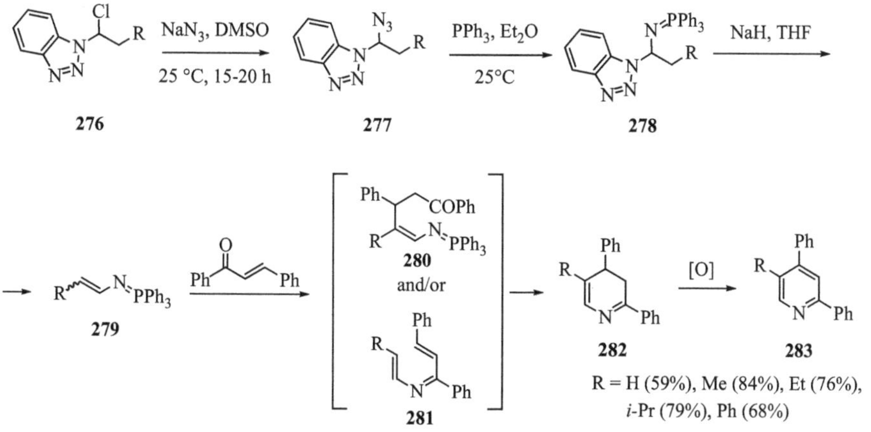

Scheme 104 Preparation of 5-alkyl-2,4-diphenylpyridines 283

reaction of phosphoranes **279** with chalcones gave 3,4-dihydropyridine intermediates **282**. The formation of intermediate **282** can be envisioned through two pathways: (1) a Michael-type addition of **279** to an enone followed by proton transfer to generate iminophosphorane **280**, which then underwent an aza-Wittig reaction, or (2) aza-Wittig reaction leading to aza-triene **281**, which then underwent a thermal 6π-electrocyclization. Final oxidation of **282** under the reaction conditions furnished 5-alkyl-2,4-diphenylpyridines **283** (Scheme 104) [196].

The Michael addition of benzotriazol-1-ylacetonitrile **264** [19] to chalcones in the presence of secondary amines gave compounds **284**, which cyclize to tetrahydropyridines **285**. Further elimination of water and benzotriazole yielded 2-aminopyridines **286** (Scheme 105) [192].

Vilsmeier-type salt **145** [121] reacted readily with conjugated β-enaminonitriles to give tetrasubstituted pyridines **288**. The reaction appeared to proceed through intermediate **287**, as such structures can be identified in the reaction mixture before their treatment with a base (Scheme 106) [197].

The reaction of α-benzotriazolyl ketones **249** with chalcones is a versatile [3+2+1] synthesis of 2,4,6-trisubstituted pyridines. The reaction pathway is believed to start with a Michael addition to give diketone **289**, which underwent cycloaddition with ammonia and subsequent elimination of benzotriazole to furnish 2,4,6-trisubstituted pyridines **290** (Scheme 107) [198].

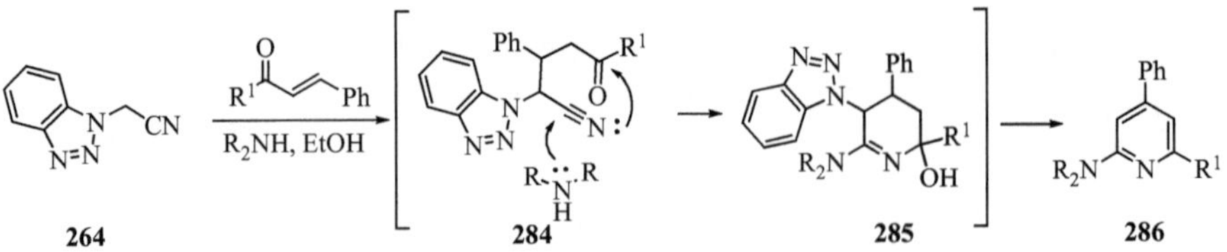

264 284 285 286

R^1 = Ph, R = Et (64%), $(CH_2)_4$ (74%), $(CH_2)_5$ (68%), $(CH_2)_2O(CH_2)_2$ (65%)
R^1 = 4-MeC_6H_4, R = Et (66%), $(CH_2)_4$ (70%), $(CH_2)_5$ (64%), $(CH_2)_2O(CH_2)_2$ (56%)
R^1 = Me, R = Et (43%), $(CH_2)_5$ (35%), $(CH_2)_2O(CH_2)_2$ (45%)

Scheme 105 Preparation of 2-aminopyridines **286**

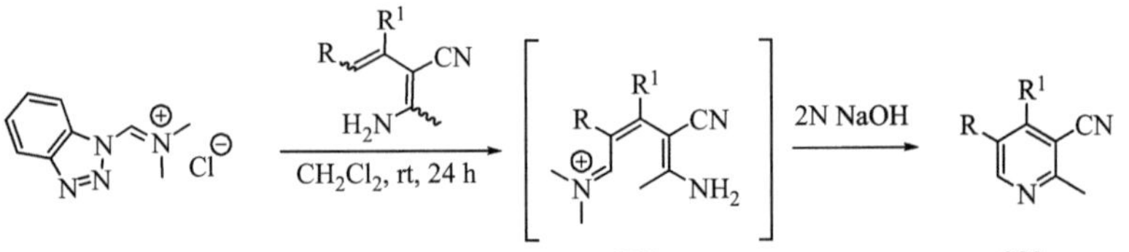

145 287 288

R = Me, R^1 = Et (75%), R = H, R^1 = *i*-Pr (57%), *c*-Pr (50%), *t*-Bu (56%), Ph (60%), R = Et, R^1 = Ph (70%)
R = R^1 = $(CH_2)_3$ (68%), $(CH_2)_4$ (74%), $(CH_2)_5$ (68%), $(CH_2)_3CHCH_3$ (72%), $C(CH_3)_3CH_2C(CH_3)_2$ (60%)

Scheme 106 Preparation of tetrasubstituted pyridines **288**

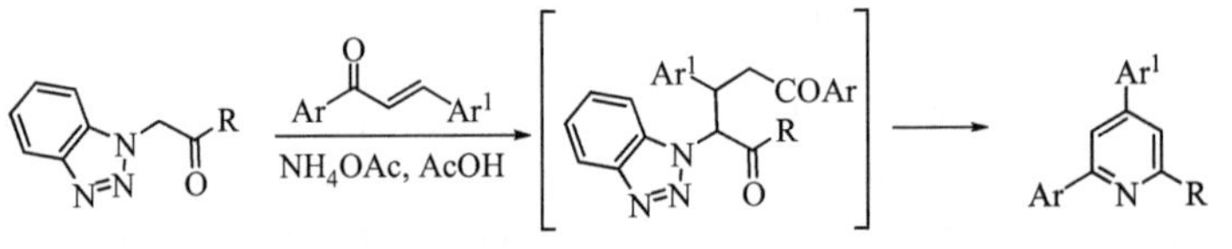

249 289 290

Ar = Ar^1 = Ph, R = Ph (81%), 4-MeC_6H_4 (78%), 4-$MeOC_6H_4$ (71%), 4-BrC_6H_4 (84%)
Ar = 4-BrC_6H_4, Ar^1 = 4-MeC_6H_4, R = 4-$MeOC_6H_4$ (85%), Ar^1 = 3,4-$(OCH_2O)C_6H_3$, R = 4-MeC_6H_4 (83%),
Ar = 2-naphthyl, Ar^1 = 3,4-$(OCH_2O)C_6H_3$, R = 4-MeC_6H_4 (87%), Ar^1 = 4-$NO_2C_6H_4$, R = Ph (62%),
Ar = Ph, Ar^1 = R = 4-ClC_6H_4 (80%)

Scheme 107 Preparation of 2,4,6-trisubstituted pyridines **290**

5.5 *Pyridazines*

Pyridazine and its saturated derivatives are of great interest, reflecting their wide range of pharmaceutical activities [199, 200]. Treatment of 2,5-bisbenzotriazolyl-*N*-aminopyrrolidines **291** with Grignard reagents resulted in ring enlargement to give 1,6-disubstituted 1,4,5,6-tetrahydropyridazines **294**. The reaction is believed to proceed with the deprotonation of **291** to produce anion **292**, in which the intramolecular attack displaced one of the benzotriazolyl moieties. The aziridine intermediate **293** then underwent ring expansion, driven by elimination of the other

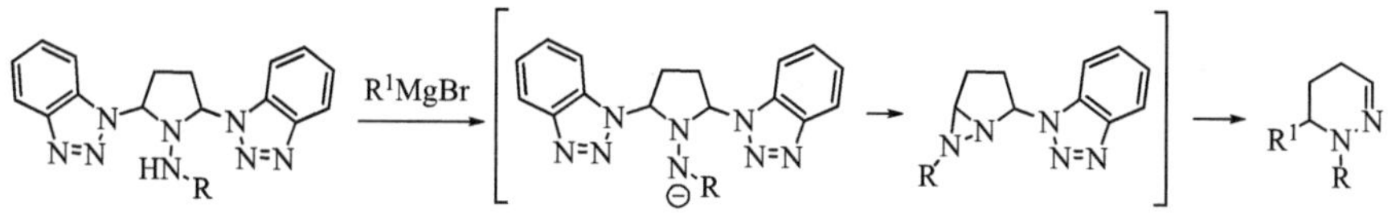

291 292 293 294

R= Ph, R^1 = Me (87%), Ph (77%), 2-anisyl (74%), ethynyl (89%)
vinyl (60%), 2,2-dimethylvinyl (67%), allyl (85%),
R = 4-MeC_6H_4, R^1 = Ph (76%), 2-anisyl (74%)
R = 4-ClC_6H_4, R^1 = Me (58%), ethynyl (55%), Ph (60%)

Scheme 108 Preparation of 1,6-disubstituted 1,4,5,6-tetrahydropyridazines **294**

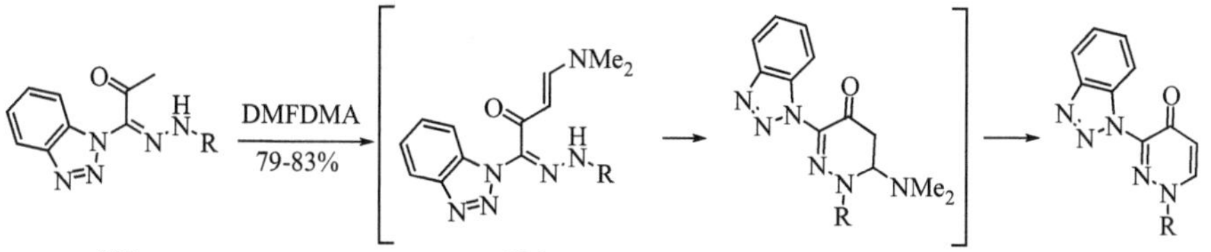

295 296 297 298

R = Ph, 4-$MeOC_6H_4$, 4-$NO_2C_6H_4$

Scheme 109 Preparation of pyridazin-4-one **298**

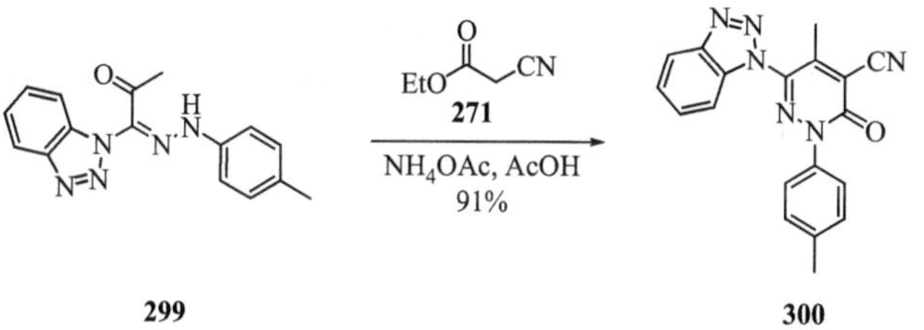

299 300

Scheme 110 Preparation of pyridazin-6-one **300**

benzotriazolyl group, to yield iminium intermediate, which was quenched by Grignard reagent to furnish 1,6-disubstituted 1,4,5,6-tetrahydropyridazines **294** (Scheme 108) [201].

Prolonged heating of hydrazones **295** with dimethylformamide dimethyl acetal resulted in the formation of pyridazin-4-ones **298**. The reaction pathway involves the initial formation of enaminones **296**, which by intramolecular cycloaddition generated tetrahydropyridazin-4-ones **297**. After elimination of dimethylamine, the final product **298** was isolated in good yields (Scheme 109) [202, 203].

Similarly, cyclization of hydrazone **299** with ethyl cyanoacetate **271** in the presence of ammonium acetate yielded pyridazin-6-one **300** (Scheme 110) [202].

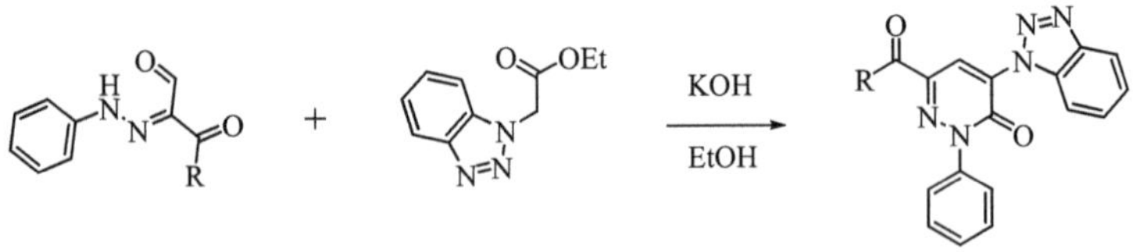

Scheme 111 Preparation of 4-disubstituted pyridazin-6-ones **302**

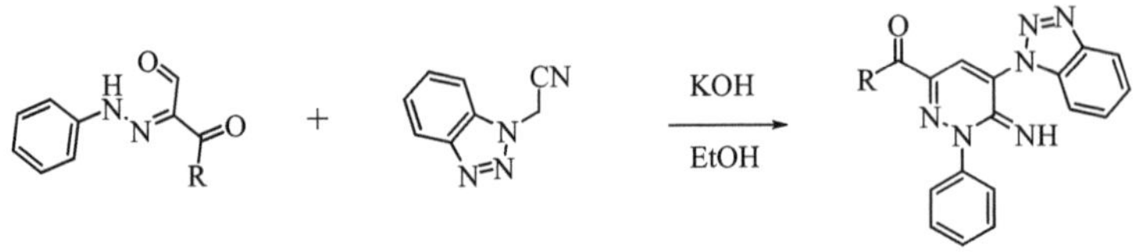

Scheme 112 Preparation of 6-iminopyridazines **303**

In a different approach, the cyclocondensation of **301** with ethyl benzotriazolylacetate in the presence of potassium hydroxide in ethanol gave 3,4-disubstituted pyridazin-6-ones **302** (Scheme 111) [202].

Similarly, when benzotriazol-1-ylacetonitrile **264** was used instead of ethyl benzotriazolylacetate in the cycloaddition with **301**, 6-iminopyridazines **303** were isolated (Scheme 112) [202, 203].

5.6 *Pyrimidines*

Pyrimidine and its saturated or partially saturated derivatives have a wide range of biological activity, which makes them desirable synthetic targets [204–207]. The condensation of propylenediamine with 3 equiv. of formaldehyde and 2 equiv. of benzotriazole provided 1,3-bis(benzotriazol-1-ylmethyl)hexahydropyrimidines **304**. The reaction of crude **304** with potassium cyanide in acetonitrile gave dinitrile **305** and with Grignard reagents in THF afforded 1,3-dialkylhexahydropyrimidines **306** (Scheme 113) [53].

In a similar manner, when monosubstituted diamines were used as starting materials, it allowed the formation of unsymmetrical hexahydropyrimidines **308** (Scheme 114) [208].

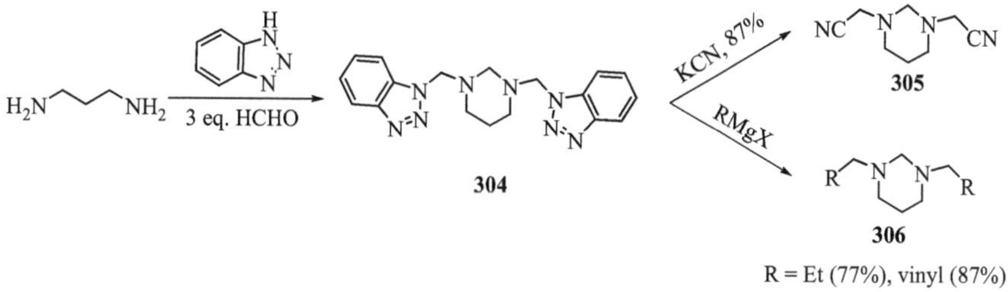

Scheme 113 Preparation of dinitrile 305 and 1,3-dialkylhexahydropyrimidines 306

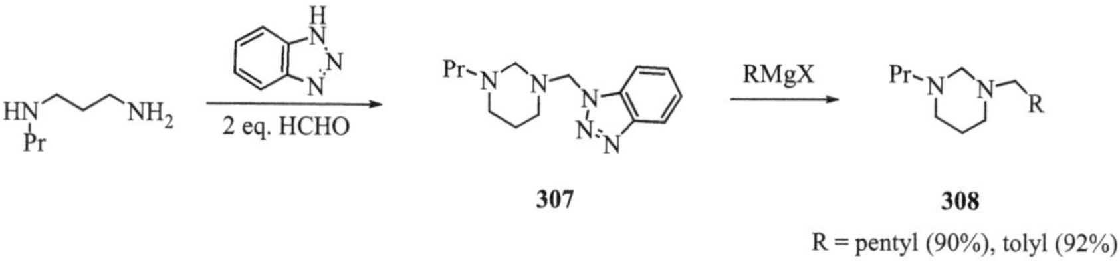

Scheme 114 Preparation of unsymmetrical hexahydropyrimidines 308

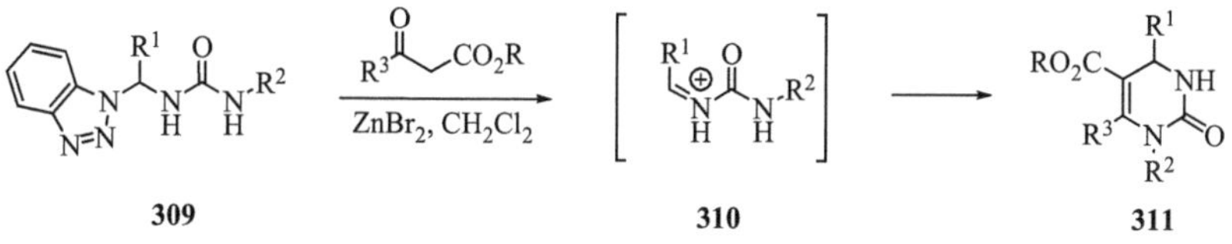

$R = R^2 = R^3 = Me$, R^1 = 1-naphtyl (84%), 2-MeO-5-BrC_6H_3 (89%), 4-ClC_6H_4 (92%), 4-CNC_6H_4 (81%), 2-thienyl (82%); R = Et, $R^2 = R^3 = Me$, R^1 = Ph (85%), 4-$NO_2C_6H_4$ (89%); R= Et, R^2 = Me, R^3 = Ph, R^1 = Ph (91%), 4-MeC_6H_4 (81%), 1-naphthyl (90%), 2-MeO-5-BrC_6H_3 (93%), 4-ClC_6H_4 (94%), 4-CNC_6H_4 (85%), 4-$NO_2C_6H_4$ (97%)

Scheme 115 Preparation of 3,4-dihydropyrimidin-2-one 311

The treatment of α-(benzotriazolyl)alkyl urea derivatives **309** [209] with β-ketoesters in the presence of zinc bromide afforded the corresponding 3,4-dihydropyrimidin-2-one **311** through intermediate **310** (Scheme 115) [210].

The reaction of bis(triphenylphosphoranylidene) derivative **205** with PhMgBr and acetylacetone provided 1,2-dihydropyrimidine **312** in 14% yield (Scheme 116) [158].

The condensation of 2-(1-benzotriazolyl)vinamidinium hexafluorophosphate salt **139** with amidines in the presence of sodium ethoxide provided pyrimidines **313** (Scheme 117) [114].

Highly substituted butadienes **314** [211] were treated with amidines in the presence of a base to furnish tetrasubstituted pyrimidines **317**. The reaction is believed to proceed by the initial substitution of the benzotriazole by the amidine to give intermediate **315** followed by an intramolecular cyclization to provide **316**.

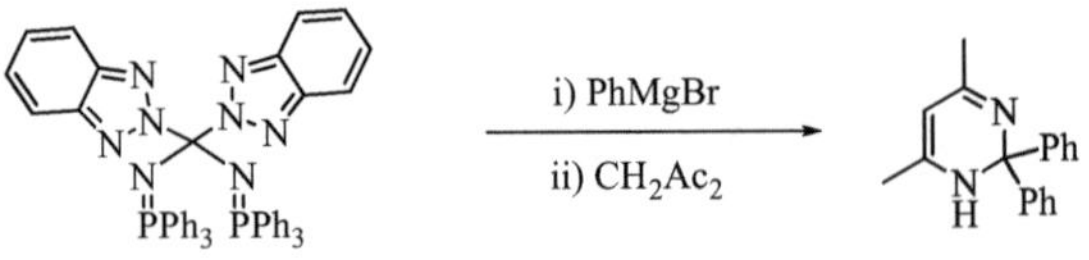

205 312

Scheme 116 Preparation of 1,2-dihydropyrimidine **312**

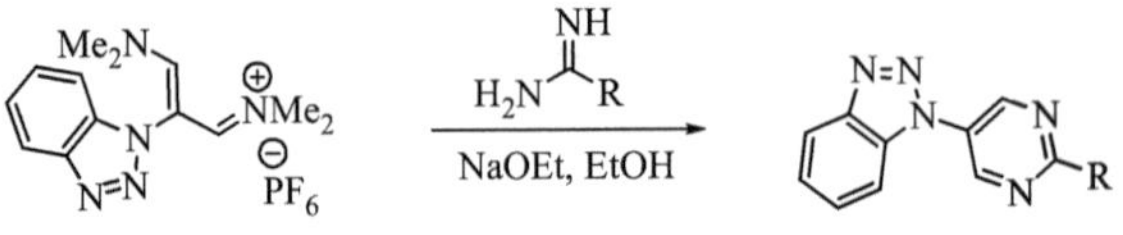

139 313

R = H (60%), Me (88%), NH_2 (85%), Ph (95%), OMe (90%), SMe (85%), NMe_2 (75%)

Scheme 117 Preparation of pyrimidines **313**

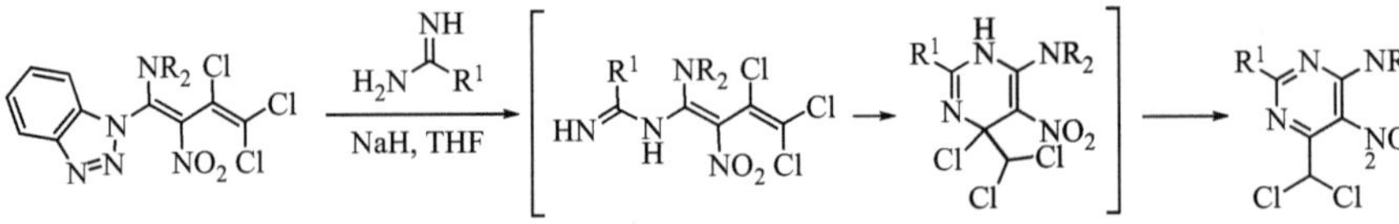

314 315 316 317

R = morpholino, R^1 = Me (56%), Ph (73%), R = piperidinyl, R^1 = Ph (76%),
R = 4-(4-fluorophenyl)piperazinyl, R^1 = Me (35%), R = 4-ClC_6H_4, R^1 = Me (85%)

Scheme 118 Preparation of tetrasubstituted pyrimidines **317**

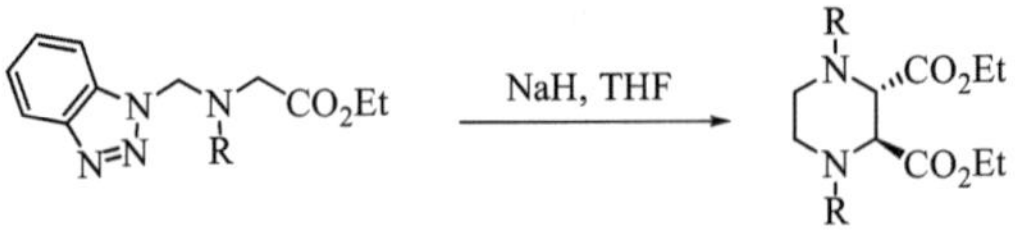

318 319 R = Me (20%), $PhCH_2$ (80%)

Scheme 119 Preparation of piperazines **319**

Elimination of a chloride and subsequent aromatization yielded tetrasubstituted pyrimidines **317** (Scheme 118) [212].

5.7 *Piperazines*

The condensation of N-substituted glycine ethyl esters with formaldehyde and benzotriazole yielded intermediates **318** as a mixture of benzotriazol-1-yl and benzotriazol-2-yl isomers. Upon treatment with sodium hydride, compounds **318** were converted into *trans*-substituted piperazines **319** (Scheme 119) [213].

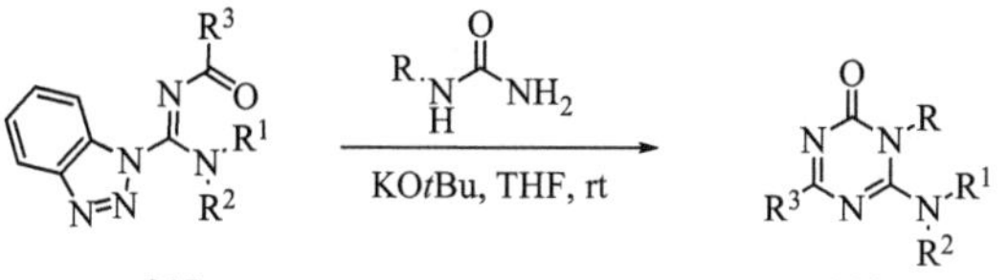

215 → 320

R = H, $R^1 = R^2$ = morpholinyl, R^3 = Ph (85%), 4-MeC_6H_4 (89%), $R^1 = R^2$ = Et, R^3 = 4-MeC_6H_4 (87%)
R = Me, $R^1 = R^2$ = morpholinyl, R^3 = Ph (85%), 2-thiophenyl (76%), 4-FC_6H_4 (91%)

Scheme 120 Preparation of 1,3,5-triazin-2-ones **320**

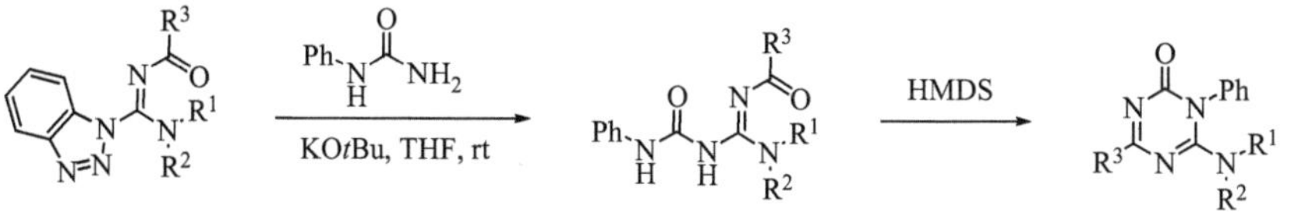

215 → 321 → 322

$R^1 = R^2$ = morpholinyl, R^3 = Ph (95%), $R^1 = R^2$ = Et, R^3 = 4-MeC_6H_4 (96%), *t*-Bu (51%),
$R^1 = R^2$ = allyl, R^3 = 4-MeC_6H_4 (88%)

Scheme 121 Preparation of triazinones **322**

5.8 Triazines

Triazine derivatives are widely used in cancer therapy [214–216]. Acyl derivatives of benzotriazole-1-carboximidamides **215** [165, 166] reacted readily with monosubstituted urea to give 1,3,5-triazin-2-ones **320** in the presence of three equivalents of potassium *tert*-butoxide (Scheme 120) [217].

In the case of *N*-phenyl urea, only the substitution of the benzotriazole occurred under the same reaction conditions giving stable intermediates **321**. Dehydration with hexamethyldisilazane (HMDS) allowed the cyclization to happen and formed triazinones **322** (Scheme 121) [217].

6 The Construction of Bicyclic Systems

6.1 (5,5)-Ring Systems

Succinaldehyde reacted with *N*-phenylethylenediamine and benzotriazole to give 1-phenyl-5-(benzotriazol-1-yl)hexahydro-1*H*-pyrrolo[1,2a]imidazole **323**, in which the benzotriazole can be replaced with nucleophiles. The reaction of **323**

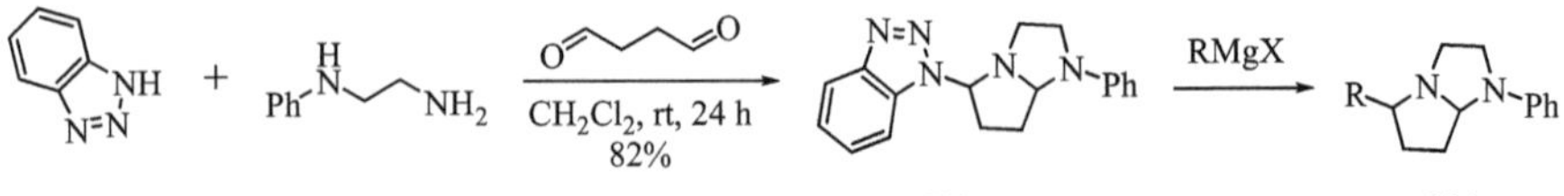

323 324

R = $PhCH_2$ (93%), pentyl (89%), allyl (91%), vinyl (86%), 4-MeC_6H_4 (88%), 3-Me-4-FC_6H_3 (90%)

Scheme 122 Preparation of 5-substituted hexahydro-1*H*-pyrrolo[1,2a]imidazole **324**

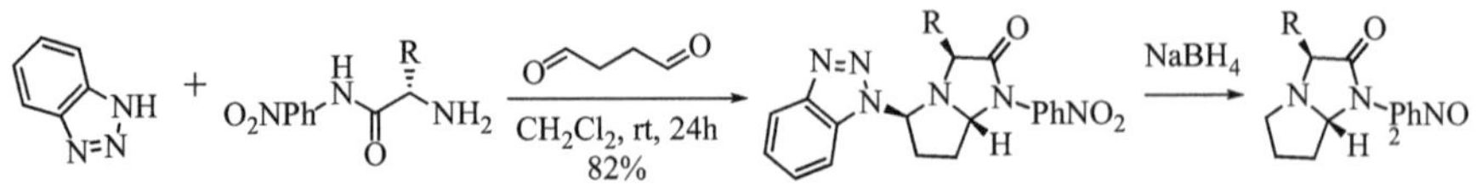

325
R = $PhCH_2$ (90%), Me (91%), *i*-Bu (84%)

326
R = $PhCH_2$ (51%), Me (48%), *i*-Bu (42%)

Scheme 123 Preparation of 1,3,5-trisubstituted hexahydro-1*H*-pyrrolo[1,2a]imidazol-2-ones **326**

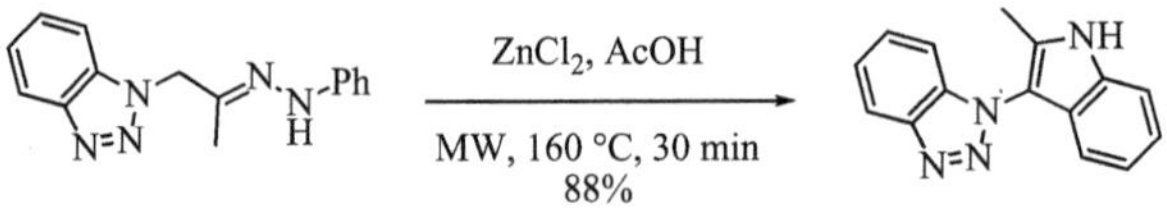

327 **328**

Scheme 124 Preparation of 3-(benzotriazolyl)-2-methylindole **328**

with Grignard reagents yielded 5-substituted hexahydro-1*H*-pyrrolo[1,2a]imidazoles **324** exclusively as *cis*-isomers (Scheme 122) [218].

Condensation of benzotriazole with succinaldehyde and amino amides derived from amino acids provided optically active 1,3,5-trisubstituted hexahydro-1*H*-pyrrolo[1,2a]imidazol-2-ones **325**. Treatment of **325** with sodium borohydride eliminated the benzotriazole group to yield major product **326**. Competing reactions are the reduction of the carbonyl group and the opening of the imidazole ring (Scheme 123) [219].

6.2 (5,6)-Ring Systems

In a Fischer-type synthesis, phenylhydrazone **327** was smoothly converted to 3-(benzotriazolyl)-2-methylindole **328** upon treatment with zinc chloride and acetic acid (Scheme 124) [220].

In the total synthesis of indoles, the heterocyclic ring is made first followed by the construction of the benzene ring. Thus, 1-propargylbenzotriazole **81** was lithiated and then reacted with various α-bromoketones and amines to form pyrroles

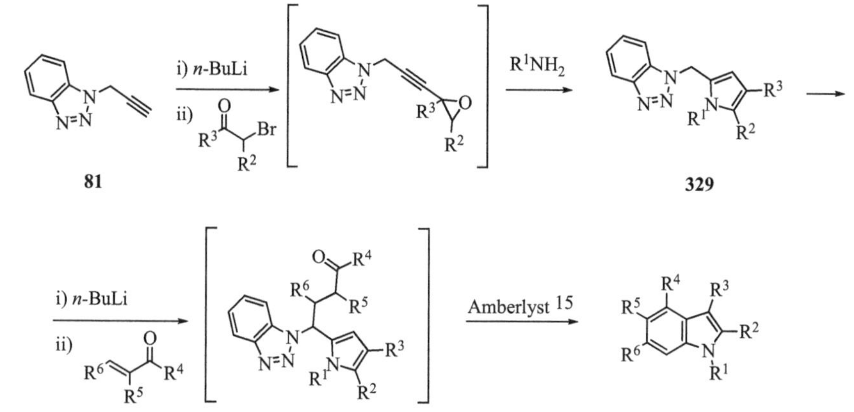

81 329 330 331

$R^1 = PhCH_2$, $R^2 = Me$, $R^3 = R^4 = R^6 = Ph$, $R^5 = H$ (56%); $R^2 = Me$, $R^3 = Ph$, $R^4 = H$, $R^5 = Ph$, $R^6 = Et$ (76%)
$R^1 = t$-Bu, $R^2 = R^5 = R^6 = H$, $R^3 = Ph$, $R^4 = Me$ (60%); $R^1 = R^2 = R^5 = R^6 = H$, $R^3 = Ph$, $R^4 = Me$ (56%)
$R^1 = MeO(CH_2)_2$, $R^2 = R^3 = Me$, $R^4 = R^6 = Ph$, $R^5 = H$ (52%); $R^2 = R^3 = Me$, $R^4 = H$, $R^5 = Ph$, $R^6 = Et$ (75%)
$R^1 = 4\text{-}MeOC_6H_4CH_2$, $R^2 = R^3 = Me$, $R^4 = R^6 = Ph$, $R^5 = H$ (51%); $R^1 = n$-Bu, $R^2 = R^3 = Me$, $R^4 = R^6 = Ph$, $R^5 = H$ (54%)

Scheme 125 Preparation of indoles **331**

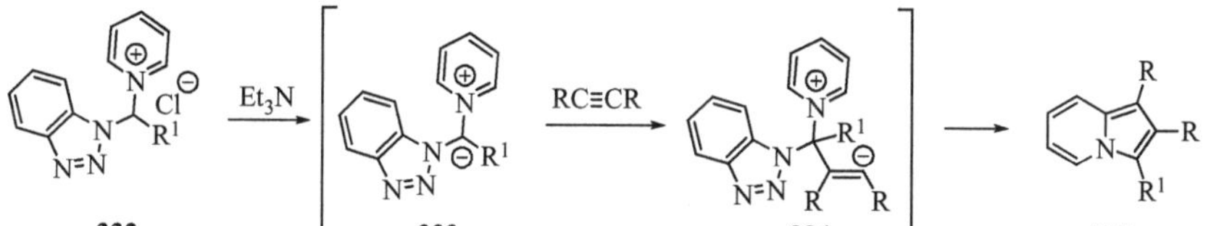

332 333 334 335

R = COOEt, $R^1 = p$-tolyl (69%), R = COOMe, R^1 = H (66%), Et (69%)

Scheme 126 Preparation of indolizine **335**

329, which were further reacted with α,β-unsaturated ketones or aldehydes in the presence of *n*-butyllithium to generate intermediates **330**. Compounds **330** were stable and can be isolated; however, they were conveniently converted to indoles **331** via the elimination of benzotriazole in cyclization reactions promoted by heating with a strongly acidic Amberlyst 15 resin (Scheme 125) [221].

Treatment of *N*-[α-(benzotriazol-1-yl)methyl]pyridinium chloride **332** [222] with triethylamine gave **333** which can be effectively trapped with activated alkynes. Thus, addition of 2-butynedioic acid esters resulted in zwitterions **334**, which cyclized and eliminated benzotriazole restoring aromaticity and producing indolizine **335** (Scheme 126) [223].

The condensation of (benzotriazol-1-yl)acetic acid **138** with *ortho*-phenylenediamine in phosphoric acid gave 2-(benzotriazol-1-ylmethyl)benzimidazole **336**. Further stepwise alkylation of the methylene group was easily accomplished using alkyl halides in the presence of *n*-butyllithium (Scheme 127) [224].

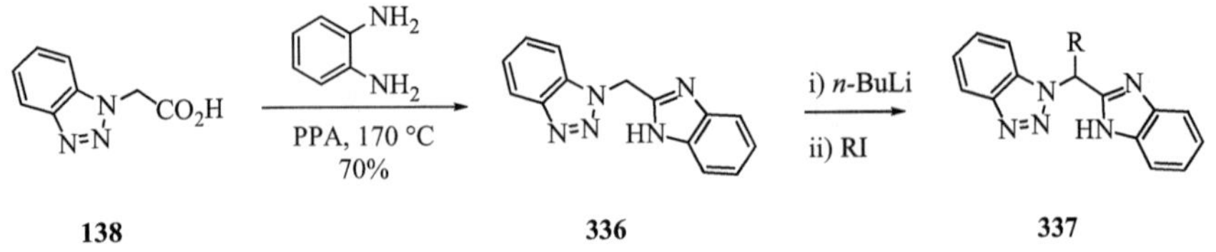

R = Me (68%), *n*-Bu (57%), octyl (62%)

Scheme 127 Preparation of benzimidazole **336**

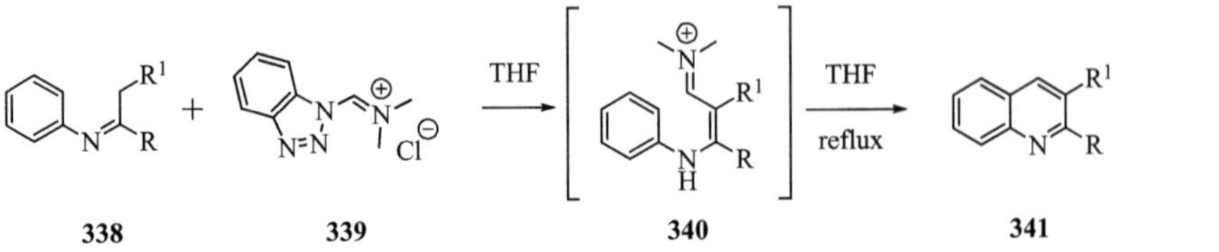

R^1 = Me, R = Et (76%), *i*-Bu (82%), Bu (78%), R^1 = Et, R = Ph (65%)

Scheme 128 Preparation of quinolines **341**

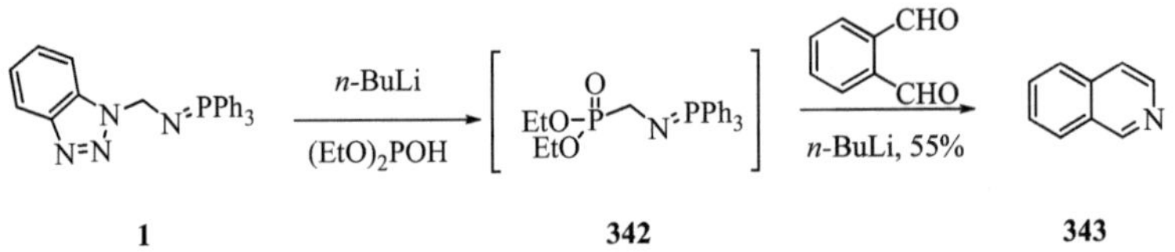

Scheme 129 Preparation of isoquinoline **343**

6.3 (6,6)-Ring Systems

The condensation of *N*-phenylimines **338** with iminium salts **339** provided vinamidinium salts **340** in quantitative yields. The reaction was remarkably regioselective, always involving only the less sterically hindered methylene group, even if the difference is small. Upon heating **340** in THF, an electrophilic attack of the iminium carbon on the aromatic ring was followed by elimination of dimethylammonium chloride to yield quinolines **341** (Scheme 128) [225].

The nucleophilic displacement of the benzotriazole group in 1-(triphenylphosphoranylideneaminomethyl)benzotriazole **1** with lithium diethyl phosphonate gave intermediate **342**. Further treatment with phthalaldehyde, in the presence of *n*-butyllithium, gave unsubstituted isoquinoline **343** (Scheme 129) [104].

References

1. Chin YW, Balunas MJ, Chai HB, Kinghorn AD (2006) AAPS J 8(2):239–253
2. Lamberth C, Dinges J (eds) (2012) Bioactive heterocyclic compound classes: agrochemicals. Wiley-VCH, New York
3. Shui L, Han N, Sun Y, Liu W et al (2002) SAE Technical Paper 1:2801
4. Kusama H, Kurashige M, Arakawa H (2005) J Photochem Photobio A Chem 169(2):169–176
5. Katritzky AR, Lan X (1994) Chem Soc Rev 23:363–373
6. Katritzky AR, Lan X, Yang JZ, Denisko OV (1998) Chem Rev 98:409–548
7. Katritzky AR, Tala SR, Singh S (2006) Synthesis 19:3231–3237
8. Katritzky AR, Lan X, Lam JN (1991) Chem Ber 124(8):1809–1817
9. Galli C, Gentili P, Lanzalunga O, Franco G, Sapienza L, Moro PA (2004) Chem Comm 20:2356–2357
10. Katritzky AR, Fali CN, Li J (1997) J Org Chem 62(12):4148–4154
11. Botuha C, Chemla F, Ferreira F, Pérez-Luna A (2011) In: Majumdar KC, Chattopadhyay SK (eds) Heterocycles in natural product synthesis. Wiley-VCH, New York, pp 1–39
12. Lowden PAS (2006) In: Yudin AK (ed) Aziridines and epoxides in organic synthesis. Wiley-VCH, New York, pp 399–442
13. McCoull W, Davis FA (2000) Synthesis 10:1347–1365
14. Tanner D (1994) Angew Chem Int Ed 33(6):599–619
15. Kim CO, Cho SJ, Park JW (2003) J Colloid Interface Sci 260(2):374–378
16. Goethals EJ, Van de Velde M, Eckhaut G, Bouquet G (2009) In: ACS symposium series, pp 219–229
17. Katritzky AR, Jiang J, Urogdi L (1990) Synthesis 7:565–567
18. Burckhalter JH, Stephens VC, Hall LAR (1952) J Am Chem Soc 74:3868–3870
19. Katritzky AR, Rachwal S, Caster KC, Mahni F, Law KW, Rubio O (1987) J Chem Soc Perkin Trans I 781–789
20. Katritzky AR, Jiang J, Urogdi L (1989) Tetrahedron Lett 30(25):3303–3306
21. Katritzky AR, Yao J, Bao W, Qi M, Steel PJ (1999) J Org Chem 64:346–350
22. Katritzky AR, Li J, Malhotra N (1992) Liebigs Ann Chem 8:843–853
23. Katritzky AR, Wu J, Kuzmierkiewicz W, Rachwal S (1994) Liebigs Ann Chem 1:1–5
24. Katritzky AR, Wu J, Kuzmierkiewicz W, Rachwal S, Balasubramanian M, Steel PJ (1994) Liebigs Ann Chem 1:7–12
25. Padwa A, Woolhouse AD (1984) In: Katritzky AR, Rees CW (eds) Comprehensive heterocyclic chemistry, vol 7. Elsevier, Amsterdam, Netherlands, pp 47–93
26. Nair V (1983) In: Hassner A (ed) Chemistry of heterocyclic compounds, vol 42. Wiley-VCH, New York, pp 215–332
27. Stapley EO, Hendlin D, Jackson M, Miller AK, Hernandez S, Mata JM (1971) J Antibiot 24 (1):42–47
28. Miller TW, Tristram EW, Wolf FJ (1971) J Antibiot 24(1):48–50
29. Molinski TF, Ireland CM (1988) J Org Chem 53:2103–2105
30. Salomon CE, Williams DH, Faulkner DJ (1995) J Nat Prod 58:1463–1466
31. Osborn HMI, Sweeney J (1997) Tetrahedron Asym 8(11):1693–1715
32. Verstappen MMH, Ariaans GJA, Zwanenburg B (1996) J Am Chem Soc 118(35):8491–8492
33. Palacios F, Ochoa de Retana AM, Gil JI, Ezpeleta JM (2000) J Org Chem 65(10):3213–3217
34. Kafarski P, Lejczak B (1991) Phos Sulf Silicon Related Elem 63:193–215
35. Katritzky AR, Zhang S, Fang Y (2000) Org Lett 2(24):3789–3791
36. Katritzky AR, Wu J (1994) Synthesis 6:597–601
37. Katritzky AR, Wang M, Wilkerson CR, Yang H (2003) J Org Chem 68(23):9105–9108
38. Filler R, Kobayashi Y, Yagupolskii LM (1993) Organofluorine compounds in medicinal chemistry and biomedical applications. Elsevier
39. Kukhar′ VP, Soloshonok VA (1993) Fluorine-containing amino acids: synthesis and properties. Wiley-VCH, Amsterdam, Netherlands

40. Cheguillaume A, Lacroix S, Marchand-Brynaert J (2003) Tetrahedron Lett 44 (11):2375–2377
41. Katritzky AR, Rachwal S, Rachwal B (1987) J Chem Soc Perkin Trans 1:799–804
42. Lacroix S, Cheguillaume A, Gérard S, Marchand-Brynaert J (2003) Synthesis 16:2483–2486
43. Ganem B (1996) Acc Chem Res 29(7):340–347
44. Bachi MD, Melman A (1997) J Org Chem 62(6):1896–1898
45. Mukai C, Sugimoto Y, Miyazawa K, Yamaguchi S, Hanaoka M (1998) J Org Chem 63 (18):6281–6287
46. Falb E, Bechor Y, Nudelman A, Hassner A, Albeck A, Gottlieb HE (1999) J Org Chem 64 (2):498–506
47. Macdonald SJF, Clarke GDE, Dowle MD, Harrison LA, Hodgson ST, Inglis GGA, Johnson MR, Shah P, Upton RJ, Walls SB (1999) J Org Chem 64(14):5166–5175
48. Lovell PJ, Bromidge SM, Dabbs S, Duckworth DM, Forbes IT, Jennings AJ, King FD, Middlemiss DN, Rahman SK, Saunders DV, Collin LL, Hagan JJ, Riley GJ, Thomas DR (2000) J Med Chem 43(3):342–345
49. Geneste H, Hesse M (1998) Tetrahedron 54(50):15199–15214
50. Katritzky AR, Luo Z, Fang Y (2000) Tetrahedron Lett 41(49):9691–9693
51. Gawley RE, Barolli G, Madan S, Saverin M, O'Connor S (2004) Tetrahedron Lett 45 (8):1759–1761
52. Katritzky AR, Cui XL, Yang B, Steel PJ (1999) J Org Chem 64(6):1979–1985
53. Katritzky AR, Pilarski B, Urogdi L (1990) J Chem Soc Perkin Trans 1:541–547
54. Katritzky AR, Fang Y, Qi M, Feng D (1998) Heterocycles 48(12):2535–2541
55. Hirai Y, Terada T, Hagiwara A, Yamazaki T (1988) Chem Pharm Bull 36(4):1343–1350
56. Brands KMJ, Meekel AAP, Pandit UK (1991) Tetrahedron 47(11):2005–2026
57. Aurrecoechea JM, Fernández-Acebes A (1996) Synlett 1:39–42
58. Aurrecoechea JM, Fernández A, Gorgojo JM, Saornil C (1999) Tetrahedron 55 (23):7345–7362
59. Bustos F, Gorgojo JM, Suero R, Aurrecoechea JM (2002) Tetrahedron 58(34):6837–6842
60. Katritzky AR, Pilarski B, Urogdi L (1989) Org Prep Proc Int 21(2):135–139
61. Padwa A, Chen YY, Chiacchio U, Dent W (1985) Tetrahedron 41(17):3529–3535
62. Padwa A, Chen YY, Dent W, Nimmesgern H (1985) J Org Chem 50(21):4006–4014
63. Noll JE, Speier JL, Daubert BF (1951) J Am Chem Soc 73(8):3867–3871
64. Katritzky AR, Köditz J, Lang H (1994) Tetrahedron 50(44):12571–12578
65. Katritzky AR, Yao J, Yang B (1999) J Org Chem 64(16):6066–6070
66. Harrison T (1995) Contemp Org Synth 2:209–224
67. Lin NH, Carrera JM Jr, Anderson DJ (1994) J Med Chem 37(21):3542–3553
68. Shih NY, Lupo AT Jr, Aslanian R, Orlando S, Piwinski JJ, Green MJ, Ganguly AK, Clark MA, Tozzi S (1995) J Med Chem 38(10):1593–1599
69. Elliott RL, Kopecka H, Lin NH, He Y, Garvey DS (1995) Synthesis 7:772–774
70. Castelhano AL, Krantz A (1984) J Am Chem Soc 106(6):1877–1879
71. Hiemstra H, Fortgens HP, Speckamp WN (1984) Tetrahedron Lett 25(29):3115–3118
72. Huang PQ, Wang SL, Ye JL, Ruan YP, Huang YQ, Zheng H, Gao JX (1998) Tetrahedron 54 (41):12547–12560
73. Baenziger M, McGarrity JF, Meul T (1993) J Org Chem 58(15):4010–4012
74. Wei ZY, Knaus EE (1993) Synlett 4:295–296
75. Katritzky AR, Szajda M, Lam JN (1993) Heterocyclic Chem 30:1261–1265
76. Katritzky AR, Mehta S, He HY, Cui X (2000) J Org Chem 65(14):4364–4369
77. Katritzky AR, Angrish P, Todadze E, Ghiviriga I (2007) Bioorg Med Chem Lett 17 (21):6000–6002
78. Katritzky AR, Zhao X, Hitchings GJ (1991) Synthesis 9:703–708
79. Katritzky AR, Hitchings GJ, Zhao X (1991) Synthesis 10:863–867
80. Katritzky AR, Verin SV (1995) J Het Chem 32(1):323–328
81. Trofimov BA (1990) Adv Het Chem 51:177–301

82. Estévez V, Villacampa M, Menéndez JC (2014) Chem Soc Rev 43:4633–4657
83. Kaiser DG, Glenn EM (1972) J Pharm Sci 61(12):1908–1911
84. Demir AS, Akhmedov IM, Sesenoglu O (2002) Tetrahedron 58(49):9793–9799
85. Meshram HM, Prasad BRV, Kumar DA (2010) Tetrahedton Lett 51(27):3477–3480
86. Davis FA, Bowen KA, Xu H, Velvadapu V (2008) Tetrahedron 64(19):4174–4182
87. Reisser M, Maas G (2004) J Org Chem 69(15):4913–4924
88. Mathew P, Asokan CV (2006) Tetrahedron 62(8):1708–1716
89. Sundberg RJ (1984) In: Katritzky AR, Rees CW (eds) Comprehensive heterocyclic chemistry. Elsevier, pp 313–376
90. Niziurski-Mann RE, Cava MP (1992) Heterocycles 34(10):2003–2027
91. Katritzky AR, Li J, Gordeev MF (1994) Synthesis 1:93–96
92. Katritzky AR, Rachwal S (1992) Pol J Chem 66:1653
93. Katritzky AR, Chang HX, Verin SV (1995) Tetrahedron Lett 36(3):343–346
94. Katritzky AR, Wu H, Xie L, Rachwal S, Rachwal B, Jiang J, Zhang G, Lang H (1995) Synthesis 10:1315–1323
95. Katritzky AR, Zhang L, Yao J, Denisko OV (2000) J Org Chem 65(23):8074–8076
96. Katritzky AR, Drewniak M, Lue P (1988) J Org Chem 53(25):5854–5856
97. Katritzky AR, Drewniak M (1988) Tetrahedron Lett 29(15):1755–1758
98. Katritzky AR, Zhu L, Lang H, Denisko O, Wang Z (1995) Tetrahedron 51(48):13271–13276
99. Katritzky AR, Yao G, Lan X, Zhao X (1993) J Org Chem 58(8):2086–2093
100. Katritzky AR, Denisko O, Lang H (1995) Tetrahedron 51(32):8703–8710
101. Katritzky AR, Huang TB, Voronkov MV, Wang M, Kolb H (2000) J Org Chem 65 (25):8819–8821
102. Katritzky AR, Jiang J, Harris PA (1990) Heterocycles 31(12):2187–2194
103. Katritzky AR, Jiang J, Harris PA (1991) Can J Chem 69(7):1153–1155
104. Katritzky AR, Jiang J, Greenhill JV (1993) J Org Chem 58(8):1987–1988
105. Katritzky AR, Sutharchanadevi M, Urogdi L (1990) J Chem Soc Perkin Trans 1:1847–1851
106. Katritzky AR, Cheng D, Musgrave RP (1997) Heterocycles 44(1):67–70
107. Katritzky AR, Wang X, Denisenko A (2001) J Org Chem 66(8):2850–2853
108. Katritzky AR, Gordeev MF, Greenhill JV, Steel PJ (1992) J Chem Soc Perkin Trans 1:1111–1117
109. Katritzky AR, Wang Z, Li J, Levell JR (1997) J Het Chem 34(4):1379–1381
110. Boger DL, Corbett WL (1993) J Org Chem 58(8):2068–2074
111. Katritzky AR, Zhang S, Wang M, Kolb HC, Steel PJ (2002) J Het Chem 39(4):759–765
112. Gupton JT, Krolikowski DA, Yu RH, Riesinger SW, Sikorski JA (1990) J Org Chem 55 (15):4735–4740
113. Angus RO Jr, Bryce MR, Keshavarz-K M, Wudl F (1988) Synthesis 9:746–748
114. Gupton JT, Hicks FA, Smith SQ, Main AD, Petrich SA, Wilkinson DR, Sikorski JA, Katritzky AR (1993) Tetrahedron 49(45):10205–10218
115. Ahn JH, Kim JA, Kim HM, Kwon HM, Huh SC, Rhee SD, Kim KR, Yang SD, Park SD, Lee JM, Kim SS, Cheon HG (2005) Bioorg Med Chem Lett 15(5):1337–1340
116. Elguero J (1984) In: Katritzky AR, Rees CW (eds) Comp Het Chem 5:167–303
117. Katritzky AR, Qiu G, Yang B (1997) J Org Chem 62(23):8210–8214
118. Ashton WT, Hutchins SM, Greenlee WJ, Doss GA, Chang RSL, Lotti VJ, Faust KA, Chen TB, Zingaro GJ (1993) J Med Chem 36(23):3595–3605
119. Flynn DL, Belliotti TR, Boctor AM, Connor DT, Kostlan CR, Nies DE, Ortwine DF, Schrier DJ, Sircar JC (1991) J Med Chem 34(2):518–525
120. Huang YR, Katzenellenbogen JA (2000) Org Lett 2(18):2833–2836
121. Katritzky AR, Cheng D, Leeming P, Ghiviriga I, Hartshorn CM, Steel PJ (1996) J Het Chem 33(6):1935–1941
122. Katritzky AR, Denisenko A, Denisenko SN, Arend M (2000) J Het Chem 37(5):1309–1314
123. Katritzky AR, Henderson SA, Yang B (1998) J Het Chem 35(5):1123–1159

124. Al-Saleh B, Abdel-Khalik MM, Darwich E, Salah OAM, Elnagdi MH (2002) Heteroatom Chem 13(2):141–145
125. Al-Omran F, El-Khair AA (2004) J Het Chem 41(3):327–333
126. Petrich SA, Main AD, Jaeger R, Smith SQ, Hicks FA, Gupton JT, Sikorski JA, Katritzky AR, Castle L (1994) Heterocycles 39(2):623–637
127. Katritzky AR, Vakulenko AV, Akue-Gedu R, Gromova AV, Witek R, Rogers JW (2007) Arkivoc I:9–21
128. Katritzky AR, Wang M, Zhang S, Voronkov MV, Steel PJ (2001) J Org Chem 66 (20):6787–6791
129. Smith JRL, Sadd JS (1975) J Chem Soc Perkin Trans 1:1181–1184
130. Katritzky AR, Lan X, Zhang Z (1993) J Het Chem 30(2):381–387
131. Katritzky AR, Fali CN, Oniciu DC (1995) Tetrahedron 51(4):1069–1076
132. Kuehne ME, Weaver SJ, Franz P (1964) J Org Chem 29(6):1582–1586
133. Katritzky AR, Musgrave RR, Breytenbach JC (1996) J Het Chem 33(6):1637–1646
134. Kim T, Kim K, Park YJ (2002) Eur J Org Chem 3:493–502
135. Sharma V, Crankshaw CL, Piwnica-Worms D (1996) J Med Chem 39(18):3483–3490
136. Sage CR, Michelitsch MD, Stout TJ, Biermann D, Nissen R, Finer-Moore J, Stroud RM (1998) Biochemistry 37(39):13893–13901
137. Olmos G, Ribera J, García-Sevilla JA (1996) Eur J Pharm 310(1):273–276
138. Fong JC, Benamour K, Szymonski B, Thomasson F, Morand JM, Cussac M (2000) Chem Pharm Bull 48(5):729–733
139. Szinai SS, Crank G, Harding DRK (1970) J Med Chem 13(6):1212–1215
140. Szinai SS, Crank G, Harding DRK (1970) J Med Chem 13(6):1215–1217
141. Katritzky AR, Suzuki K, He HY (2002) J Org Chem 67(9):3109–3114
142. Katritzky AR, Xu YJ, He HY, Steel PJ (2001) J Chem Soc Perkin Trans 1:1767–1770
143. Malawska B (2005) Curr Top Med Chem 5(1):69–85
144. Fiallo MML, Kozlowski H, Garnier-Suillerot A (2001) Eur J Pharm Sci 12(4):487–494
145. Gregoriou M, Noble MEN, Watson KA, Garman EF, Johnson LN, Krulle TM, De La Fuetene C, Fleet GWJ, Oikonomakos NG (1998) Protein Sci 7(4):915–927
146. Wu S, Lee S, Beak P (1996) J Am Chem Soc 118(4):715–721
147. Katritzky AR, Luo Z, Fang Y, Steel PJ (2001) J Org Chem 66(8):2858–2861
148. Katritzky AR, Fali CN, Li J, Ager DJ, Prakash I (1997) Synth Commun 27(9):1623–1630
149. Butula I, Jadrijević-Mladar Takač M (2000) Croat Chem Acta 73(2):569–574
150. Kalčić I, Zovko M, Jadrijević-Mladar Takač M, Zorc B, Butula I (2003) Croat Chem Acta 76 (3):217–228
151. Rajic Z, Zorc B, Raic-Malic S, Ester K, Kralj M, Pavelic K, Balzarini J, De Clercq E, Mintas M (2006) Molecules 11(11):837–848
152. Opačić N, Barbarić M, Zorc B, Cetina M, Nagl A, Frković D, Kralj M, Pavelić K, Balzarini J, Andrei G, Snoeck R, De Clercq E, Raić-Malić S, Mintas M (2005) J Med Chem 48 (2):475–482
153. Ujjinamatada RK, Baier A, Borowski P, Hosmane RS (2007) Bioorg Med Chem Lett 17 (8):2285–2288
154. Shingalapur RV, Hosamani KM, Keri RS (2009) Eur J Med Chem 44(10):4244–4248
155. Sharma D, Narasimhan B, Kumar P, Judge V, Narang R, De Clercq E, Balzarini J (2009) Eur J Med Chem 44(6):2347–2353
156. Olender D, Żwawiak J, Lukianchuk V, Lesyk R, Kropacz A, Fojutowski A, Zaprutko L (2009) Eur J Med Chem 44(2):645–652
157. Congiu C, Cocco MT, Onnis V (2008) Bioorg Med Chem Lett 18(3):989–993
158. Katritzky AR, Wrobel L, Savage GD (1990) J Chem Res Synop 1990:330
159. Sanghvi YS, Bhattacharya BK, Kini GD, Matsumoto SS, Larson SB, Jolley WB, Robins RK, Revankar GR (1990) J Med Chem 33(1):336–344
160. Moltzen EK, Pedersen H, Boegesoe KP, Meier E, Frederiksen K, Sanchez C, Lemboel HL (1994) J Med Chem 37(24):4085–4099

161. Norris P, Horton D, Levine BR (1996) Heterocycles 43(12):2643–2655
162. Katritzky AR, Zhang Y, Singh SK, Steel PJ (2003) Arkivoc XV:47–64
163. Thambidurai S, Jeyasubramanian K, Ramalingam SK (1996) Polyhedron 15(22):4011–4014
164. Katritzky AR, Rogovoy BV, Chassaing C, Vvedensky V (2000) J Org Chem 65:8080–8082
165. Katritzky AR, Rogovoy BV, Vvedensky VY, Kovalenko K, Steel PJ, Markov VI, Forood B (2001) Synthesis 6:897–903
166. Boojamra SG, Burow KM, Thompson LA, Ellman JA (1997) J Org Chem 62:1240–1256
167. Makara GM, Ma Y, Margarida L (2002) Org Lett 4(10):1751–1754
168. Katritzky AR, Vvedensky V, Cai X, Rogovoy B, Steel PJ (2002) Arkivoc VI:82–90
169. Herr JR (2002) Bioorg Med Chem 10(11):3379–3393
170. Venkatraman S, Lim J, Cramer M, Gardner MF, James J, Alves K, Lingham RB, Mumford RA, Munoz B (2005) Bioorg Med Chem Lett 15(18):4053–4056
171. Mayer J, Umkehrer M, Kalinski C, Ross G, Kolb J, Burdack C, Hiller W (2005) Tetrahedron Lett 46(43):7393–7396
172. Patani GA, LaVoie EJ (1996) Chem Rev 96(8):3147–3176
173. Dömling A, Beck B, Magnin-Lachaux M (2006) Tetrahedron Lett 47(25):4289–4291
174. Katritzky AR, Cai C, Meher NK (2007) Synlett 8:1204
175. Katritzky AR, Belyakov SA, Lam JN, Drust HD, Karpenko DV (1994) Heterocycles 39 (1):73–80
176. Watson PS, Jiang B, Scott B (2000) Org Lett 2:3679–3681
177. Okitsu O, Suzuki R, Kobayashi S (2001) J Org Chem 66(3):809–823
178. O'Hagan D (2000) Nat Prod Rep 17:435–446
179. O'Hagan D (1997) Nat Prod Res 14:637–651
180. Katritzky AR, Fan W (1990) J Org Chem 55(10):3205–3209
181. Veerman JJN, Klein J, Aben RWM, Scheeren HW, Kruse CG, van Maarseveen JH, Rutjes FPJT, Hiemstra H (2002) Eur J Org Chem 18:3133–3139
182. Wand X, Dong Y, Sun J, Xu X, Li R, Hu Y (2005) J Org Chem 70(5):1897–1900
183. Katritzky AR, Qui G, Yang B, Steel PJ (1998) J Org Chem 63(19):6699–6703
184. Zheng JF, Jin LR, Huang PQ (2004) Org Lett 6(7):1139–1142
185. Katritzky AR, Luo Z, Cui XL (1999) J Org Chem 64:3328–3331
186. Katritzky AR, Luo Z, Fang Y, Feng D, Ghiviriga I (2000) J Chem Soc Perkin Trans 2:1375–1380
187. Katritzky AR, Gupta V, Gordeev M (1993) J Het Chem 30(4):1073–1077
188. Katritzky AR, Gordeev MF (1993) J Org Chem 58(15):4049–4053
189. Katritzky AR, Shcherbakova IV (1996) J Het Chem 33(6):2031–2036
190. Alnajjar AA, Abdelkhalik MM, Al-Enezi A, Elnagdi MH (2009) Molecules 14(1):68–77
191. Al-Saleh B, El-Apasery MA, Abdelkhalik MM, Elnagdi MH (2003) J Het Chem 40 (1):171–175
192. Katritzky AR, Belyakov SA, Sorochinsky AE, Henderson SA, Chen J (1997) J Org Chem 62 (18):6210–6214
193. Basak S, Ho YH, Tsai EW, Rajeshwar K (1989) Chem Commun 54:462–464
194. Katritzky AR (1980) Tetrahedron 36(6):679–699
195. Katritzky AR, Kuzmierkiewicz W, Rachwal B, Rachwal S, Thomson J (1987) J Chem Soc Perkin Trans 1:811–817
196. Katritzky AR, Mazurkiewicz R, Stevens CV, Gordeev MF (1994) J Org Chem 59 (10):2740–2742
197. Katritzky AR, Denisenko A, Arend M (1999) J Org Chem 64(16):6076–6079
198. Katritzky AR, Ashraf AAAF, Tymoshenko DO, Samy AE (1999) Synthesis 12:2114–2118
199. Heinisch G, Kopelent-Frank H (1992) Prog Med Chem 29:141–183
200. Heinisch G, Frank H (1990) Prog Med Chem 27:1–49
201. Katritzky AR, Pleynet DPM, Yang B, Yao J (1998) Synthesis 11:1627–1631
202. Al-Omran F, Al-Awadl N, Yousef O, Elnagdi MH (2000) J Het Chem 37:167–170

203. Al-Mousawi S, Elassar A, El-Apasery MA (2006) Phosphorus, sulfur, silicon and the related elements. 181:1755–1771
204. Drandarov K, Guggisberg A, Hesse M (1999) Helv Chem Acta 82:229–237
205. Siddiqui, AQ, Merson-Davies L, Cullis PM (1999) J Chem Soc Perkin Trans I 3243–3252
206. Horvath D (1997) J Med Chem 40:2412–2423
207. Grandos R, Alverez M, Vails N, Salas M (1983) J Med Chem 20:1271–1275
208. Katritzky AR, Singh SK, He H (2002) J Org Chem 67:3115–3117
209. Katritzky AR, Toader D, Jiang J (1995) Org Prep Proc Int 27:179–184
210. Abdel-Fattah AAA (2003) Synthesis 15:2358–2363
211. Zapol'skii VA, Potkin VI, Nechai NI, Kaberdin RV, Pevzner MS (1997) Russian J Org Chem 33:1461–1468
212. Zapol'skii VA, Namyslo JC, Altug C, Gjikaj M, Kaufmann DE (2008) Synthesis 304–311
213. Katritzky AR, Wang Z, Lang H (1996) Het Commun 2:109–115
214. Steuber CP, Krischer J, Holbrook T, Camitta B, Land V, Sexauer C, Mahoney D, Weinstein H (1996) J Clin Oncol 14:1521–1525
215. Momparler RM, Bovenzi V (2000) J Cell Physiol 183:145–154
216. Sacchi S, Kantarjian HM, O'Brien S, Cortes J, Rios MB, Giles FJ, Beran M, Koller CA, Keating MJ, Talpaz M (1999) Cancer 86:2632–2641
217. Katritzky AR, Rogovoy BV, Vvedensky VY, Herbert N, Forood B (2001) J Org Chem 66:6797–6799
218. Katritzky AR, Qiu G, He H, Yang B (2000) J Org Chem 65:3683–3689
219. Katritzky AR, He H, Wang J (2002) J Org Chem 67:4951–4956
220. Al-Mousawi SM, El-Apasery MA (2009) Molecules 14:2976–2984
221. Katritzky AR, Levell JR, Li J (1996) Tetrahedron Lett 37:5641–5644
222. Katritzky AR, Watson CH, Dega-Szafran Z, Eyler JR (1990) J Am Chem Soc 112:2471–2478
223. Katritzky AR, Qiu G, Yang B, He H (1999) J Org Chem 64:7618–7621
224. Katritzky AR, Asian D, Shcherbakova IV, Chen J, Belyakov SA (1996) J Het Chem 33:1107–1114
225. Katritzky AR, Arend M (1998) J Org Chem 63:9989–9991

Top Heterocycl Chem (2016) 43: 235–284
DOI: 10.1007/7081_2015_198

Published online: 14 December 2015

Benzotriazole: Much More Than Just Synthetic Heterocyclic Chemistry

Kiran Bajaj and Rajeev Sakhuja

Abstract Benzotriazole could be considered as a premium synthetic auxiliary that has been extensively explored for the synthesis of natural and synthetic-based molecules of varied biological and pharmaceutical importance. Much more than heterocyclic compounds, its derivatives have exhibited outstanding properties in medicinal chemistry including anticancer, antifungal, antibacterial, antiviral, antiparasitic, and antioxidative activities. In addition, benzotriazole derivatives have found profound applications as corrosion inhibitors, UV filters, and materials for solar and photovoltaic cells. The aim of this review is to provide an overview of the applications of benzotriazole derivatives in medicinal chemistry and material sciences.

Keywords Anticancer • Anticorrosive • Anti-inflammatory • Antimicrobial • Antipsychotic • Benzotriazole • Material science • Medicinal chemistry • Photovoltaic cells • UV ray absorber

Contents

K. Bajaj (✉) and R. Sakhuja
Department of Chemistry, Birla Institute of Technology and Science, Pilani 333031, Rajasthan, India
e-mail: kbajaj26@gmail.com

1 Introduction

Benzotriazole (BtH), an extremely useful and synthetically explored auxiliary, has occasionally been described as a "tame halogen." Various heterocyclic compounds were synthesized by using this synthetic scaffold including benzoxazines, quinazolines, triazine triones, hydantoins, oxadiazines, diazepanes, and many other heterocyclic compounds. It also served for the edition of libraries of carbamates, ureas, semicarbazides, carbazides, sulfonylureas, sulfonylcarbazides, and oligopeptides. Benzotriazoles are also useful synthons for the Graebe–Ullmann reaction [1–9].

An enormous contribution toward the development of benzotriazole chemistry has been laid by Katritzky and coworkers. By developing a systematic benzotriazole methodology, Katritzky demonstrated the huge chemical versatility of benzotriazole for the synthesis of heterocyclic and non-heterocyclic frameworks such that it would be justified as an ideal synthetic auxiliary [10–12].

The medicinal importance of benzotriazole derivatives has been also deeply explored over the last decade. Benzotriazole derivatives have proved their potency to treat various kinds of conditions such as cancers, microbial infections, psychotropic disorders, and many more. The combined presence of a large conjugated system capable of forming π–π stacking interactions and hydrogen bond acceptors makes benzotriazole derivatives susceptible to bind with enzymes and receptors in biological systems. Such diverse non-covalent interactions endow benzotriazole derivatives with a broad spectrum of biological properties including plant growth regulator [13], choleretic [14], antibacterial [15], antiprotozoal [16], and antiviral activity [17].

Benzotriazole metal complexes possess the bioactivities of both the benzotriazole nucleus and metal supramolecular complexes and exert double action mechanism to overcome drug resistances [18]. For the above reasons, the benzotriazole moiety has been used to construct innovative drug molecules [19].

Apart from its medicinal importance, benzotriazole derivatives show various applications in industrial processes as well as in households, e.g., as corrosion inhibitors in dishwasher detergents and deicing/anti-icing fluids, as UV stabilizers in plastics, and as antifogging agents in photography [20, 21]. As UV stabilizers, benzotriazole derivatives have been added to various industrial products such as construction materials, automobile components, wax and painting products, adhesive agents, films, glasses, and sportswear to prevent yellowing and degradation

[22]. Details on the application of benzotriazole in medicinal chemistry and material sciences are discussed below.

2 Applications of Benzotriazole Derivatives in Medicinal Chemistry

The benzotriazole fragment is a constituent of many bioactive heterocyclic compounds endowed with diverse clinical applications. A variety of benzotriazole derivatives showed diverse biological activities such as antitumor, antibacterial, antifungal, antimalarial, antiviral, and anti-inflammatory activities. Being an important pharmacophore, the benzotriazole nucleus plays a critical role in medicinal agents that act as agonists for many proteins. Many benzotriazole-based biologically active heterocycles are under clinical trial such as vorozole **1** [23], a nonsteroidal aromatase inhibitor; alizapride **2** [24], a dopamine antagonist having prokinetic and antiemetic effects; and 1-isopropylbenzotriazole-5-carboxylic acid **3**, a selective small-molecule agonists of the human orphan G-protein-coupled receptor GPR109b (HM74) (Fig. 1) [25]. Among these, vorozole has been withdrawn from clinical development in the US Phase III trial despite having similar efficacy to other aromatase inhibitors [26].

2.1 Benzotriazole Derivatives as Antimicrobial Agents

One of the major advances in medical science over the two decades has been the emergence of novel antimicrobial agents. However, the widespread and often misuse of these antimicrobial agents has led to the emergence of drug-resistant microorganisms. Infection by such drug-resistant pathogens has become an important cause of morbidity and mortality worldwide. Therefore, a new class of antimicrobials is needed to fight against these microbes. Benzotriazoles have been extensively explored in this area. A certain number of benzotriazole derivatives having antimicrobial activity are in clinical trial and are discussed in the following subsections.

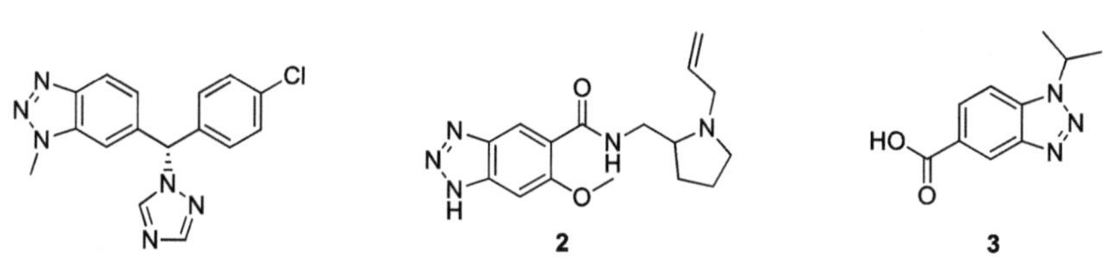

1 2 3

Fig. 1 Benzotriazole-based drugs under clinical trial

2.1.1 Benzotriazole Derivatives as Antibacterial Agents

The discovery of antibacterial agents to clinical medicine was one of the greatest medical achievements of the twentieth century. However, the gradual emergence of antibacterial-resistant pathogenic bacteria has become a major global health concern nowadays. Despite the diversity of clinically available antibiotics and chemotherapeutics such as tetracyclines, aminoglycosides, β-lactams, macrolides, polyenes, and other synthetic drugs including sulfonamides, quinolones, oxazolidinones, allylamines, etc., the development of new drugs for the treatment of bacterial infectious diseases still remains a challenging task [27]. With the advance knowledge on multidrug-resistant microbial pathogens such as methicillin-resistant *Staphylococcus aureus* (MRSA) and carbapenem-resistant *Enterobacteriaceae*, a thrust toward the development of new antibacterial agents acting through distinct mechanisms has been observed over the last few years. In this context, the antibacterial activity of benzotriazole derivatives has been extensively investigated and become an active highlight. Some benzotriazole derivatives are in clinical trials, among which levofloxacin acylbenzotriazole derivative **4** was found effective on Gram-positive bacteria (MRSA) and Gram-negative bacteria (*Proteus* spp.) with similar inhibition zones as compared to ofloxacine. The introduction of a 2-aminobenzotriazole group in levofloxacin improved its antibacterial behavior via interactions with the target enzymes or direct penetration into bacteria (Fig. 2) [28]. Another promising benzotriazole derivative is berberine benzotriazole derivative **5** that possesses a good inhibitory activity against Gram-negative bacteria *Shigella dysenteriae* ATCC51252. The minimum inhibitory concentration for **5** was equivalent to that of the standard drug chloromycin (MIC = 32 μg/mL) and was fourfold higher than that of berberine itself (MIC = 256 μg/mL) (Fig. 2) [29].

Linezolid **6**, a synthetic antibiotic, has been derivatized with a benzotriazole moiety to improve the efficacy of linezolid on some enterococcal microbes. The benzotriazole derivatives **7a–k** were tested against numerous MDR Gram-positive organisms (Fig. 3). Most notably, compounds **7a–k** were active against problematic

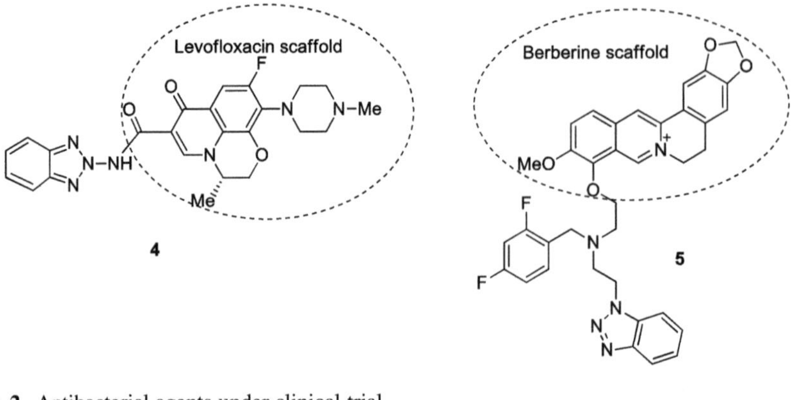

Fig. 2 Antibacterial agents under clinical trial

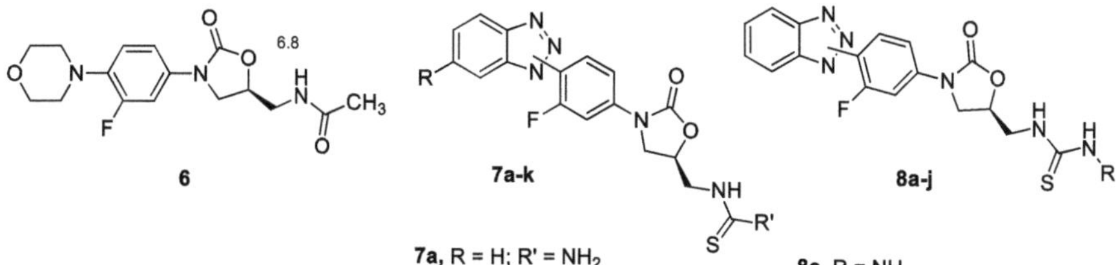

7a, R = H; R' = NH_2
7b, R = H; R' = $NH(CH_2)_2OH$
7c, R = H; R' = $NH(CH_2)_3OH$
7d, R = H; R' = $NH(CH_2)_2NEt_2$
7e, R = H; R' = OMe
7f, R = H; R' = OCH_2CF_3
7g, R = H; R' = $OCH_2CF_2CHF_2$
7h, R = H; R' = OCH_2CH_2OH
7i, R = H; R' = $OCH_2CH_2NMe_2$
7j, R = COOMe; R' = NH_2
7k, R = COOMe; R' = OMe

8a, R = NH_2
8b, R = NHCO-4-pyridyl
8c, R = $CH(Et)CH_2OH$
8d, R = $CH_2CH(OEt)_2$
8e, R = $CH_2CH_2NH_2$
8f, R = CH_2CH_2-2-pyridyl
8g, R = cyclopropyl
8h, R = CH_2CH_2-pyrrolidinyl
8i, R = CH_2CH_2-piperidinyl
8j, R = CH_2CH_2-morpholinyl

Fig. 3 Linezolid benzotriazole derivatives as antibacterial agents

pathogens including methicillin- and vancomycin-resistant *Staphylococcus aureus* (MRSA), vancomycin-resistant *Enterococci* (VRE), and penicillin- and cephalosporin-resistant *Streptococcus pneumoniae*. Compounds **7a–k** inhibit protein synthesis in the initial stage. Most of these analogues exhibited good to excellent antibacterial activity against sensitive and drug-resistant Gram-positive bacteria. Some of these benzotriazole homologues of linezolid such as the unsubstituted derivative **7a** possessed a superior in vitro activity in comparison to the parent molecule [30]. Similarly, 1-[3-(4-benzotriazol-1/2-yl-3-fluoro-phenyl)-2-oxo-oxazolidin-5-ylmethyl]-3-substituted-thiourea derivatives **8a–j** (Fig. 3) have been evaluated for their antimycobacterial activity against *Mycobacterium tuberculosis*, *M. avium*, and *M. intracellulare*. The in vitro activity of some derivatives was found to be equivalent or better than linezolid and isoniazid against sensitive and resistant strains of *M. tuberculosis*. Compound **8g** possesses excellent in vitro antimycobacterial activity against drug-sensitive and drug-resistant clinical isolates of *M. tuberculosis*, with MIC value equivalent to linezolid and superior to isoniazid against all these strains [31].

Besides the above useful benzotriazole agents, several other benzotriazole derivatives have been evaluated for their antibacterial activity by testing them against different bacterial strains. Major attention has been focused on the synthesis of substituted benzotriazoles using bioactive pharmacophores to develop new categories of antibacterial agents. For example, N^1-acredine-substituted benzotriazoles **9a–b** were reported to show remarkable inhibition of bacterial cultures and found less reactive than the standard drug ampicillin (Fig. 4) [32]. Benzotriazole substituted with β-lactams **10a–b** possessing nitro substitution at *ortho* and *para* positions and thiazolidinone-substituted benzotriazole derivatives **11a–b** displayed an antibacterial activity against *B. subtilis*, *E. coli*, *K. Pneumoniae*, *and S. aureus* bacterial strains comparable to the standard drug streptomycin (Fig. 4) [33, 34].

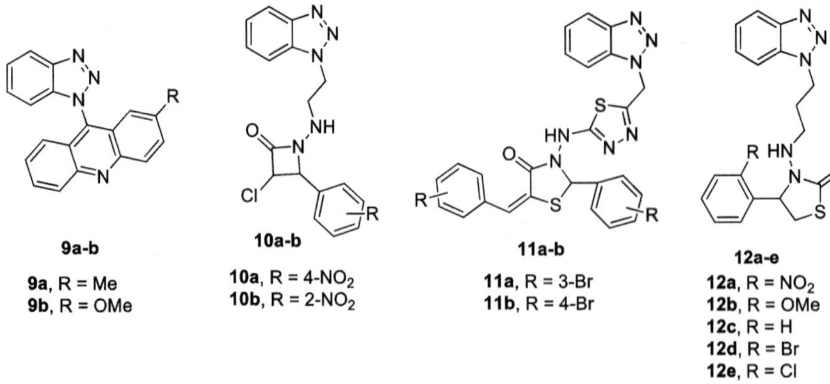

Fig. 4 Benzotriazole-derivatized heterocycles as antibacterial agents

Thiazolidinone benzotriazole derivatives **12a** and **12b** were also reported to have good antibacterial activities. The nitro-substituted derivative **12a** showed an equivalent activity against bacterial culture *S. aureus* comparable with the standard drug ofloxacin (MIC = 0.1 μg/mL), while the methoxy-substituted derivative **12b** showed a comparable activity to the standard drug miconazole (MIC = 0.1 μg/mL) against *A. niger* (Fig. 4) [35]. Unsubstituted aryl derivative **12c** came with a strongly decreased antibacterial activity as compared to substituted ones.

Another series of thiazole-substituted benzotriazoles **13a–y** was synthesized and tested against the standard strains of *Bacillus subtilis* (2250), *Staphylococcus aureus* (2079), *Escherichia coli* (2109), and *Pseudomonas aeruginosa* (2036) (MIC = 16–128 μg/mL) and for their antifungal activity against *Candida albicans* (3471) and *Aspergillus niger* (Fig. 5). Most of these compounds exhibited moderate to good antibacterial activities. An enhanced biological activity of few compounds was observed, particularly in the presence of F, Cl, and Br substituents on the aryl system, irrespective of their position in the molecule [36]. Similarly, triazol-substituted benzotriazoles **14** displayed comparable antibacterial activity (MIC = 1.56–6.25 μg/mL) against three Gram-positive bacterial strains (*Bacillus subtilis*, *Staphylococcus aureus*, and *Streptococcus faecalis*) and three Gram-negative bacterial strains (*Escherichia coli*, *Pseudomonas aeruginosa*, and *Enterobacter cloacae*) to that of the reference drugs kanamycin and penicillin (Fig. 5) [37]. The replacement of the bromo group in compound **14** by a methyl group reduced its anti-*B. subtilis* activity. The authors concluded that the introduction of bromo substituent may increase the hydrophobicity of the synthesized compounds, hence leading to an enhanced antibacterial activity.

The introduction of a benzotriazole fragment at C-7 position of a coumarin ring via an ether linkage afforded conjugated compounds **15** with a broad antibacterial spectrum against both Gram-positive and Gram-negative bacteria. Notably, compound **15a** displayed a twofold more active inhibition (MIC = 8 μg/mL) than the reference drug chloromycin (MIC = 16 μg/mL) against *Proteus vulgaris* ATCC 6896 as well as a similar antibacterial efficacy against *S. aureus* ATCC 25923 and

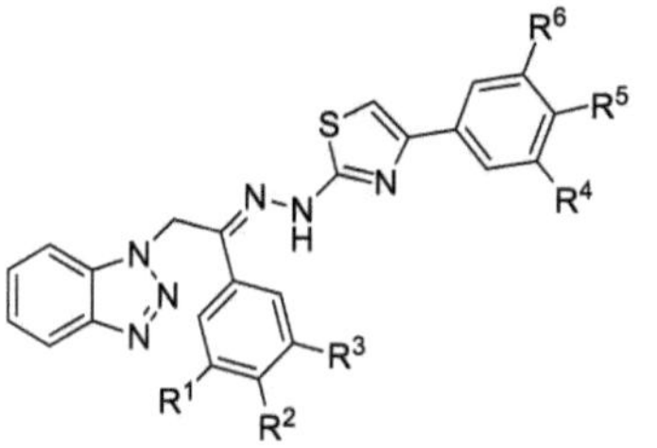

13a-y

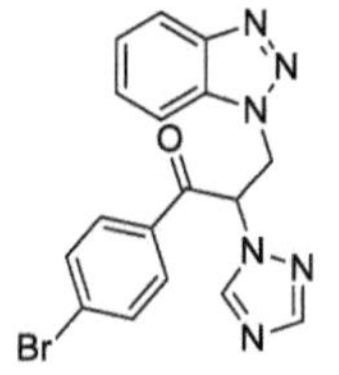

14

13a, R^1 = H, R^2 = F, R^3 = H, R^4 = H, R^5 = F, R^6 = H
13b, R^1 = H, R^2 = F, R^3 = H, R^4 = H, R^5 = Cl, R^6 = H
13c, R^1 = H, R^2 = F, R^3 = H, R^4 = H, R^5 = Br, R^6 = H
13d, R^1 = H, R^2 = F, R^3 = H, R^4 = H, R^5 = NO_2, R^6 = H
13e, R^1 = H, R^2 = F, R^3 = H, R^4 = CF_3, R^5 = H, R^6 = CF_3
13f, R^1 = H, R^2 = Cl, R^3 = H, R^4 = H, R^5 = F, R^6 = H
13g, R^1 = H, R^2 = Cl, R^3 = H, R^4 = H, R^5 = Cl, R^6 = H
13h, R^1 = H, R^2 = Cl, R^3 = H, R^4 = H, R^5 = Br, R^6 = H
13i, R^1 = H, R^2 = Cl, R^3 = H, R^4 = H, R^5 = NO_2, R_6 = H
13j, R^1 = H, R^2 = Cl, R^3 = H, R^4 = CF_3, R^5 = H, R^6 = CF_3
13k, R^1 = H, R^2 = Br, R^3 = H, R^4 = H, R^5 = F, R^6 = H
13l, R^1 = H, R^2 = Br, R^3 = H, R^4 = H, R^5 = Cl, R^6 = H
13m, R^1 = H, R^2 = Br, R^3 = H, R^4 = H, R^5 = Br, R^6 = H
13n, R^1 = H, R^2 = Br, R^3 = H, R^4 = H, R^5 = NO_2, R^6 = H
13o, R^1 = H, R^2 = Br, R^3 = H, R^4 = CF_3, R^5 = H, R^6 = CF_3
13p, R^1 = H, R^2 = NO_2, R^3 = H, R^4 = H, R^5 = F, R^6 = H
13q, R^1 = H, R^2 = NO_2, R^3 = H, R^4 = H, R^5 = Cl, R^6 = H
13r, R^1 = H, R^2 = NO_2, R^3 = H, R^4 = H, R^5 = Br, R^6 = H
13s, R^1 = H, R^2 = NO_2, R^3 = H, R^4 = H, R^5 = NO_2, R^6 = H
13t, R^1 = H, R^2 = NO_2, R^3 = H, R^4 = CF_3, R^5 = H, R^6 = CF_3
13u, R^1 = CF_3, R^2 = H, R^3 = CF_3, R^4 = H, R^5 = F, R^6 = H
13v, R^1 = CF_3, R^2 = H, R^3 = CF_3, R^4 = H, R^5 = Cl, R^6 = H
13w, R^1 = CF_3, R^2 = H, R^3 = CF_3, R^4 = H, R^5 = Br, R^6 = H
13x, R^1 = CF_3, R^2 = H, R^3 = CF_3, R^4 = H, R^5 = NO_2, R^6 = H
13y, R^1 = CF_3, R^2 = H, R^3 = CF_3, R^4 = CF_3, R^5 = H, R^6 = CF_3

Fig. 5 Thiazole- and triazole-substituted benzotriazoles **13** and **14** as antibacterial agents

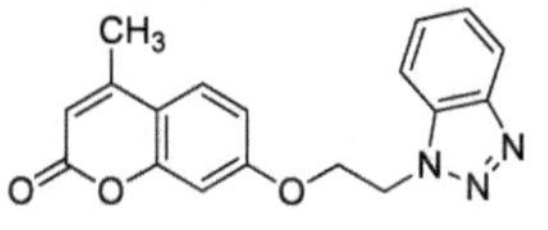

15a

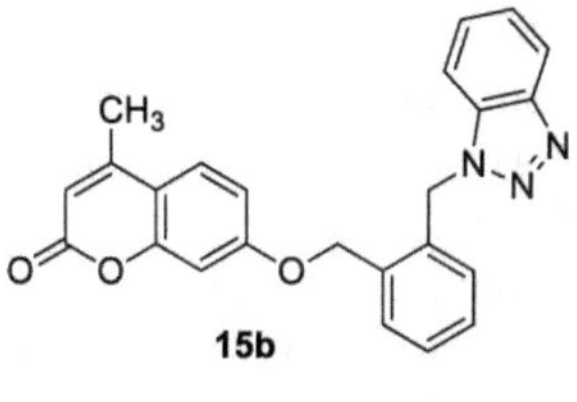

15b

Fig. 6 Coumaryl-substituted benzotriazoles **15a,b** as antibacterial agents

Micrococcus luteus ATCC 4698 to chloromycin (Fig. 6). Increasing the length of the linker led to compounds with decreased bacterial inhibitory activity. The replacement of the alkyl chain linker by a benzene ring yielded another coumaryl benzotriazole derivative **15b** that showed a decreased antibacterial activity as well. The combination of the most active compound **15a** and fluconazole could effectively inhibit the growth of *C. albicans*, *S. cerevisiae*, and *A. fumigatus* with a MIC value of 0.25 μg/mL that was eightfold or fourfold higher than the isolated compounds, as a result of synergistic effects (Fig. 6) [38].

Recently, *N*-aminoalkylated benzotriazoles **16a–j**, containing heterocyclic and aliphatic amines at the end of the lipophilic chain, were synthesized and tested for in vitro antimicrobial activities against three Gram-positive bacteria (*Staphylococcus aureus*, *Bacillus subtilis*, and *Staphylococcus epidermis*) and four Gram-

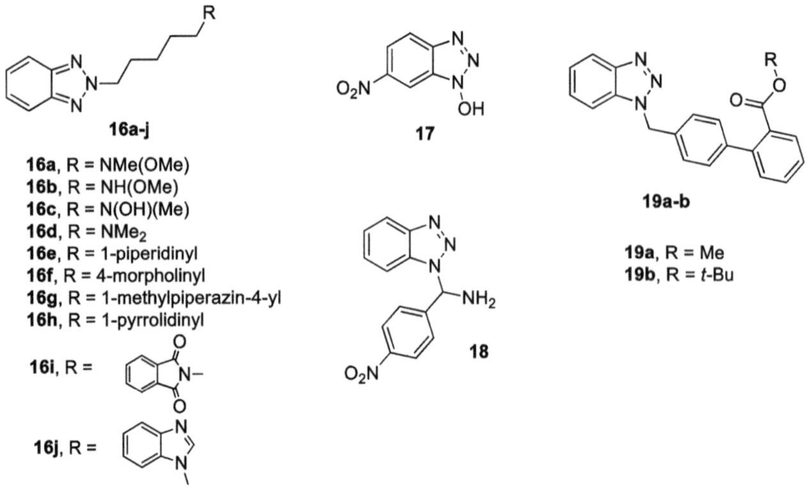

Fig. 7 *N*-substituted benzotriazoles **16–19** as antibacterial agents

negative bacteria (*Escherichia coli, Pseudomonas aeruginosa, Staphylococcus typhi, and Klebsiella pneumoniae*) (Fig. 7). Most of these compounds showed moderate to good activity against *Staphylococcus epidermis, Staphylococcus aureus, Pseudomonas aeruginosa, and Salmonella typhi* [39]. The antibacterial activity of the nitro-substituted *N*-hydroxybenzotriazole **17** was reported (Fig. 7). The clinical trials included six methicillin-resistant (MRSA), two methicillin-sensitive (MSSA) *S. aureus* strains, and one methicillin-resistant *S. epidermidis*, three *E. faecalis*, and two *E. faecium* strains. The range of MIC value for compound **17** observed for the aforementioned strains was ranging from 50 to 400 μg/mL [40]. Moreover, benzotriazolyl-substituted phenylmethanamines **18** exhibited a better inhibition activity against *B. subtilis* than the standard drug ampicillin. The exchange of the amino group by methyl or methoxy group reduced its antibacterial activity (Fig. 7) [41]. Other biphenyl-substituted benzotriazoles **19a–b** were found to possess strong antibacterial activity against various Gram-positive and Gram-negative microorganisms (Fig. 7) [42].

The introduction of hydrophilic groups such as amino, hydrazo, and azo groups usually enhances the antibacterial property of these benzotriazole derivatives since they enable the formation of hydrogen bonds. Besides, such structural modifications can also improve the solubility of the corresponding derivatives. Compounds **20a–p** displayed exceptionally high antibacterial activities. The most active compounds **20j** were those with a nitro substitution on phenyl groups (Fig. 8). The nitro group was believed to affect the charge distribution and to interact with some intercellular target, hence improving the antibacterial activity. Chlorophenyl-substituted compound **20o** also showed a better activity against *Bacillus cereus* (MTCC – 430) compared to the reference co-trimoxazole, whereas all other synthesized compounds have an antibacterial activity comparable to co-trimoxazole [43]. *N*-substituted 2-(1*H*-benzotriazol-1-yl)-acetohydrazide **21a–b** bearing a

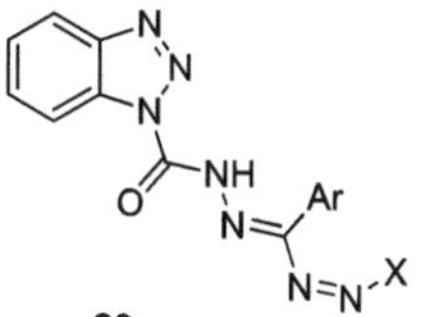

20a-p

20a, Ar = Ph, X = Ph
20b, Ar = Ph, X = $C_6H_4NO_2$
20c, Ar = Ph, X = C_6H_4Cl
20d, Ar = Ph, X = C_6H_4Br
20e, Ar = C_4H_3O, X = Ph
20f, Ar = C_4H_3O, X = $C_6H_4NO_2$
20g, Ar = C_4H_3O, X = C_6H_4Cl
20h, Ar = C_4H_3O, X = C_6H_4Br
20i, Ar = $C_6H_4NO_2$, X = Ph
20j, Ar = $C_6H_4NO_2$, X = $C_6H_4NO_2$
20k, Ar = $C_6H_4NO_2$, X = C_6H_4Cl
20l, Ar = $C_6H_4NO_2$, X = C_6H_4Br
20m, Ar = C_6H_4Cl, X = Ph
20n, Ar = C_6H_4Cl, X = $C_6H_4NO_2$
20o, Ar = C_6H_4Cl, X = C_6H_4Cl
20p, Ar = C_6H_4Cl, X = C_6H_4Br

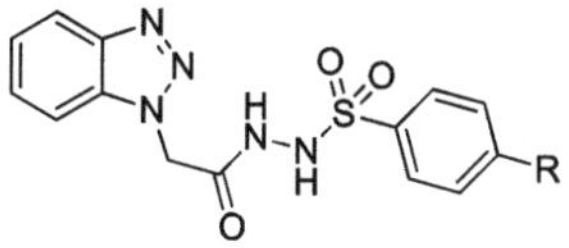

21a-b

21a, R = Me
21b, R = Cl

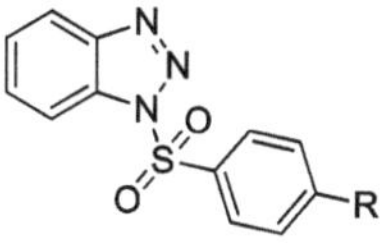

22a-b

22a, R = H
22b, R = $NHCOCH_3$

Fig. 8 N^1-substituted benzotriazoles as antibacterial agents

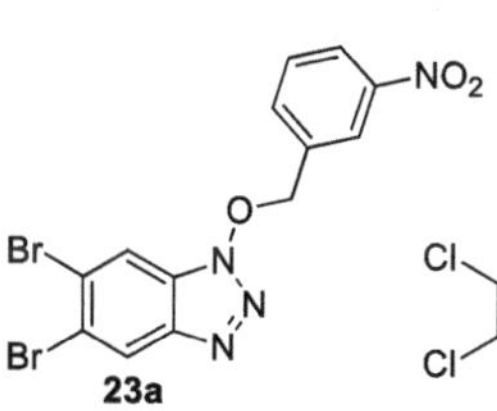

23a

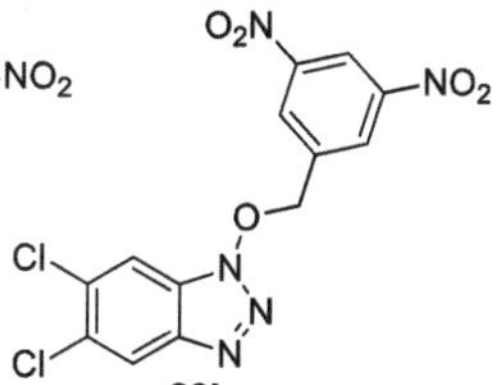

23b

Fig. 9 Nitrobenzyloxy-substituted 1*H*-benzotriazole derivatives **23a,b** as antibacterial agents

sulfonylhydrazino moiety displayed good antibacterial activity, although they were less potent than reference drug sulfacetamide (Fig. 8) [44]. Benzenesulfonyl-substituted benzotriazoles **22a–b** have also been reported to have high antibacterial activity against *E. coli* and *Staphylococcus aureus* compared to sulfathiazole. At the tested concentration, these derivatives did not inhibit the growth of the dermatophytes strains (250 μg/mL) (Fig. 8) [45].

Nitrobenzyloxy-substituted benzotriazoles **23a–b** were tested against four *Mycobacterium* strains. These 5,6-dibromo- and 5,6-dichlorosubstituted 1*H*-benzotriazole derivatives **23a–b** bearing nitrobenzyloxy groups were found considerably active against the reference strain H37Rv, their MIC values being similar to that of the reference isoniazid (Fig. 9) [46].

Few benzotriazole-based metal complexes were also reported with antibacterial effects. The Cu(II) complex of a 5-aminosalicylic acid benzotriazole derivative **24** exerted moderate inhibitions against both Gram-positive bacteria (*Bacillus subtilis* and *Staphylococcus aureus*) and Gram-negative bacteria (*Escherichia coli* and *Salmonella typhi*). Results showed that variation in structure on coordination

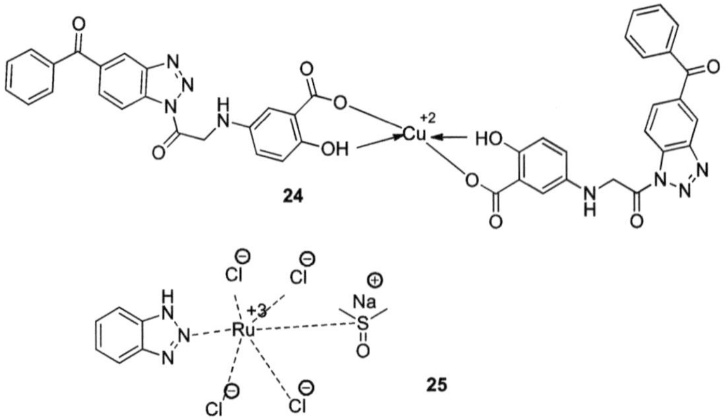

24

25

Fig. 10 Benzotriazole-based metal complexes as antibacterial agents

affects the growth of microorganism by reducing the toxicity of metal ions toward some organisms (Fig. 10) [47]. Benzotriazole-ruthenium (III) complex **25** possessed an efficient activity against Gram-negative *Escherichia coli* in comparison to either benzotriazole or the parent ruthenium precursors (Fig. 10) [48]. The increased lipophilicity of this complex reduced the permeability barriers of the cells and retarded the normal cell process of bacteria, thus resulting in enhanced antibacterial activity.

2.1.2 Benzotriazole Derivatives as Antifungal Agents

Fungal infections are quiet common diseases, and the incidence of systemic fungal infections has been increased gradually in immunocompromised hosts such as patients who have undergone complex organ transplants surgeries. During anticancer chemotherapy, organ transplants, or long treatment with antimicrobial agents, patients become immune suppressed and susceptible to life-threatening systemic fungal infections such as candidiasis, cryptococcosis, and aspergillosis. There is an increasing need to develop strong antifungal agents with an improved efficiency, less toxicity, and an increased bioavailability, with the growing emergence of the intrinsic and acquired antifungal resistance caused by the abuse of available drugs.

Among the different kinds of antifungal agents, a number of azole-based molecules are used as clinical drugs including fluconazole, ketoconazole, and clotrimazole, which effectively inhibit the growth of fungal infection. Their mode of action involves replacing lanosterol from cytochrome P45014αDM and blocking the biosynthesis of ergosterol. This further destroys the integrity of the fungal cell wall and inhibits the growth and breeding of fungi [49]. These antifungal drugs act through a mechanism in which the heterocyclic nitrogen atom (*N*-3 of imidazole or *N*-4 of triazole) binds to the heme iron atom in cytochrome P450 enzymes. Particularly, triazole derivatives possess a greater affinity for fungal cytochrome

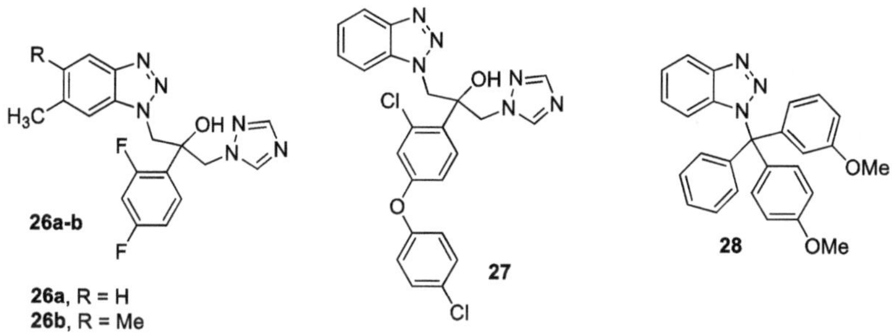

Fig. 11 Benzotriazole derivatives of fluconazole and clotrimazole

P450 enzymes than for the mammalian ones, thus contributing to a favorable safety profile [50].

Similarly to triazoles or imidazole, benzotriazole has a conjugated system that could bind more readily with various receptors in organisms. Thus, a benzotriazole ring has been introduced into fluconazole to improve its bioactivity. Fluconazole benzotriazole derivative **26** showed an increased antifungal activity against *Candida glabrata* (MIC value = 25 μg/mL) and was found more potent than fluconazole (Fig. 11) [51]. Structure–activity relationships suggested that a small hydrophobic such as a methyl group on the benzotriazole ring at the 5- or 6-position enhanced its activity. Another fluconazole benzotriazole derivative **27** inhibited the growth of *C. arachidicola* with an improved efficiency (62.7%) by comparison with the commercial fungicide difenoconazole. However, the replacement of the benzotriazole fragment by other molecular fragments bearing smaller groups such as alkylamino, alkoxy, triazolyl, or substituted benzyl substituents also inhibited the oxidative removal of sterol C(*14*) methyl groups by the cytochrome P450 enzyme, thus enhancing the antifungal activity (Fig. 11) [52, 53].

Clotrimazole is another clinically used antifungal agent as an effective inhibitor of lanosterol 14α-demethylase (cytochrome P45014αDM). Benzotriazole-substituted clotrimazole analogue **28** displayed equivalent anti-*Trichophyton rubrum* activity to the reference fluconazole (MIC = 32 μg/mL). However, the removal of the methoxy group resulted in the decreased antifungal efficiency (Fig. 11) [54].

The combination of multiple functional groups with different mode of action onto one single molecular scaffold could produce better antifungal activity. Thus, a series of 1-(benzoyloxy)benzotriazoles **29a–h**, 1-(benzoyloxy)-6-nitrobenzotriazoles **29r-z**, and 1-(benzoyloxy)-6-chloro benzotriazoles **29j–q** have been evaluated for their antifungal activity by Dabhade and Jain (Fig. 12). It was observed that the nitro-substituted derivatives **29o**, **29w**, and **29z** showed an improved activity comparable to ketoconazole on *E. floccosum* [55]. However, none of the derivatives were found to be effective against *Malassezia furfur*. Excellent antifungal activity has been exhibited by 1-(1*H*-benzo[*d*][1,2,3]triazole-1-carbonyl) derivatives **20a–p** against *Candida albicans* (MTCCe 3018) (Fig. 12). The compounds were inactive against *Aspergillus niger*, *Aspergillus flavus*, and *Saccharomyces*

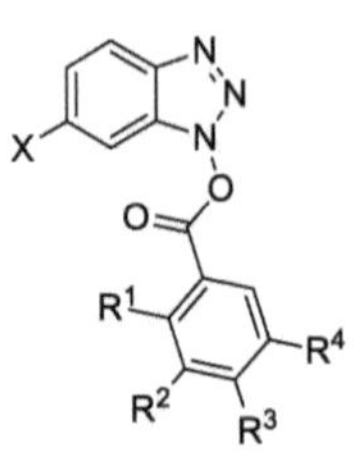

29a-z

29a, X = H, R^1, R^2, R^3, R^4 = H
29b, X = H, R^1, R^2, R^4 = H, R^3 = NO_2
29c, X = H, R^1, R^3, R^4 = H, R^2 = NO_2
29d, X = H, R^1 = Cl, R^2, R^3, R^4 = H
29e, X = H, R^1, R^2, R^4 = H, R^3 = Cl
29f, X = H, R^1, R^3 = H, R^2, R^4 = NO_2
29g, X = H, R^1, R^3, R^4 = H, R^2 = Me
29h, X = H, R^1, R^3, R^4 = H, R^2 = OMe
29i, X = Cl, R^1, R^2, R^3, R^4 = H
29j, X = Cl, R^1 = Me, R^2, R^3, R^4 = H
29k, X = Cl, R^1, R^2, R^4 = H, R^3 = NO_2
29l, X = Cl, R^1, R^3, R^4 = H, R^2 = NO_2
29m, X = Cl, R^1 = Cl, R^2, R^3, R^4 = H
29n, X = Cl, R^1, R^2, R^4 = H, R^3= Cl
29o, X = Cl, R^1, R^3 = H, R^2, R^4 = NO_2
29p, X = Cl, R^1, R^3, R^4 = H, R^2 = Me
29q, X = Cl, R^1, R^3, R^4 = H, R^2 = OMe
29r, X = NO_2, R^1, R^2, R^3, R^4 = H
29s, X = NO_2, R^1, R^2, R^4 = H, R^3 = NO_2
29t, X = NO_2, R^1, R^3, R^4 = H, R^2 = NO_2
29u, X = NO_2, R^1 = Cl, R^2, R^3, R^4 = H
29v, X = NO_2, R^1, R^2, R^4 = H, R^3 = Cl
29w, X = NO_2, R^1, R^3 = H, R^2, R^4 = NO_2
29x, X = NO_2, R^1 = Me, R^2, R^3, R^4 = H
29y, X = NO_2, R^1, R^3, R^4 = H, R^2 = Me
29z, X = NO_2, R^1, R^3, R^4 = H, R^2 = OMe

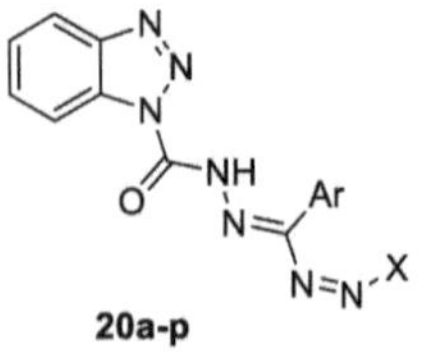

20a-p

20a, Ar = Ph, X = Ph
20b, Ar = Ph, X = $C_6H_4NO_2$
20c, Ar = Ph, X = C_6H_4Cl
20d, Ar = Ph, X = C_6H_4Br
20e, Ar = C_4H_3O, X = Ph
20f, Ar = C_4H_3O, X = $C_6H_4NO_2$
20g, Ar = C_4H_3O, X = C_6H_4Cl
20h, Ar = C_4H_3O, X = C_6H_4Br
20i, Ar = $C_6H_4NO_2$, X = Ph
20j, Ar = $C_6H_4NO_2$, X = $C_6H_4NO_2$
20k, Ar = $C_6H_4NO_2$, X = C_6H_4Cl
20l, Ar = $C_6H_4NO_2$, X = C_6H_4Br
20m, Ar = C_6H_4Cl, X = Ph
20n, Ar = C_6H_4Cl, X = $C_6H_4NO_2$
20o, Ar = C_6H_4Cl, X = C_6H_4Cl
20p, Ar = C_6H_4Cl, X = C_6H_4Br

Fig. 12 Benzotriazole-based antifungal agents **20, 29**

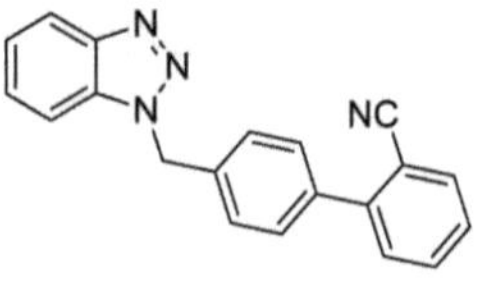

30

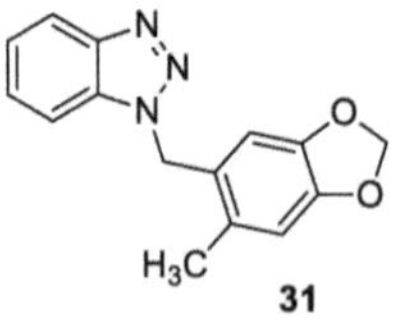

31

Fig. 13 *N*-alkylated benzotriazoles **30, 31** as antifungal agents

cerevisiae. A typical MIC value of these compounds was 62.5 μg/mL. A substantial improvement in their antifungal activity was observed by anchoring phenyl groups or phenyl groups bearing electron-withdrawing substituents [56].

N-Alkylated benzotriazole derivatives **30–31** bearing biologically active pharmacophores such as cyano-biphenyl, 6-methyl-benzo[1,3]-dioxolyl exhibited better antifungal activity against *Aspergillus niger*, *Aspergillus flavus*, *Fusarium oxysporum*, and *Trichoderma* species with reference to the tested drug nystatin. The presence of bulky hydrophobic groups such as a cyano-biphenyl moiety in **30** and a benzodioxole moiety in **31** accounts for the inhibitory activity of these compounds (Fig. 13) [57].

The fusion of two biologically active parts may enhance the bioactivity of a given compound. Thus several heterocycles were attached to the benzotriazole fragment and tested for their bioactivity such as pyridine benzotriazole derivatives

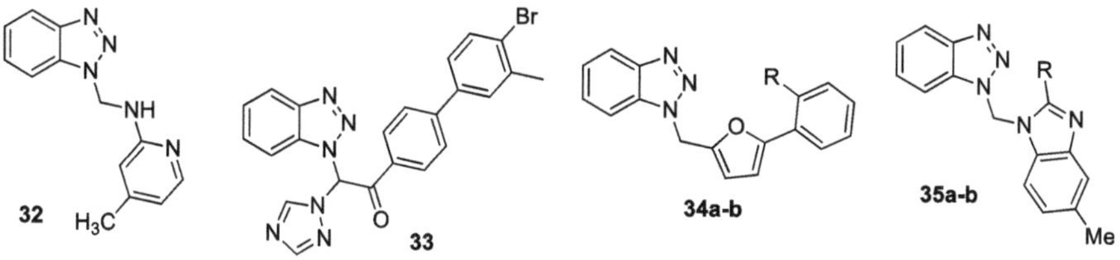

32 33 34a-b 35a-b

34a, R = H
34b, R = Cl

35a, R = Et
35b, R = 3-pyridyl

Fig. 14 Benzotriazole-derivatized heterocycles as antifungal agents

32, which displayed a strong antifungal action on plant pathogenic fungi such as *Botryodiplodia* spp. *Alternaria tennis*, *Heleminthosporium turicum*, *Fusarium oxysporum*, and *F. moniliforme*. The same study revealed that the derivative bearing a methyl group at 4th position on the pyridine ring gave an improved fungal inhibition than the derivative substituted at the 3rd position (Fig. 14) [58]. Moreover, triazole-substituted benzotriazole derivative **33** was reported with antifungal and plant growth regulatory activities (Fig. 14) [59]. Oxadiazole benzotriazole derivatives **34a–b** exhibited good antifungal activity against *C. albicans*, *A. niger*, and *A. flavus* with larger zones of inhibition than the standard drug streptomycin (Fig. 14). The chloro-substituted derivative **34b** showed better antifungal efficiency against all tested strains [60]. Benzimidazole benzotriazole derivatives **35a–b** were reported to show good inhibition profile against a broad range of fungi, such as *P. oryzae*, *B. cinerea*, *A. niger*, *C. albicans*, and *T. rubrum*. The ethyl-substituted derivative **35a** displayed a better inhibition against *P. oryzae* than the commercial fungicide griseofulvin. The replacement of the ethyl group by a 3-pyridyl moiety gave compound **35b**, which displayed an antifungal activity against *B. cinerea* comparable to griseofulvin [61].

Hydantoin are widely used to treat fungous infections due to their low cytotoxicity and advanced antifungal activity. Benzotriazole-modified hydantoin **36** was reported with antifungal activities equivalent to the standard drug griseofulvin and with comparable zone inhibitions against the tested strains *C. albicans* and *A. Niger* (Fig. 15) [62]. Further, pyrazilidin-3,5-dione-substituted benzotriazole derivatives **37** also exhibited antifungal activity against *C. albicans* comparable to the standard drug clotrimazole (Fig. 15). The combination of a pyrazolidine-3,5-dione moiety with a chlorobenzotriazole fragment contributed significantly to the antifungal activity [63].

Triazole-based derivatives are versatile ligands for a variety of metal cations as they could provide multi-coordinated sites to link more metal centers. They also show excellent π–π stacking interactions with other aromatic rings and are thus capable of generating multinuclear complexes or polymers. The Ag(I)-(triazol-1-ylmethylbenzotriazole complex **38** displayed 86.1% inhibitory effect against *Physalospora piricola* as well as good antifungal activity against other strains including *Gibberella zeae*, *Fusarium oxysporum*, *Cercospora arachidicola*, and *Alternaria solani* (Fig. 16) [64]. It was proposed that the benzotriazole ring affected the coordination and antifungal activity of this metal complex. The substituted

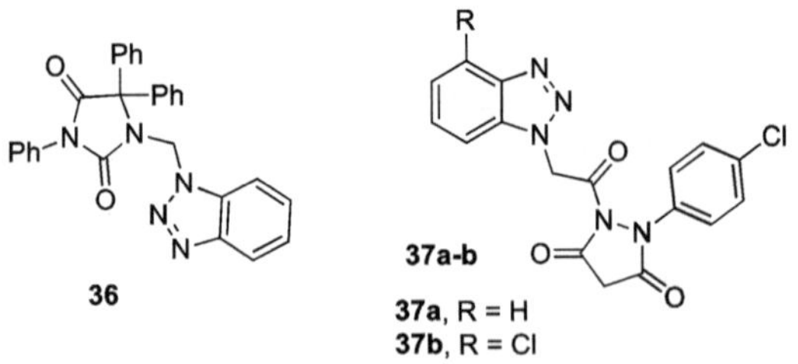

Fig. 15 Hydantoin- and pyrazolidinone-derivatized benzotriazole **36**, **37** as antifungal agents

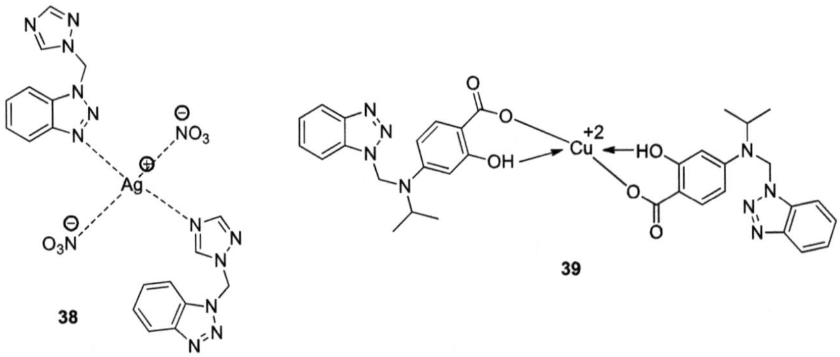

Fig. 16 Benzotriazole-based metal complexes **38–39** endowed with antifungal activity

4-aminosalicylic acid benzotriazole derivative Cu(II) complex **39** displayed antifungal activities against *Penicillium expansum*, *Botryodiplodia theobromae*, *Nigrospora* sp., *Trichothecium* sp., and *Rhizopus nigricans*. The fungal inhibitory action was reduced when the Cu^{2+} ion was replaced with other transition metal ions such as Mn^{2+}, Co^{2+}, Ni^{2+}, or Zn^{2+} ions (Fig. 16) [65].

2.1.3 Benzotriazole Derivatives as Antiparasitic Agents

Parasitosis is a kind of epidemic disease which is considered to be associated with infectious parasite present in multiform, including helminthiasis and protozoiasis. Amebiasis, caused by the protozoan parasite *Entamoeba histolytica*, is responsible for large numbers of deaths and infections in humans worldwide.

5-Chlorobenzotriazole **40** possesses a low micromolar activity ($IC_{50} = 0.339$ μg/mL) and is thus more active than the reference metronidazole for the treatment of amebiasis [16]. Similarly, *N*-benzenesulfonyl benzotriazole **41** showed a good inhibitory activity ($IC_{50} = 21.56$ μg/mL) against epimastigotes of *Trypanosoma cruzi*, whereas the standard benzotriazole exhibited no inhibition on the growth of this parasite form (Fig. 17) [66].

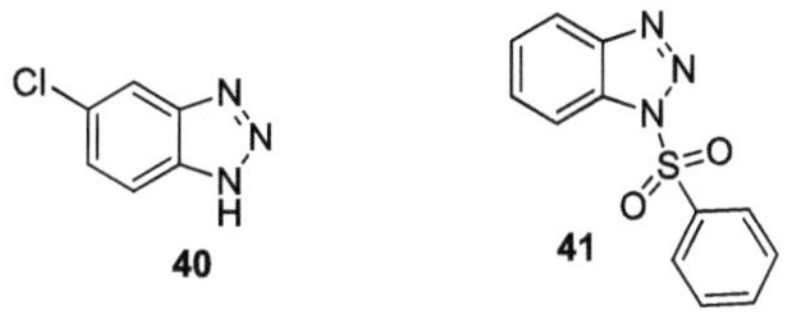

Fig. 17 Benzotriazole derivatives as antiparasitic agents

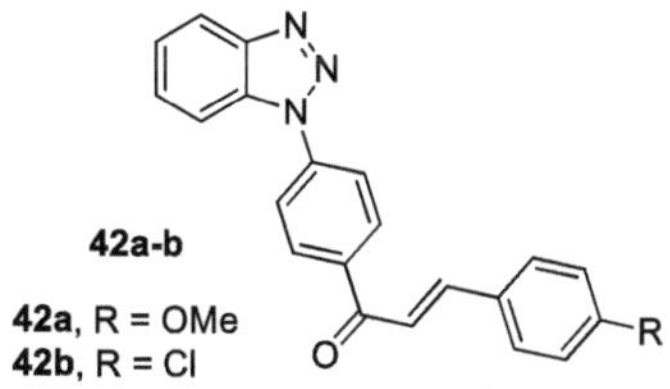

Fig. 18 N^1-substituted benzotriazoles as active antiparasitic agents

The absence of any safe and effective drug to combat adult human filarial worms has made filarial parasite a major medicinal challenge. Due to the presence of strong antioxidative system in filarial parasites, these parasites get protected from the reactive oxygen species (ROS) produced by the immune cells of the hosts. The combination of a chalcone pattern with a benzotriazole fragment such as in compounds **42a,b** showed a significant suppression in antifilarial activity on *Setaria cervi* using glutathione-*S*-transferase (GST) as a drug target against adult female parasite at a concentration of 3 μg/mL. Compounds **42** exhibited major irreversible effects on bioactivity, thus resulting in the death of parasite. They inhibited the GST activity with the percentage of 84–100% in vitro (Fig. 18) [67].

2.1.4 Benzotriazole Derivatives as Antiviral Agents

Viral infections are serious threat to human health as these are responsible for 60% of epidemic infectious diseases. Traditional nucleosides are used as drugs to treat viral infections. However, the structural modifications of nucleosides face major challenges such as poor solubility in common organic solvents [68]. Recently discovered antiviral agents can not only inhibit the growth of virus instead of directly destroying and killing them but also damage the host cell. A large number of investigations have been therefore focused on the design and development of non-nucleoside compounds as antiviral agents.

Hepatitis C is a severe liver disease that could lead to cirrhosis, liver cancer, or liver failure. This is caused by the hepatitis C virus (HCV). Thiourea-substituted benzotriazole derivative **43** inhibited HCV subgenomic replication with a moderate efficiency ($IC_{50} > 50$ μg/mL) (Fig. 19). This compound showed lower cytotoxicity and better pharmacokinetic activities than other aryl thiourea derivatives [69].

Sakthi et al. reported examples of 4-(3*H*)-quinazoline that revealed potential antiviral characteristics, especially against HIV-1 (IIIB) and HIV-2 (ROD) in MT-4 cells. The benzotriazole-substituted quinazoline derivative **44** exhibited an improved antiviral activity against IIIB ($CC_{50} = 61.33$ μg/mL) by comparison

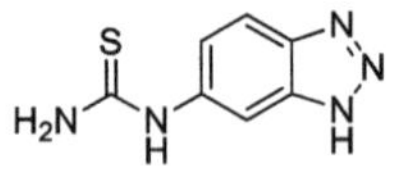

43

Fig. 19 Thiourea-based benzotriazole derivative as powerful agent against hepatitis C virus

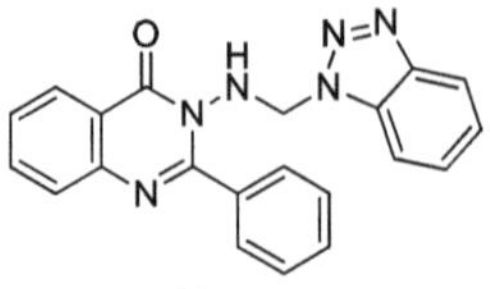

44

Fig. 20 Quinazolinone benzotriazole derivative **44** as HIV inhibitor

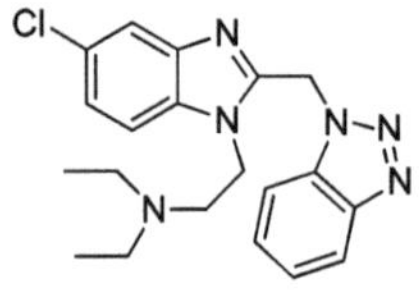

45

Fig. 21 Benzimidazole-derivatized benzotriazole as RSV inhibitor

with the standard drug AZT ($CC_{50} = 72.00$ μg/mL) (Fig. 20). Compound **44** also exhibited good cytotoxic effect against other viruses such as HSV-I, HSV-II, parainfluenza-3, Coxsackievirus B4, and Punta Toro virus [70].

Benzimidazole-substituted benzotriazole **45** was found to have a significant antiviral effect on respiratory syncytial virus (RSV) with an EC_{50} value of 0.1 μg/mL and was more effective than the reference drug azauridine ($EC_{50} = 1.2$ μg/mL) (Fig. 21) [71]. This compound was proved to be a potent RSV inhibitor. The presence of benzotriazole was assumed to reduce the cytotoxicity and contribute to its high selectivity index.

N,N'-Bis[4-(1*H*(2*H*)-benzotriazol-1(2)-yl)phenyl]alkyldicarboxamides **46** and **47** have been screened for in vitro cytotoxicity and antiviral activity against various viruses representative of *Picornaviridae*, such as *Enterovirus* Coxsackie B2 (CVB-2), and Polio (Sb-1) and few genera of the *Flaviviridae* such as bovine viral diarrhea virus (BVDV) and yellow fever virus (YFV). Compound **46** was found to be inactive, whereas compound **47** exhibited a good activity against the *Enteroviruses* with EC_{50} ranging from 7 to 11 μM against CVB-2 and 19–52 against Sb-1 (Fig. 22). Interestingly, selective activity against CVB-2 ($EC_{50} = 4$–11 μM) was observed for some bis-5,6-dichloro-benzotriazol-2-yl substituted derivatives, while these derivatives were found inactive against all the other viruses screened [72].

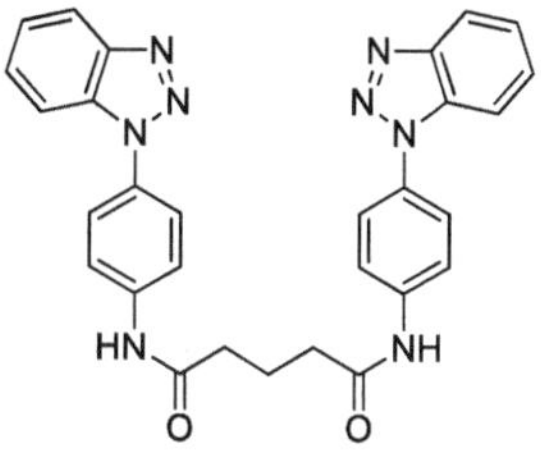

46

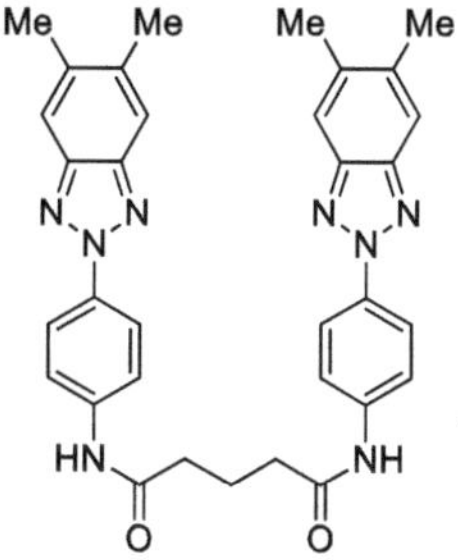

47

Fig. 22 Antiviral *N,N'*-bis[4-(1*H*(2*H*)-benzotriazol-1(2)-yl)phenyl]alkyldicarboxamide derivatives

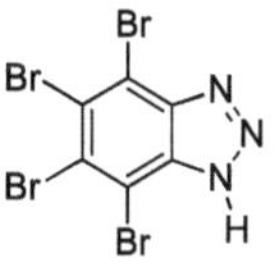

48

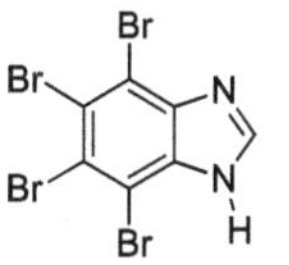

49

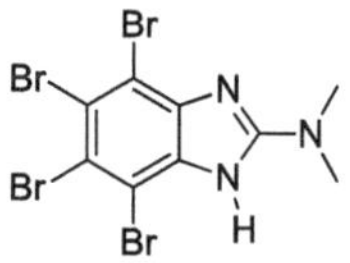

50

Fig. 23 CK2 inhibitors

2.2 *Benzotriazole Derivatives as Anticancer Agents*

Cancer is the second leading cause of death after cardiovascular diseases, accounting for about 8 million deaths per year worldwide [73]. There are different therapeutic approaches to treat cancer including surgical treatment, radiation therapy, immunotherapy, or chemotherapy. Polychemotherapy implies the use of several different drugs, which would ideally operate selectively toward cancerous cells, thus saving the host healthy cells. Nowadays, a variety of anticancer drugs are in clinical use such as alkylating agents, platinum complexes, porphyrin drugs, and azole agents [74]. However, most of the clinical anticancer drugs are often toxic to normal tissues and cause various severe side effects, which, in turn, limit the treatment efficacy. Long-term effectiveness is also limited by dose-related cumulative cardiotoxicity as well as drug resistance [75, 76].

Several benzotriazole derivatives have been found to possess potent anticancer activity, for example, the antineoplastic agent vorozole **1** (Fig. 1), and 4,5,6,7-tetrabromobenzotriazole (TBB) **48** (Fig. 23) is a selective inhibitor of protein kinase CK2 and thus a potent anticancer agent [77].

A variety of cellular functions in gene expression, signal transduction, proliferation, and cell survival are implicated by CK2 (casein kinase 2) type of protein kinases; these kinase inhibition is one of the important strategies to treat cancers [78]. Several classes of CK2 inhibitors are known to be effective in low micromolar ranges. Among these compounds, 4,5,6,7-tetrabromo-1*H*-1,2,3-benzotriazole (TBB) **48**, 4,5,6,7-tetrabromo-1*H*-benzimidazole (TBBi) **49** and 4,5,6,7-tetrabromo-1*H*-benzimidazole-2-*N,N*-dimethylamine (DMAT) **50** are well-known CK2 inhibitors (Fig. 23) [77, 79, 80].

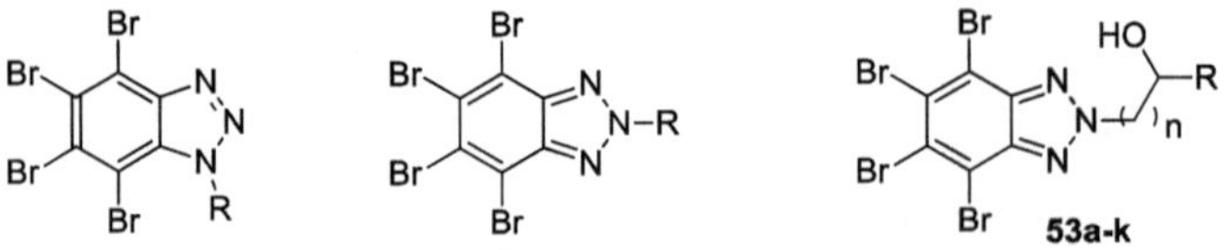

51a-e

51a, R = C_2H_4OH
51b, R = C_2H_4Br
51c, R = C_3H_6OH
51d, R = C_3H_6Cl
51e, R = C_4H_8OH

52a-e

52a, R = C_2H_4OH
52b, R = C_2H_4Br
52c, R = C_3H_6OH
52d, R = C_3H_6Cl
52e, R = C_4H_8OH

53a-k

53a, n = 2, R = Ph
53b, n = 2, R = 4-BrC_6H_4
53c, n = 2, R = 4-ClC_6H_4
53d, n = 2, R = 4-FC_6H_4
53e, n = 2, R = 3-BrC_6H_4
53f, n = 2, R = 4-ClC_6H_4
53g, n = 2, R = 4-$OMeC_6H_4$
53h, n = 1, R = Me
53i, n = 2, R = Me
53j, n = 1, R = Et
53k, n = 1, R = CH_2OH

Fig. 24 TBB-based CK2 inhibitors

Numerous TBB derivatives **51–52** have been synthesized and examined for their inhibitory effects on human protein kinase CK2 (Fig. 24). It was observed that instead of bromo-substitution, chloro- and methyl-substituted derivatives showed a lower inhibitory effect. In compounds **51** and **52**, the inhibitory action also strongly depends upon the length of the alkyl substituents. Compound **51c** (IC_{50} 0.32 ± 0.15 μM) and **52c** (0.34 ± 0.12 μM) exhibited higher potency against the catalytic subunit of CK2a than TBB (**48**, 0.50 ± 0.07 μM) [81]. Other TBB derivatives such as compounds **53a–g** bearing a single stereogenic center have also been reported and examined for their effect on human protein kinase CK2a (Fig. 24). The aliphatic chiral derivatives **53h–j** and **53k** showed IC_{50} values within the range 0.8–7.2 mM. Compound **53k** was almost equipotent with (*S*)-**53i** and **53j** (IC_{50} in 2.17–2.50 mM range). Considerably lower inhibitory activity, with an IC_{50} values in the range 100–300 μM was observed for all aromatic-substituted derivatives **53a–g**. The lowest IC_{50} value (0.80 mM) was obtained for enantiopure (*R*)-**53i** and found fourfold more active than its (*S*)-enantiomer (*S*)-**53i** [82].

The same group further reported other analogues of TBB (compounds **54a–g** and **55** in Fig. 25) by substituting the hydrophobic bromine atoms with ethyl, carboxyl, or ethylphenylcarboxamide fragments. Compounds with carboxyl substituent at C4 or C5 displayed a major decrease in affinity for CK2, while the methyl-substituted derivatives **54c** showed only a minor decrease in inhibitory potency in comparison to TBB **48**. C4 ethyl-substituted derivative **54e** showed a small improvement in CK2 inhibitory activity (IC_{50} 0.16 μM) [83].

A series of triazolyl-linked TBB derivatives **56a–e** and **57a–i** have also been designed to bind with ATP and to interact simultaneously with the Mg^{2+}-chelating residues and the protein substrate binding residues (Fig. 25). These derivatives were tested against CK2 inhibitor action by molecular docking studies. It showed that the presence of an amine functionality in some derivatives induced additional polar interactions with aspartic acid units and that the presence of a carboxylic functionality interacted with lysine and asparagine units. In a few cases, a hydrogen bond between a serine residue from the phosphate interacting loop and one of the

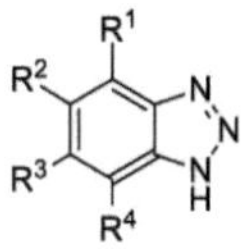

54a-h

54a, R^1 = Br, R^2= Me, R^3, R^4 = Br
54b, R^1 = Br, R^2 = COOH, R^3, R^4 = Br
54c, R^1 = Me, R^2, R^3, R^4 = Br
54d, R^1 = COOH, R^2, R^3, R^4 = Br
54e, R^1 = Et, R^2, R^3, R^4 = Br
54f, R^1 = Me, R^2= H, R^3 = Br, R^4 = H
54g, R^1 = $CONHC_6H_3$(2-Et), R^2 = H, R^3 = Br, R^4 = H

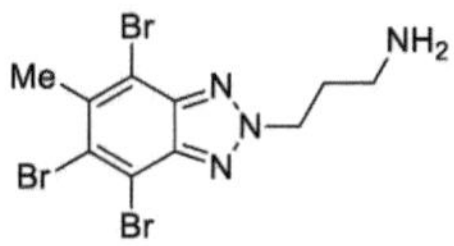

55

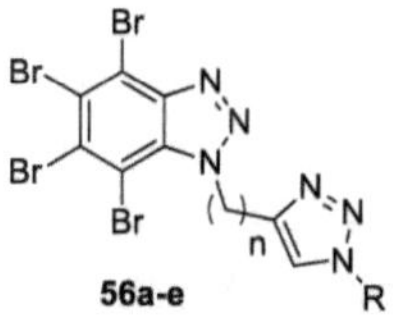

56a-e

56a, n= 1, R = $(CH_2)_3CONH(CH_2)_2CH(CH_3)_2$
56b, n= 1, R = $(CH_2)_3CONHCH_2C_6H_5$
56c, n= 1, R = $(CH_2)_3CONHCh(CH_2CH_2)_2NCH_2C_6H_5$
56d, n= 2, R = $(CH_2)_3CONH(CH_2)_2CH(CH_3)_2$
56e, n= 1, R = $(CH_2)_3CONHCH_2C_6H_5$

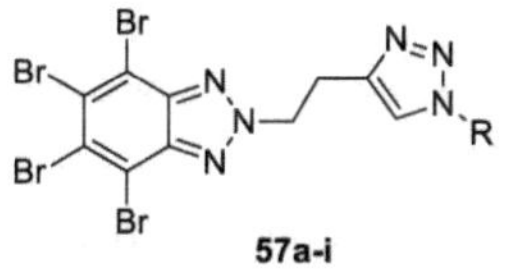

57a-i

57a, R = $(CH_2)_3CONH(CH_2)_2CH(CH_3)_2$
57b, R = $(CH_2)_3CONHCH_2C_6H_5$
57c, R = $(CH_2)_3CONHCH(CH_2CH_2)_2NCH_2C_6H_5$
57d, R = $(CH_2)_9COOH$
57e, R = $(CH_2)_7COOH$
57f, R = $(CH_2)_4NHCOOC(CH_3)_3$
57g, R = $(CH_2)_2CH(NHCOOC(CH_3)_3)CONH(CH_2)_2CH(CH_3)_2$
57h, R = $(CH_2)_4NH_2$
57i, R = $(CH_2)_2CH(NH_2)CONH(CH_2)_2CH(CH_3)_2$

Fig. 25 Tetrasubstituted benzotriazole-based CK2 inhibitors

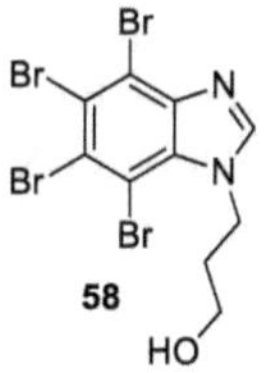

58

59

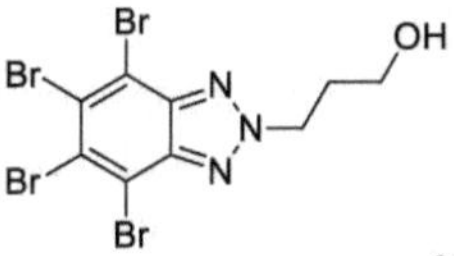

60

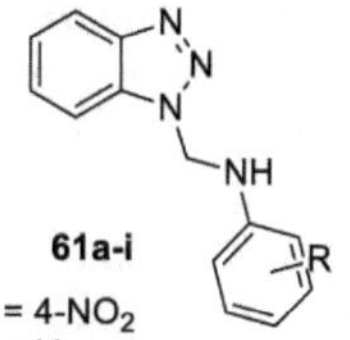

61a-i

61a, R = 4-NO_2
61b, R = H
61c, R = 4-OH
61d, R = 4-COOEt
61e, R = 4-OMe
61f, R = 4-Cl
61g, R = 2,3-2C
61h, R = 3,4-2Cl
61i, R = 4-morpholinyl

Fig. 26 TBB- and TBI-based CK2 and RSK2 inhibitors

nitrogen atoms in the triazole ring was also observed. These interactions make these derivatives multisite-directed inhibitors [84].

Genetic alterations in malignant gliomas decrease their susceptibility toward cell death due to the elevation of CK2 kinase in tumors. The inhibition of CK2 expression by treatment with inhibitors of CK2 affects the survival rates by inducing apoptosis in various cancer cells. Kaminska et al. compared the cytotoxic effects of well-known CK2 inhibitors TBB **48**, TBBi **49**, and DMAT **50** and new derivatives such as 3-(4,5,6,7-tetrabromo-1*H*-benzimidazol-1-yl)propan-1-ol **58**, 3-(4,5,6,7-tetrabromo-1*H*-1,2,3-benzotriazol-1-yl)propan-1-ol **59**, 3-(4,5,6,7-tetrabromo-2*H*-1,2,3-benzotriazol-2-yl)propan-1-ol **60** on cultured malignant glioma cells (Fig. 26). TBBi **49** and new derivatives **58–60** were more effective than

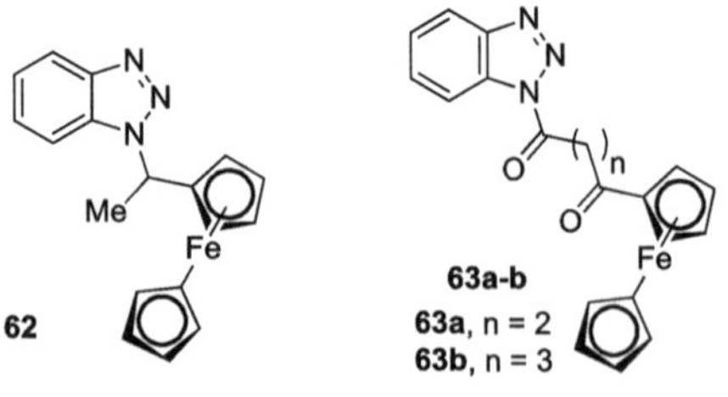

Fig. 27 Ferrocenyl benzotriazoles as anticancer agents

TBB **48** in growth arrest and cell death in glioma cells. These derivatives also have an improved efficiency toward malignant glioblastoma cells than TBB **48** [85]. Other substituted benzotriazoles of the type **61a–i** have been tested as RSK2 inhibitors for the treatment of cancer [86].

Ferrocenium salts are well known for their various bioactivities such as antitumor and antianemic activities. Ferrocenyl benzotriazole derivative **62** showed good inhibition against different human cancers including non-small cell lung cancer, endometrial cancer, and esophageal cancer and found even better than the standard drug cisplatin (Fig. 27). The benzotriazole moiety contributed to its high bioactivity since it provided a transport for the lipophilic ferrocenyl moiety to ensure the membrane permeability. Moreover, the plane hydrophilic structure of benzotriazole could intercalate into the planes of DNA nucleic bases and form hydrogen bonds with phosphate groups at cleavage points of DNA, thus enhancing its anticancer activity [87]. Additional ferrocenyl benzotriazole derivatives **63a–b** were also found to have inhibitory activity of non-small cell lung cancer, namely, H1299 cell tumor, and could be used as lead compounds in screening drug against non-small cell lung cancer (Fig. 27) [88].

A series of 1,3,4-oxadiazole derivatives containing a benzotriazole moiety **64a–s** were reported for their anticancer activity against MCF-7 (human breast cancer) and HT29 (human colorectal cancer) cell lines and compared with cisplatin (Fig. 28). Fluoro-substituted compound **64a** showed the best activity against MCF-7 cells with a IC_{50} value of 5.68 μg/mL. The latter was thus much better than reference drug cisplatin with a IC_{50} value of 11.20 μg/mL. Biological activity against HT29 cells was found only in fluoro-substituted compound **64a** with an IC_{50} value of 10.21 μg/mL. It was also observed that compounds bearing an electron-withdrawing groups showed a stronger anticancer potency than those with electron-donating groups as per the following order: $F > Cl > Br > NO_2 > OCH_3 > CH_3$ [89]. Different benzotriazole–oxadiazole derivatives **65a–h** were also found to be active against human cervical cancer cells (Hela) and hepatoma cell (HepG2). Compound with *m*-methoxyphenyl substitution exhibited the IC_{50} values of 4.83 and 3.84 mg/mL against HepG2 and Hela cell lines, respectively (Fig. 28) [90].

Benzotriazole-substituted benzoate derivatives such as 3-(1*H*-benzo[*d*][1,2,3] triazol-1-yl)-1-(4-methoxyphenyl)-1-oxopropan-2-yl benzoate (BmOB) **66a** effectively inhibited the proliferation of hepatocarcinoma BEL-7402 cell with low IC_{50} value of 0.082 mg/mL (Fig. 29) [91]. Both derivatives **66a,b** exhibited potent

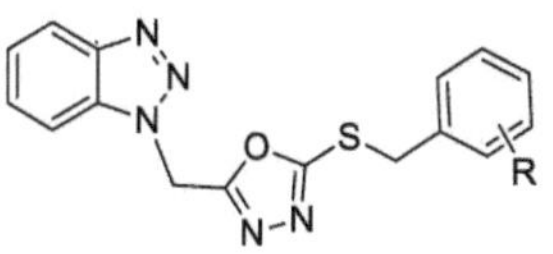

64a-s

64a, R = 2-F
64b, R = 2-Cl
64c, R = 2-Br
64d, R = 2-Me
64e, R = 2-OMe
64f, R = 2-NO_2
64g, R = 3-F
64h, R = 3-Cl
64i, R = 3-Br
64j, R = 3-Me
64k, R = 3-OMe
64l, R = 3-NO_2
64m, R = 4-F
64n, R = 4-Me
64o, R = 4-Br
64p, R = 4-OMe
64q, R = 4-NO_2
64r, R = 3,4-2Cl
64s, R = 3,4-2F

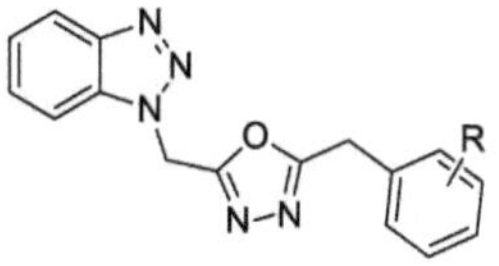

65a-h

65a, R = H
65b, R = 3-Br
65c, R = 4-Br
65d, R = 3-Cl
65e, R = 3-F
65f, R = 2-OMe
65g R = 3-OMe
65h, R = 4-OMe

Fig. 28 Oxadiazole-substituted benzotriazoles as anticancer agents

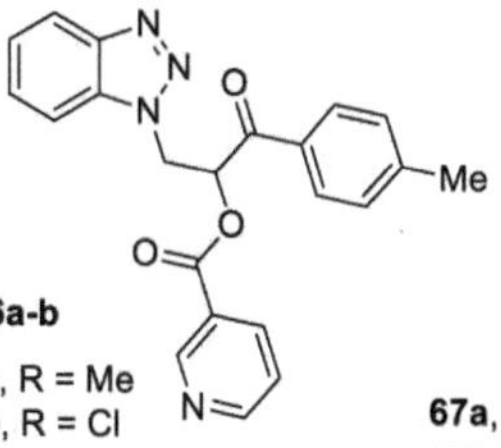

66a-b

66a, R = Me
66b, R = Cl

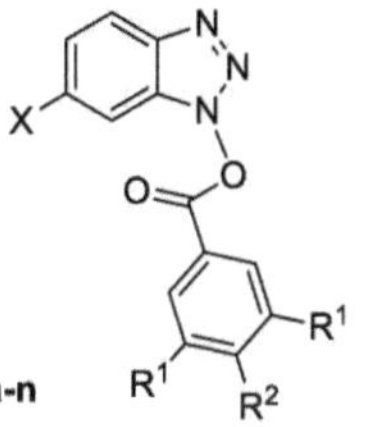

67a-n

67a, X = H, R^1, R^2 = H
67b, X = H, R^1= H, R^2 = OH
67c, X = H, R^1= OH, R^2 = H
67d, X = H, R^1, R^2 = OH
67e, X = H, R^1 = H,R^2 = OMe
67f, X = H, R^1 = OMe, R^2 = H
67g, X = H, R^1, R^2 = OMe
67h, X = OMe, R^1, R^2 = H
67i, X = OMe, R^1 = H, R^2 = OH
67j, X = OMe, R^1= OH, R^2 = H
67k, X = OMe, R^1, R^2 = OH
67l, X = OMe, R^1 = H, R^2 = OMe
67m, X = OMe, R^1 = OMe, R^2 = H
67n, X = OMe, R^1, R^2 = OMe

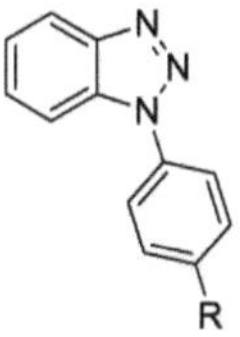

68a-e

68a, R = COOEt
68b, R = CN
68c, R = Cl
68d, R = H
68e, R = F

Fig. 29 N^1-substituted benzotriazoles as anticancer agents

inhibition in liver and galactophore cancer cells and possess good antitumor activity for hepatoma 7,402 cells and breast cancer 4T-1 cells [92]. Another series of benzotriazole derivatives bearing substituted benzoic acids linked via an ester linkage **67a–n** was tested and showed considerable activity against three human cancer cell lines with the half maximal inhibitory concentration values of 1.2–2.4 nM quite comparable to doxorubicin [93]. Compound **67f** showed IC_{50} values of 42 nM against oral epithelial carcinoma cells (KB), 53 nM against H460 lung cancer cells, and 38 nM against gastric cancer cells [94]. *N*-Aryl-substituted benzotriazoles **68a–e** were reported for their inhibitory activity toward indoleamine 2,3-dioxygenase (Fig. 29). Among compounds **68a–e**, compounds bearing an ethyl carboxylate substituent **68a** showed the highest inhibitory activity. This particular type of inhibitory activity is an important therapeutic approach for the treatment of cancer [95].

2.3 *Benzotriazole Derivatives as Antioxidants*

Reducing agents that stabilize free radicals produced by cellular metabolism are termed as antioxidants. The stabilization or quenching of these free radicals or reactive oxygen species (ROS) is necessary to avoid cell destruction. Thus, there is a need to develop antioxidant agents which eliminate the excessive free radicals and protect the cells against oxidative stress. Prophylactic agents are known to be promising antioxidants in pathogenesis [96]. It has been demonstrated that some benzotriazole derivatives are potential candidates with a promising antioxidative activity.

Primaquine is a well-known antimalarial drug having prooxidant effects in blood, and its benzotriazole derivative **69** is an interesting molecule with potential antioxidative activity. Benzotriazole-substituted primaquine **69** showed a higher antioxidative interaction (73.8%) than the parent compound primaquine (31%). In addition, it also exhibited a good lipoxygenase inhibitory (LOX) inhibition ($IC_{50} = 260$ μg/mL) (Fig. 30) [97].

Ketoprofen, a nonsteroidal anti-inflammatory drug (NSAID), has analgesic and antipyretic activities. Ketoprofen benzotriazole derivative **70** possessed a good interaction with 1,1-dipheny-l,2-picrylhydrazyl (DPPH), which is a stable free radical with spared electron delocalization over the whole molecule (Fig. 30). The interaction between compound **70** and DPPH indicated its radical scavenging ability in an iron-free system, as well as its reductant character. This derivative also

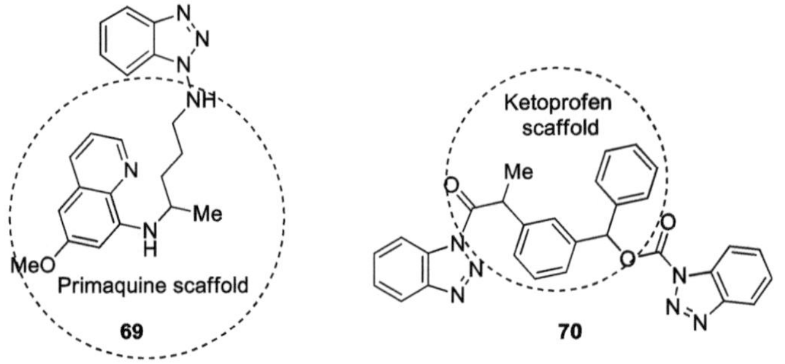

Fig. 30 Benzotriazole derivatized drugs as antioxidants

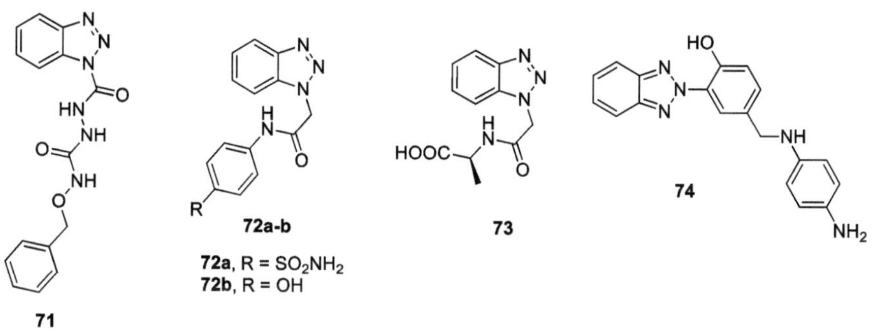

Fig. 31 Benzotriazole-based antioxidants

showed a high soybean lipoxygenase inhibition activity (95%) [98]. Substitution of benzotriazole with pyrrolyl or piperidyl groups reduced the antioxidant activity, which indicated that the presence of benzotriazole was beneficial to its antioxidant property.

N^1-Carbonyl-substituted benzotriazole derivative **71** showed good DPPH interaction value (85%) as compared to the reference compound nordihydroguaiaretic acid (91%) at the same concentration (Fig. 31). This compound also displayed a good lipid peroxidation (LP) inhibition (31%) activity [99]. Other compounds such as 2-(1*H*-1,2,3-benzotriazol-1-yl)-*N*-phenylacetamide derivatives **72a,b** and [(1*H*-benzotriazol-1-ylacetyl)amino]acetic acid derivative **73** also showed antioxidant activity compared with the reference ascorbic acid. The presence of both carbonyl and amine functionalities was believed to interact with nitric oxide, resulting in a reduced production of the nitrite anion (Fig. 31), and makes them nitric oxide scavengers [100]. These results highlighted the promising activity of the benzotriazole group as a new scaffold in the rational design of new antioxidant compounds. Other benzotriazole derivatives bearing a free phenolic and amine groups such as compound **74** were also reported with a pronounced antioxidant and antiozonant activity (Fig. 31) [101].

2.4 *Benzotriazole Derivatives as Antipsychotics*

The search for newer antipsychotic drugs with higher therapeutic efficiency and a wider spectrum of action on positive, negative, and cognitive symptoms of schizophrenia is one of the biggest challenges of modern psychopharmacology. Several dopamine and serotonin agonists/antagonists have been developed for the treatment of schizophrenia and Parkinson's diseases. Tomic et al. reported various heterocyclic arylpiperazines bearing a specific heteroaryl group such as benzimidazole, substituted benzimidazoles, benzotriazoles, and 1,4-dihydroquinoxaline-2,3-diones that mimics the catechol moiety of dopamine. Several compounds were tested, from which benzotriazole-substituted aryl piperazines **75** and **76** showed a higher affinity

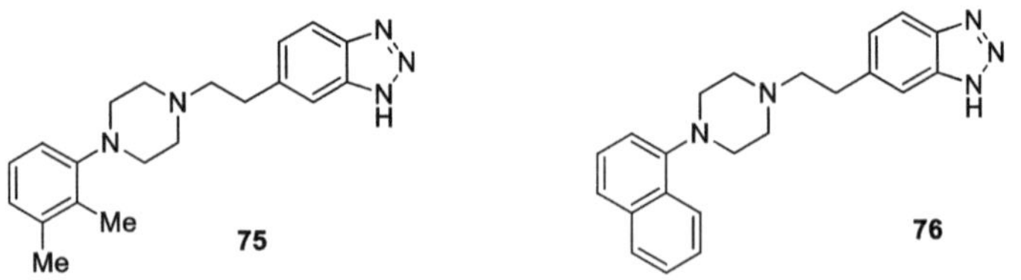

Fig. 32 Piperazine-derivatized benzotriazole as antipsychotic agents

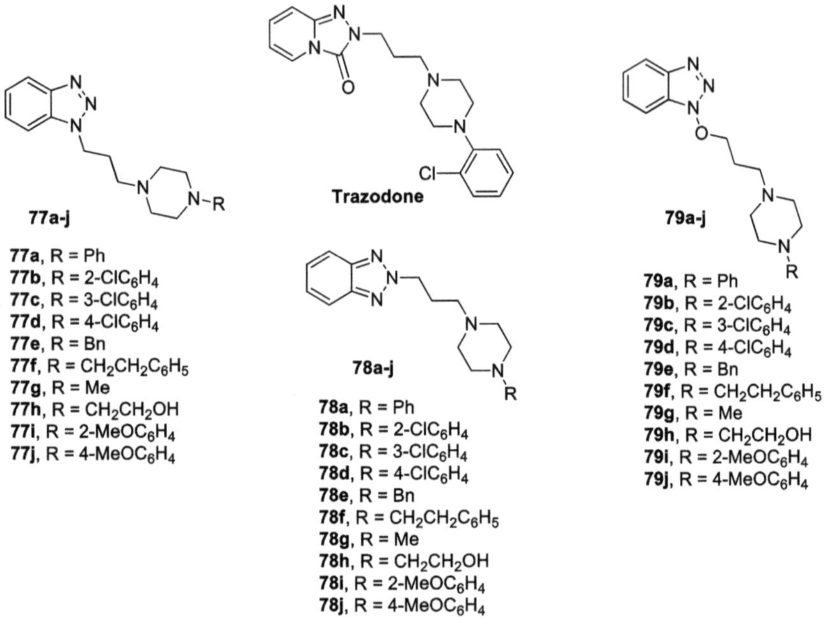

Fig. 33 Trazodone type-benzotriazole-based antipsychotics

for 5-HT_{2A} than for D_2 receptors. Compound **75** expressed a comparable affinity to the reference clozapine for the D_2, 5-HT_{2A}, and α1-adrenergic receptors and a lower affinity for the 5-HT_{2C} receptors. It also exhibited a poor affinity for the D_1 receptor and no binding potency for the 5-HT_3 receptor. On the other hand, benzotriazole derivative **76** showed a higher affinity for the 5-HT_{1A} receptor than clozapine (K_i values of 90.2 and 415 nM, respectively) (Fig. 32) [102].

A series of 1/2-[3-(4-(*R*)-1-piperazinyl)alkyl]-benzotriazole derivatives **77a–j** and **78a–j** have been synthesized as structural analogues of trazodone (Fig. 33). The presence of either an unsubstituted phenyl ring or a 2- or 3-chloro phenyl moiety at the 4-piperazine nitrogen showed a good pharmacological profile similar to that of the antidepressant trazodone [103]. The binding assays with radiolabeled ligand assays of **77a–j**, **78a–j**, and other derivatives **79a–j** for the recombinant human receptor subtypes 5-HT_0, 5-HT_{2A}, 5-HT_{2C}, and 5-$HT_{1D\beta}$ (Fig. 33) suggested a high affinity for the 5-HT_{2A} receptor and no affinity for subtype 5-$HT_{1D\beta}$. The presence

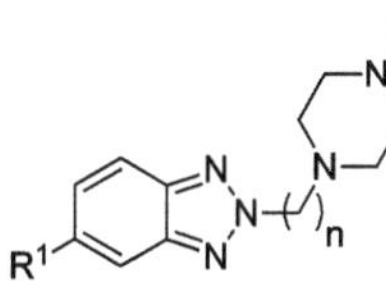
80a-l

80a, R^1 = OMe, R^2 = Ph, n =2
80b, R^1 = OMe, R^2 = Ph, n = 3
80c, R^1 = OMe, R^2 = 2-MeOC$_6$H$_4$, n =2
80d, R^1 = OMe, R^2 = 3-MeOC$_6$H$_4$, n = 3
80e, R^1 = Cl, R^2 = Ph, n =2
80f, R^1 = Cl, R^2 = Ph, n = 3
80g, R^1 = Cl, R^2 = 2-MeOC$_6$H$_4$, n = 2
80h, R^1 = Cl, R^2 = 3-MeOC$_6$H$_4$, n = 3
80i, R^1 = H, R^2 = 2-pyrimidinyl, n = 2
80j, R^1 = H, R^2 = 2-pyrimidinyl, n = 3
80k, R^1 = H, R^2 = 2-pyrimidinyl, n = 4
80l, R^1 = H, R^2 = 3-CF$_3$C$_6$H$_4$, n = 3

Fig. 34 N^2-derivatized benzotriazoles as antipsychotics

of phenyl, 3-chlorophenyl or 4-chlorophenyl, or 2-methoxyphenyl at the *N*-4 position of the piperazine ring increased the affinity toward 5-HT$_{2A}$ receptor [104].

Another series of piperazinyl-substituted benzotriazoles **80a–l** were prepared and evaluated, by modification of the benzotriazole moiety (introduction of substituents such as Cl and OCH$_3$) or the aryl side chain (introduction of 2-pyrimidinyl or 3-trifluoromethylphenyl groups). The modified compounds demonstrated moderate to good affinity for serotonin receptor 5-HT$_{1A}$ and poor affinity to the dopamine D$_2$ receptor as compared to buspirone (Fig. 34) [105].

2.5 Benzotriazole Derivatives as Anti-inflammatory Agents

Inflammation is majorly caused by tissue injury, infection of trauma, or biochemical stimulation. An increasing number of nonsteroidal anti-inflammatory drugs (NSAIDs) are available on the market for the treatment of pain and inflammation. However, drawbacks such as gastrointestinal ulcers and hemorrhages are commonly associated with the extensive use of most NSAIDs. There is therefore a need for developing of newer anti-inflammatory agents with an improved activity and fewer side effects. Among various heterocycles, benzotriazole-6-carboxylic acid **81** displayed good cPLA2α inhibition (IC$_{50}$ = 0.016 μmol/L) and is thus a potent anti-inflammatory drug candidate (Fig. 35). The replacement of the carboxyl benzotriazole scaffold by a carboxyl indole or a carboxyl benzimidazole moiety resulted in decreased inhibitory activities (IC$_{50}$ = 0.035 and 0.085 μmol/L, respectively), thereby suggesting the importance of the benzotriazole ring for enhancing the anti-inflammatory ability of the drug candidate [106]. Another tetrazole-linked sulfanilamide benzotriazole derivative **82** possessed superior anti-inflammatory efficiency as compared to the standard drug paracetamol, with 47% inhibition of paw edema of albino rats. Compound **82** had comparable anti-nociceptive activity to the standard drug pentazocine. The introduction of substituted sulfonyl moiety and benzotriazole may enhance the anti-inflammatory property (Fig. 35) [107].

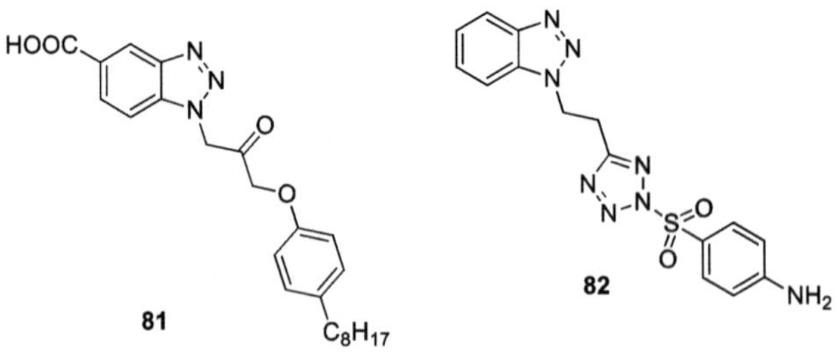

Fig. 35 Benzotriazole-based anti-inflammatory agents

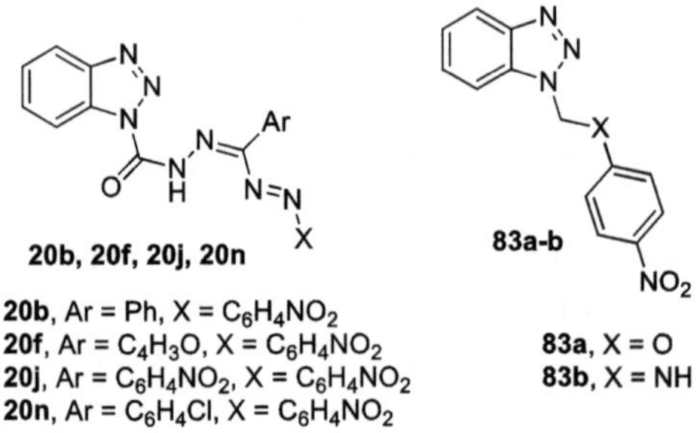

Fig. 36 Benzotriazole-based anthelmintic agents

2.6 *Benzotriazole Derivatives as Anthelmintic Agents*

Infections by different worms constitute a major cause for numerous chronic ill-health diseases worldwide especially for the people living in tropical and subtropical areas [108]. Benzotriazole-formazans **20b**, **20f**, **20j**, and **20n** were screened for their anthelmintic activity against *Pheretima posthuma* using albendazole and mebendazole as a reference drug. These compounds **20b, 20f, 20j, and 20n** exhibited equal or comparable anthelmintic activity with albendazole (Fig. 36). Among the abovementioned derivatives, compound **20j** showed superior activity [109]. Similarly, N^1-(*p*-nitrophenyl) aminomethylenebenzotriazole **83a** and N^1-benzyloxymethylenebenzotriazole **83b** showed an excellent anthelmintic activity and required less time for causing paralysis and death of the earthworms by comparison with the reference drug albendazole (Fig. 36) [110].

2.7 *Benzotriazole Derivatives as Antihyperglycemic Agents*

Diabetes has now become a worldwide disease with an increase in the number of diabetic patients at an alarming rate. After a long-term treatment, a large number of type 2 diabetic mellitus (T2DM) patients lose response with most of the oral antihyperglycemic agents. For the treatment of T2DM, the inhibition of protein

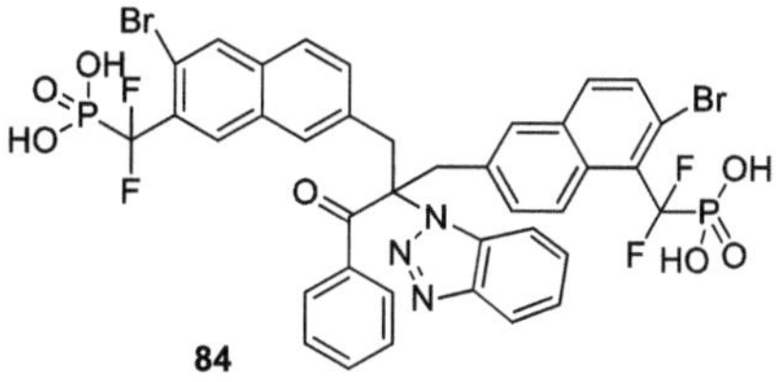

Fig. 37 Benzotriazole-based antihyperglycemic agent

tyrosine phosphatase 1B (PTP1B) has been considered as the best validated biological target. PTPIB acts as a negative regulator in insulin signaling pathways and dephosphorylates key tyrosine residues within the regulatory domain of the subunit of the insulin receptor. Benzotriazole-based PTP1B inhibitor **84** showed remarkable antihyperglycemic effects in animal models, along with an improved oral bioavailability. Compound **84** also displayed high selectivity and improved inhibitory activity against PTP1B and T-cell protein tyrosine phosphatase (TCPTP) with IC_{50} values of 5 and 589 nmol/L, respectively (Fig. 37) [111].

3 Application of Benzotriazole Derivatives in Material Sciences

3.1 *Benzotriazole Derivatives as Corrosion Inhibitors*

Copper alloys are widely used in various household and industrial applications. Good machinability, high thermal and electrical conductivity, and resistance to corrosion enforce these alloys to be used in heat exchangers, water distribution systems, water treatment units, and multistage flash (MSF) in the desalination plants. However, alloys of copper are susceptible to corrosion in aggressive media. Benzotriazole derivatives are effective inhibitors of corrosion for copper alloys in both immersed conditions and atmospheric environment [112] due to their chelating action and the formation of an insoluble physical diffusion barrier on the metal surface through strong π interactions with the surface of the material. This could also be attributed to the fact that benzotriazole nucleus binds to the copper surface through its nitrogen lone pairs, which makes benzotriazole derivatives efficient corrosion inhibitors for copper and its alloys in aqueous media [113, 114]. For example, *N*-[(benzylidenehydrazino)-propyl]-benzotriazole (BPBT) **85** and *N*-[(4-oxo-2-phenyl-1,3-thiazolidineimino)-propyl]-benzotriazole (OPBT) **86** have been reported as good corrosion inhibitors (Fig. 38). Compound OPBT **86** exhibited the best performance and showed 94% inhibition efficiency in artificial sea water [115].

A library of imidazoline-containing benzotriazoles **87** and **88** were reported as efficient corrosion inhibitors by Chen et al. (Fig. 39). It was observed that the incorporation of an imidazoline ring increases the number of *N* atoms, providing more chelating sites to coordinate with a metal center and therefore increases metal

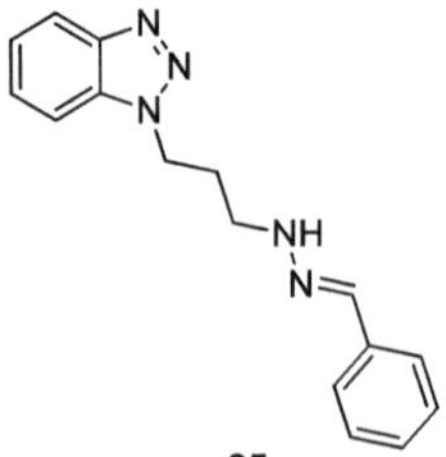

85

BPBT

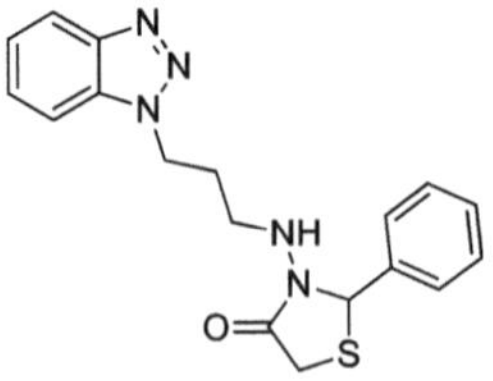

86

OPBT

Fig. 38 BPBT and OPBT as corrosion inhibitors

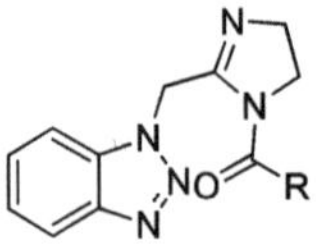

87

R = $C_{15\text{-}18}$ saturated and unsaturated carboxylic acids

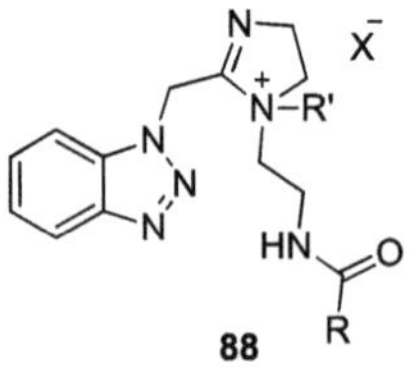

88

R = $C_{15\text{-}18}$ saturated and unsaturated carboxylic acids
R' = Me, Et, Pr, Bu, substituted Bn
X = Cl, Br

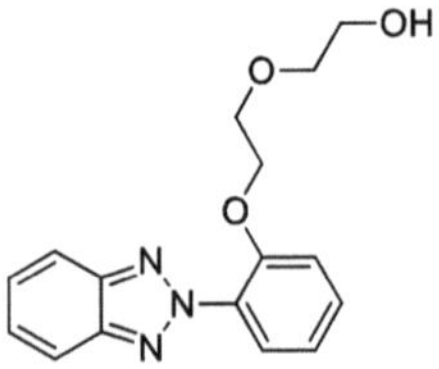

89

Fig. 39 Imidazoline benzotriazole derivatives as corrosion inhibitors

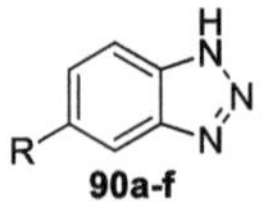

90a-f

90a, R = hexyl
90b, R = dodecyl
90c, R = (1-undecyl)dodecyl
90d, R = Cl
90e, R = Me
90f, R = *n*-Bu

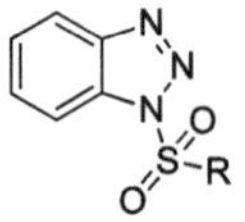

91a-c

91a, R = Ph
91b, R = 3-pyridyl
91c, R = 4-pyridyl

Fig. 40 Some corrosion inhibitors based on a benzotriazole scaffold

resistance to corrosion [116]. Another 2*H*-benzotriazole derivative **89** bearing a diethylene glycol fragment also displayed excellent inhibition against copper corrosion [117].

Some alkyl derivatives of benzotriazole such as 5-hexyl-1,2,3-benzotriazole **90a**, 5-dodecyl-1,2,3-benzotriazole **90b,** and [5-(1-undecyl)dodecyl]-1,2,3-benzotriazole **90c** were synthesized. The influence exerted by the aliphatic chain on the inhibiting properties of the benzotriazole toward bronze corrosion was investigated (Fig. 40). The results showed that benzotriazole fragments bearing long aliphatic chains acted as very effective corrosion inhibitors, with an increased efficiency than the parent nucleus [118]. Various substituents at 5th position on the benzotriazole scaffold were studied and compared. For instance, compounds **90d,e** were compared for the inhibition of Cu corrosion in an aerated HCl solution (0.1 mol/L) using electrochemical polarization in the presence of different concentrations of benzotriazole and its two derivatives **90d,e**. The inhibition efficiencies were obtained from cathodic Tafel plots and showed markedly increased efficiency with an increase in the additive concentration. Compound **90e** behaved as a cathodic corrosion inhibitor for concentrations $>10^{-1}$ mmol/L. Compound **90d** behaved as a cathodic corrosion inhibitor for concentration up to 1 mmol/L and as an anodic type inhibitor above it. The inhibition efficiencies depended on the inhibitor concentration and follow the order: 5-chlorobenzotriazole >5-methylbenzotriazole >1*H*-benzotriazole. The corrosion inhibition efficiency of N^1-butyl benzotriazole **90f** was slightly higher (73.0%) than benzotriazole itself (60.2%). Similarly to derivatives **90d,e**, an increasing concentration led to an significantly higher corrosion inhibition. For instance, when a 10 mmol/L concentration of **90f** was used, the corrosion inhibitor efficiency increased up to 93.5% [119].

Sulfonyl benzotriazoles such as 1-(phenylsulfonyl)-1*H*-benzotriazole (PSB) **91a**, 1-(3-pyridinylsulfonyl)-1*H*-benzotriazole (3PSB) **91b**, and 1-(2-pyridinylsulfonyl)-1*H*-benzotriazole (2PSB) **91c** have also been investigated for the corrosion inhibition of copper in 1 M HNO_3 at different concentrations at 25°C using chemical (weight loss) and electrochemical (Tafel polarization method) measurements (Fig. 40). The results indicated that compounds **91a–c** were firmly adsorbed on the copper surface through the benzotriazole ring and heteroatoms and had excellent corrosion inhibition performances [120].

Borate ester **92** was studied for its tribological properties and used as a multifunctional additive in rapeseed oil. Compound **92** possessed good load-carrying, antiwear, and excellent friction-reducing capacity (Fig. 41). In addition, compound **92** also showed good anticorrosion performance [121]. Further polysulfide-bounded benzotriazole derivatives **93** were explored for their anticorrosive action. Heterocyclic polysulfides **93** exhibited better corrosion-inhibiting properties and thermal stability by comparison with diisobutylpolysulfides (DIBPS) (Fig. 41). An improvement in the extreme pressure properties of rapeseed oil as well as antiwear and friction-reducing abilities was observed by using benzotriazole-substituted polysulfide additives **93** [122].

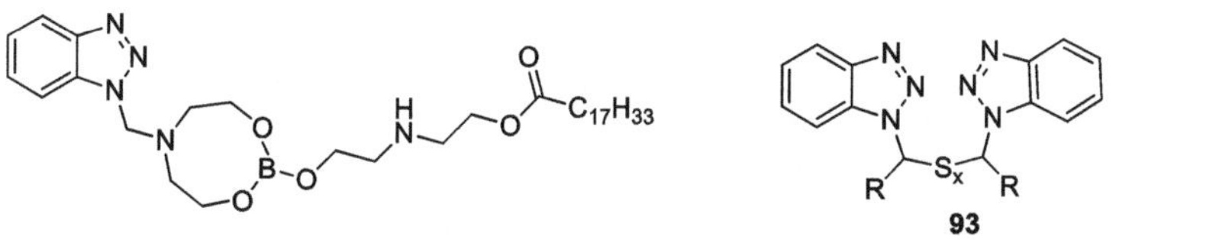

92

93

R - *n*-Pr, *n*-hex; X = integer or fraction from 2-5

Fig. 41 Borate ester **92** and polysulfide benzotriazole derivatives **93** as corrosion inhibitors

3.2 *Benzotriazole Derivatives as UV Absorbers*

Most plastic, polymer and wood products upon absorbing ultraviolet radiation undergo a rapid photolytic degradation. Several efforts have been devoted to improve the durability of these materials. To protect these materials from the harmful effects of UV radiation, many UV-absorbing agents or UV filters have been developed and tested, such as inorganic and organic materials possessing high absorption coefficients in the UV range. These compounds transform the absorbed energy into less harmful energy before reaching to the substrate, therefore preventing its photodegradation. Among the various UV absorbers, hindered amine light stabilizers (HALS) are effective UV absorbers that have been explored to a great extent. The presence of acid and basic groups in a reasonably close proximity is a common requirement of the UV-absorbing molecules [123]. These UV filters exhibit a pronounced difference between excited state and ground state acidity due to the change in their electronic environment. Usually, the acidic group in UV filters is a phenolic hydroxyl group, and the basic acceptor is a heteroatom present in an adjacent heterocycle or a carbonyl group.

To be considered as an efficient UV stabilizer, a molecular scaffold should combine strong UV absorption bands, short excited state lifetimes, and high photostability. For instance, 2-(2′-hydroxy-5′-methylphenyl)benzotriazole (TIN, trade name Tinuvin P) **94** and its silylated analogue drometrizole trisiloxane (trade name Mexoryl XL) **95** are commonly used as UV stabilizer (Fig. 42) [124]. Photoinduced excited state intramolecular proton transfer (ESIPT) is believed to explain the photostability of these compounds, thereby leading to an efficient (more than 99%) and rapid non-radiative dissipation of the harmful UV energy.

Modified Mexoryl XL-type compounds **96** and **97** have also been reported and studied for their photophysical and photochemical behaviors in order to assess their ability to act as UV filters (Fig. 43). Both compounds (**96–97**) exhibited potential applicability as UV filters due to their UVA absorption capability and photostability [125].

Some more UV filters based on 2-(hydroxyphenyl)benzotriazole derivatives were extensively explored and reported in the literature and possessed various applications. These compounds are summarized in Table 1.

Other benzotriazole-based UV absorbers include compounds **117a–c** (Fig. 44). These derivatives were used as UV absorbers for paints. Results showed that paints

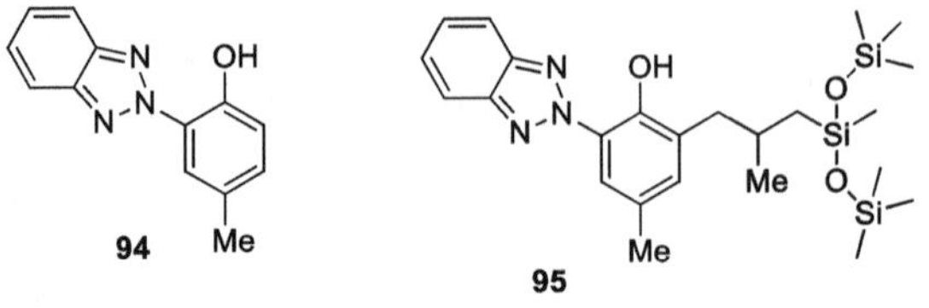

Fig. 42 Tinuvin P and Mexoryl XL as UV filters

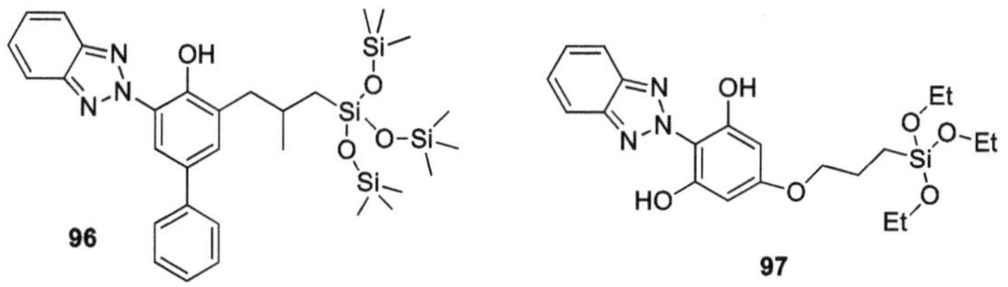

Fig. 43 Modified Mexoryl XL-type UV filters

containing **117a–c** were endowed with an improved stability profile against photo-aging/oxidation than standard paints available on the market [145].

Coumarin-based benzotriazole derivatives were also reported to improve the photostability of various industrial products. For example, compound **118** containing a coumarin and a triazine scaffold was prepared and tested as a thermo-stable fluorescent brightener and yellowing inhibitor in the paper industry (Fig. 45). This synthetic yellowing inhibitor showed UV absorption peaks at 345 nm and at 285 nm and was reported with an improved UV absorption effect and behaved as a superior fluorescent brightening agent [146]. Another coumarin-containing benzotriazole derivative **119** was also studied for its photoluminescent behavior when included in PMMA. The results showed that compound **119** exhibited a strong blue emission under UV light excitation (Fig. 45) [147]. 7-*N*,*N*-diethylamino-3-(benzotriazol-1-yl)coumarin **120** and 7-*N*,*N*-diethylamino-3-(benzotriazol-2-yl)coumarin **121** also displayed strong blue and blue-green emissions under ultraviolet light excitation (Fig. 45) [148]. 3-(1-Benzotriazole)-4-methyl-coumarin **122** was also found to possess interesting photoluminescent characteristics (Fig. 45) [149].

Dimers of 2-(hydroxyphenyl)benzotriazole such as compounds **123, 124** [150], and **125** [151] were also reported in literature as useful UV absorber agents in oil-in-water cosmetic emulsions, mostly in sun creams (Fig. 46). These compounds showed good dispersion stability of even in the presence of electrolytes.

Two yellow-green emitting derivatives **126**, **127** of 2-(2-hydroxyphenyl)-benzotriazole and benzo[*de*]-isoquinoline-1,3-dione have been reported. These compounds combined a benzotriazole UV absorber unit bearing a 1,8-naphthalimide fluorescent unit within a single molecular framework (Fig. 47). These compounds **126** and **127** showed good photostabilizing efficiency and were found suitable for preparing PMMA-based copolymers with intensive yellow-green fluorescence and high photostability [152].

A library of N^2-alkyl or amino alkyl-substituted benzotriazole derivatives **128** have also been used as UV absorbers or for the photostabilization of macromolecular architectures such as foils, fibers, and coating compositions. These compounds were also for solar cells and photovoltaic modules (Fig. 48) [153].

Table 1 2-(Hydroxyphenyl)benzotriazole-type UV filters

No.	Compound	Substitution	Application	Reference
98	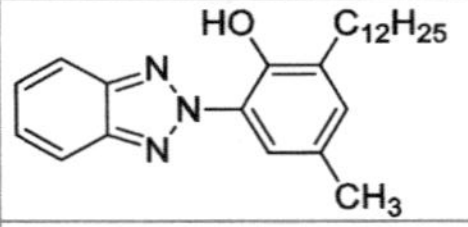	–	Used in cosmetics and dermatological compounds	[126]
99	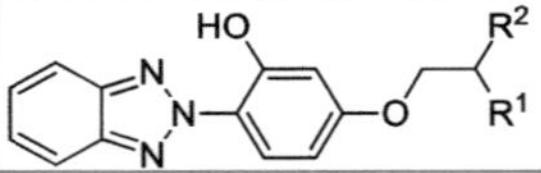	$R^1=C_{1\text{-}6}$ linear alkyl; $R^2=C_{1\text{-}3}$ linear alkyl	Used in cosmetics and sunscreen lotions	[127]
100	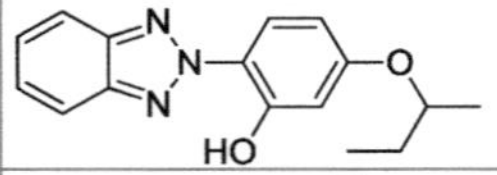	–	Used as water-in-oil sunscreen emulsions to provide synergistic effects	[128]
101	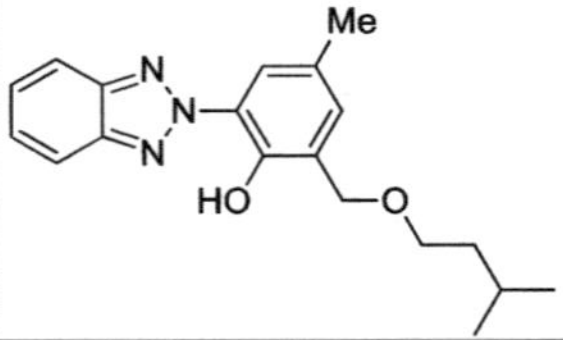	–	Used in sunscreen formulations comprising oil-in-water emulsion	[129]
102	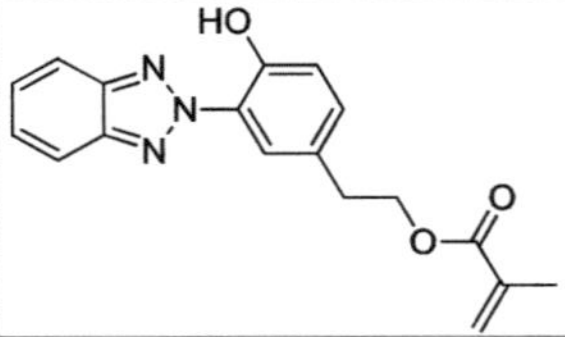	–	Used as UV stabilizers in coating materials for wood	[130]
103	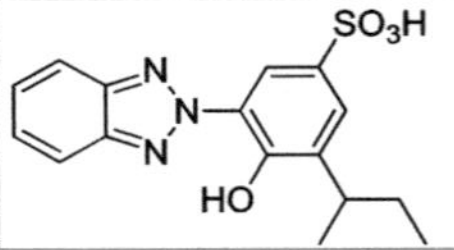	–	Present in fragrance compositions as UVA-type sunscreen agent	[131]

104	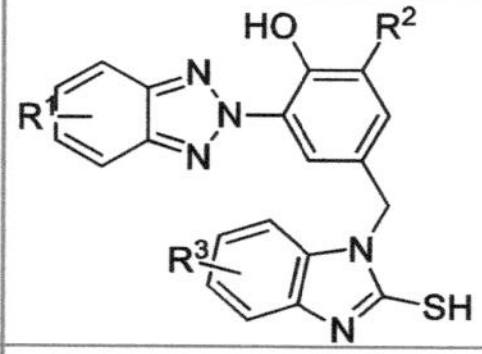	R^1=H, halo, $C_{1\text{-}12}$ alkyl, alkoxy; R^2=H, $C_{1\text{-}8}$ alkyl, cyclopentyl, cyclohexyl, cumyl; R^3=H, Me, Et	Antioxidant and UV absorber	[132]
105	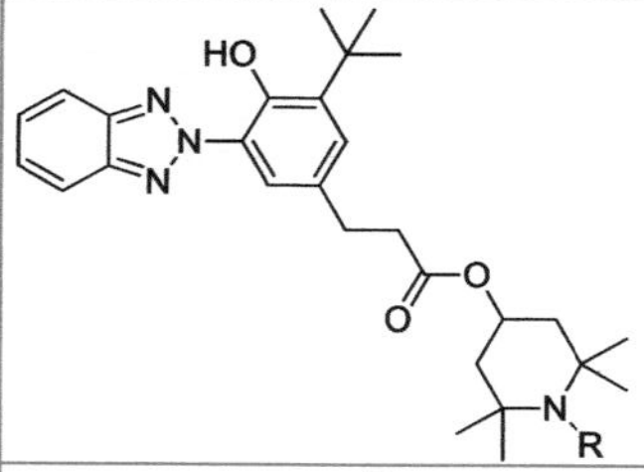	R=H, Me	UV absorbers and radical scavengers	[133]
106	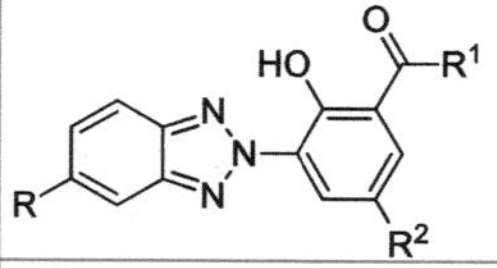	R=H, halo, alkyl, alkoxy; R^1=aryl, alkyl; R^2=alkyl, alkoxy, alkylaryl	Used as protective agents and stabilizers in coatings, plastics, and topically applied products	[134]
107	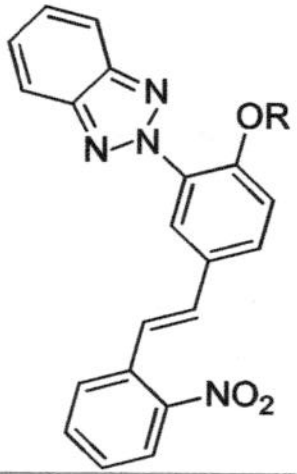	R=H, Me	Uses as photosensitizer in near-UV polymerization of acrylic monomers in solution	[135]

(continued)

Table 1 (continued)

No.	Compound	Substitution	Application	Reference
108	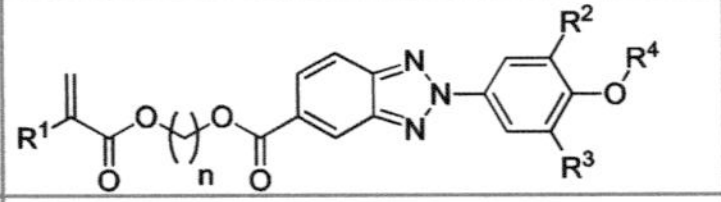	R^1=H, Me; R^2, R^3=H, Halo, $C_{1\text{-}8}$ alkyl, NO_2, CN, acylsulfo R^4=$C_{1\text{-}8}$ alkyl	Coating material and shows excellent compatibility to other polymers and device components	[136]
109	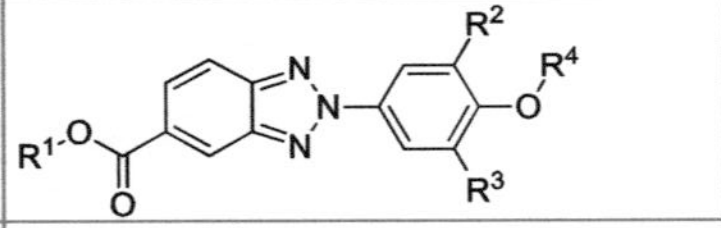	R^1=H, $C_{1\text{-}8}$ alkyl; R^2, R^3=H, Halo, $C_{1\text{-}8}$ alkyl, aryl, NO_2, CN, etc.; R^4=$C_{1\text{-}8}$ alkyl	Uses in organic fluorescent dyes	[137]
110	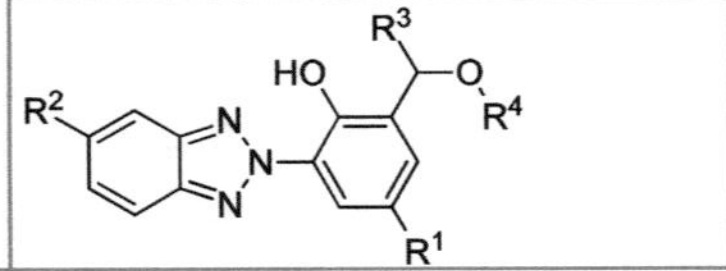	R^1=H, alkyl, alkoxy, alkoxycarbonyl, cycloalkyl, aryl, or aralkyl; R^2=H, alkyl, alkoxy or halo, H, Cl; R^3=H or alkyl; R^4=alkyl, cycloalkyl	As photostabilizer and solubilizer for dibenzoylmethane derivatives	[138]
111	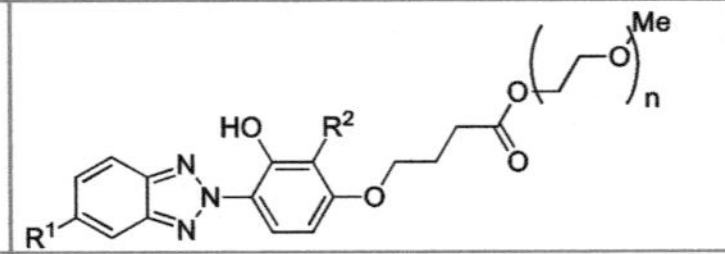	R^1=H, Cl; R^2=H, Me; $n=8$–22	Water-soluble UV absorbers for cosmetics	[139]
112	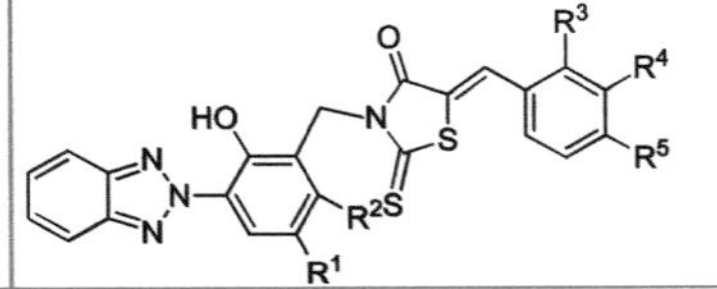	R^1-R^3, R^5, R^6=H, halo, OH, NH_2, $C_{1\text{-}4}$ disubstituted amino, NO_2, $C_{1\text{-}20}$ alkyl,$C_{1\text{-}20}$ alkyloxycarbonyl	Useful as organic pigments for optical recording medium	[140]
113	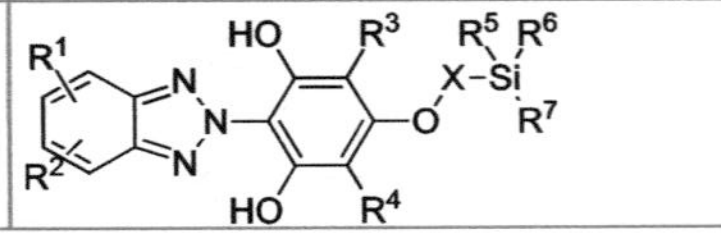	R^1-R^4=H, OH, SH, halo, NH_2, $C_{1\text{-}10}$ alkyl (oxy); R^5-R^7=H, $C_{1\text{-}4}$ alkyl(oxy), aryl(oxy), aralkyl, aryloxyalkyl, oligo(siloxanyloxy), X=$C_{1\text{-}6}$ alkylene	Useful as photoactive UV-absorbing components for skin photoprotecting applications	[141]

114	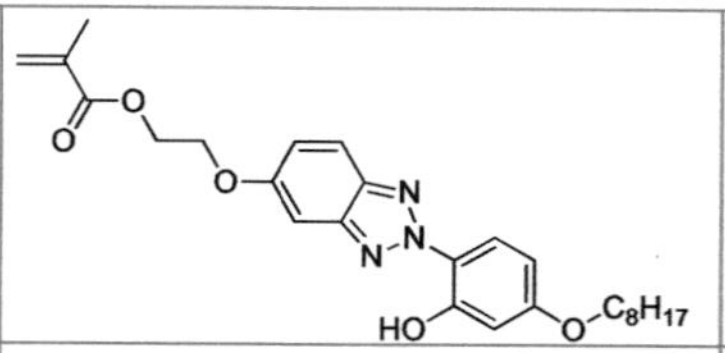		As copolymer for optical films, plastic lenses, packaging materials, agrochemical films, dye-sensitized solar cell, etc.	[142]
115	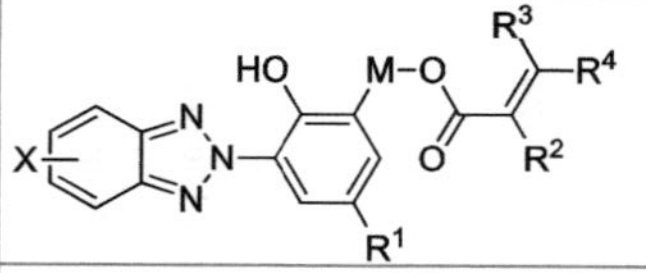	X=H, halo, alkyl, alkoxyl; R=alkyl, cycloalkyl, carboxyl, aryl, R^1, R^2, and R^3=H, alkyl, alkenyl; M=alkylene	Uses as photostabilizer for antiaging polymers	[143]
116	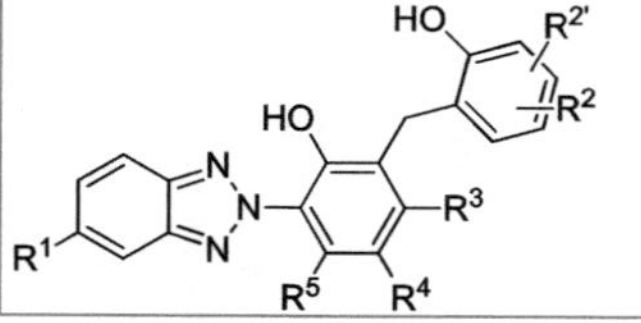	R^1=H, halo, alkyl, alkoxy; R^2 , $R^{2'}$=H, OH, halo, alkyl, alkoxy, carboxy alkyl, carboxyl, R^3–R^5=H, alkyl group, alkylaryl, alkoxy, Ar, etc.	Use as photostabilizer in polymers and/or carriers	[144]

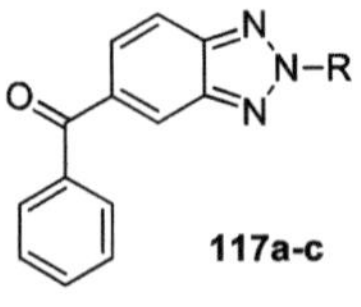

117a-c

117a, R = β-naphthol
117b, R = p-cresol
117c, R = 2,6-xylenol

Fig. 44 N^2-substituted benzotriazole derivatives as UV filters

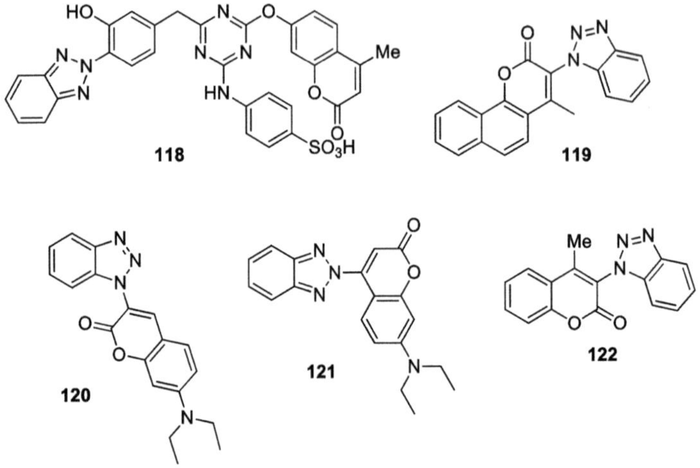

118 **119** **120** **121** **122**

Fig. 45 Coumarin-derivatized benzotriazoles as UV absorbers

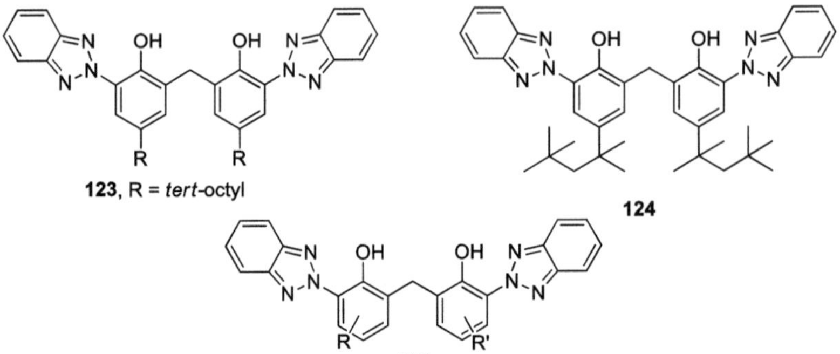

123, R = *tert*-octyl

124

125

R, R' = C_{1-18} alkyl, C_{5-12} cycloalkyl, aryl, monoalylate of polyglycerine.

Fig. 46 2*H*-Benzotriazole derivatives as component of sunscreen agents

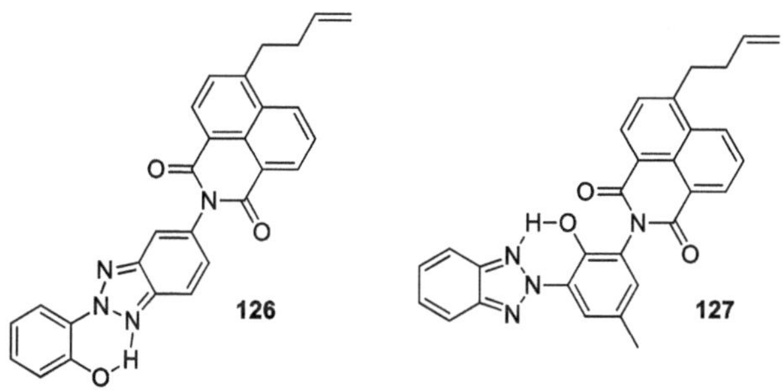

Fig. 47 2*H*-Benzotriazole-based photostabilizers

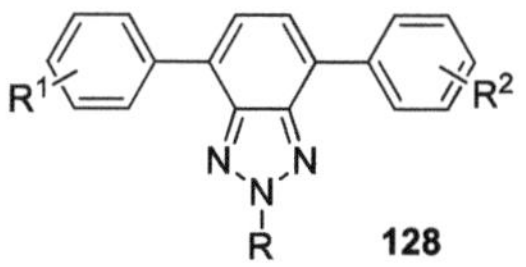

R = substituted alkyl, alkenyl, heteroaryl, aryl, amino, amido, cycloamido, carboxy and carbonyls.
R^{1-2} = substituted and unsubstituted alkoxy.

Fig. 48 Substituted 2*H*-benzotriazoles as photostabilizers

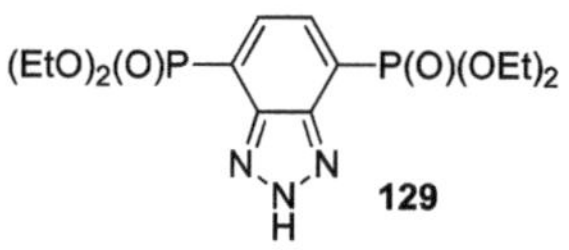

Fig. 49 Benzotriazole diphosphonate derivative **129** as component of PEM

3.3 Uses of Benzotriazole Derivatives in Fuel Cells

Fuel cells are known to convert the chemical energy directly into electrical energy with a high efficiency and a low environmental impact. The performance of these cells depends on the properties of the component materials. Proton exchange membrane fuel cells (PEMFCs) constitute a very useful power source, exhibiting high power density and high power-to-weight ratio by comparison with other kinds of fuel cells. The proton transport properties of the proton exchange membrane (PEM) strongly depend on their water content and, consequently, limit their operation temperatures up to 90°C. In this context, Teixeira et al. synthesized benzotriazole diphosphonate derivative **129** as a precursor for novel membrane materials for PEMFCs with modified properties (Fig. 49). Novel membrane materials with potentially high proton conductivity for intermediate temperature PEMFCs were obtained by incorporating these benzotriazole derivatives into proton-conductive inorganic–organic hybrid membranes of mesoporous silica [154].

3.4 Uses of Benzotriazole Derivatives in Solar and Photovoltaic Cells

Bulk heterojunction (BHJ)-based organic photovoltaic cells are ideal candidates for low-cost, light weight, and flexible solar cells. Such junctions include an electron rich π-conjugated polymer donor capable of exhibiting broad absorption with high absorption coefficient in the visible region and an electron-deficient fullerene-based acceptor, yielding a nanoscale bicontinuous interpenetrating network. Benzotriazole-based donor–acceptor copolymers containing a dithienosilole (DTS) donor unit and a benzotriazole acceptor unit such as PDTS-BTA **130** and with an optional thiophene bridge such as PDTS-DTBTA **131** have been reported by Li et al. (Fig. 50). The addition of a thiophene bridge in PDTS-DTBTA improved the planarity by improving the effective conjugation between polymer chains, hence leading to a broader and stronger absorption, a higher hole mobility, and a better photovoltaic performance [155].

Another benzotriazole-based polymer **132** has been used as a photovoltaic material (Fig. 51). Compound **132** was soluble in the most common organic solvents, such as chloroform, tetrahydrofuran, and chlorobenzene and showed excellent film-forming properties. PCDTBTz polymers (**132**) exhibited good

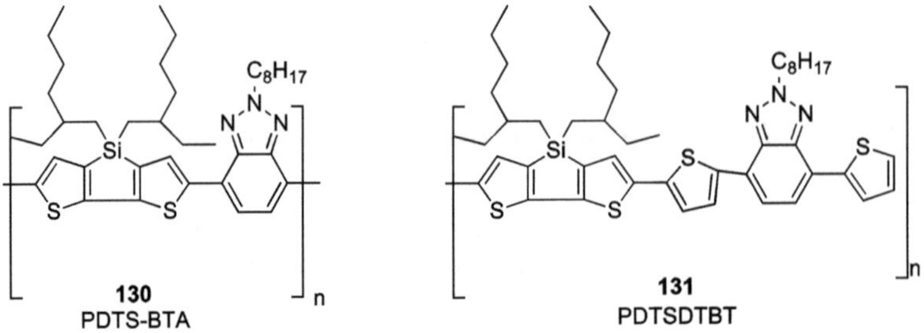

130 PDTS-BTA **131** PDTSDTBT

Fig. 50 Benzotriazole-based copolymers for OPV

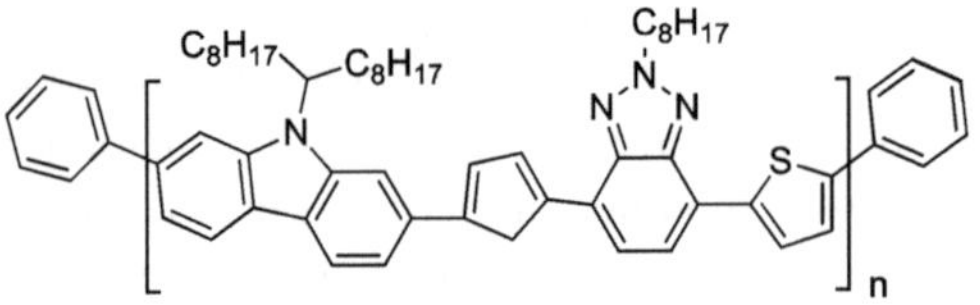

132 PCDTBTz

Fig. 51 Benzotriazole-based polymer for photovoltaic applications

Fig. 52 Benzotriazole-based polymer for photovoltaic cell

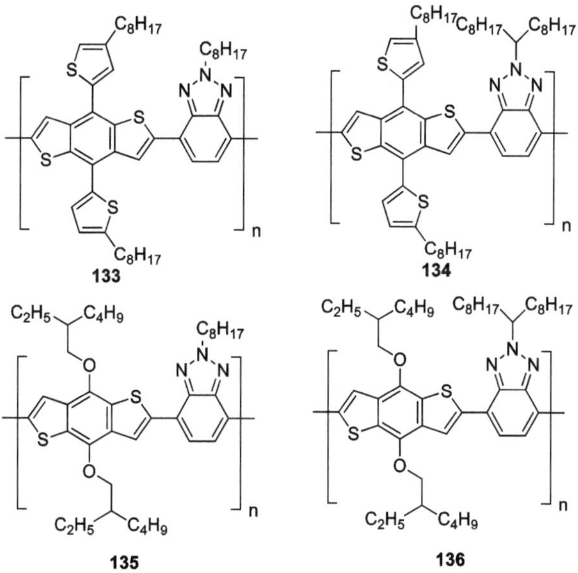

thermal and air stability. An absorption band between 300 and 610 nm was reported for polymer films made from **132**. An open-circuit voltage of 0.92 V, a power conversion efficiency of 2.2%, and a short circuit current of 5.33 mA/cm^2 were observed for the photovoltaic cells that are composites of PCDTBTz with indium tin oxide [156].

Benzotriazole derivatives **133–136** have been synthesized as low band polymers by coupling electron-accepting benzotriazole-based monomer units with electron-donating benzodithiophene-based counter-monomer units (Fig. 52). The introduction of either a linear *n*-octyl or a branched 1-octylnonyl group on the benzotriazole moiety, as well as the substitution of the benzodithiophene ring by an alkylthiophene group, drastically improved the solubility of the corresponding polymers, making them suitable to form smooth and uniform thin films by spin-casting. These polymers also exhibited a good thermal stability and lost less than 5% of their weight upon heating to approximately 300°C. Optical band gap energies in the range 2.03–1.90 eV and an induction of a broad absorption from 300 to 650 nm were observed by intramolecular charge transfers between the electron-donating and electron-accepting groups in the polymer backbone [157].

Few other benzotriazole-based monomers, such as 1-benzyl-4,7-di(thiophen-2-yl)-2*H*-benzo[*d*][1,2,3]triazole (BBTA) **137** and 2-benzyl-4,7-di(thiophen-2-yl)-2*H*-benzo[*d*][1,2,3]triazole (BBTS) **138**, have been synthesized by Toppare et al. (Fig. 53). Altering the position of phenyl group changed the electronic structure of the polymer, thereby enabling different optical and electrochemical behaviors. The polymer obtained from BBTS possesses multicolored property in its different oxidized and reduced states. A maximum absorption at 390 nm in its neutral state and multicolored electrochromic property upon stepwise oxidation was exhibited

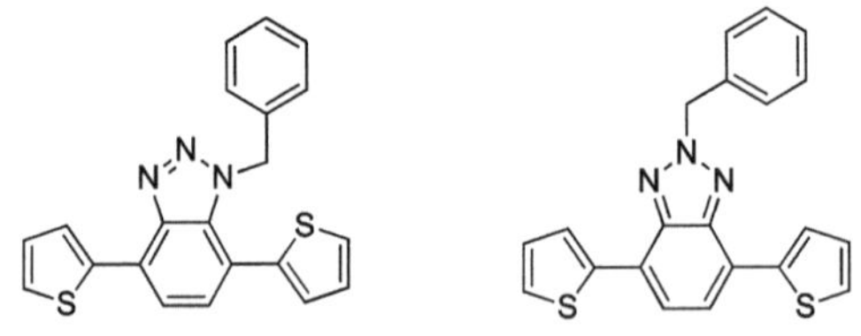

Fig. 53 Thiophene-substituted benzotriazole derivatives

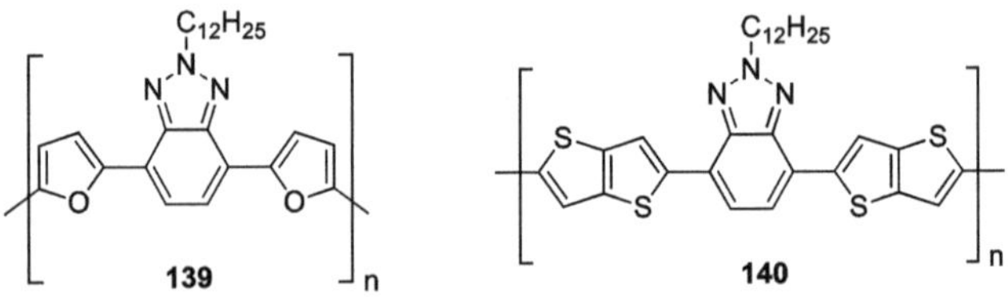

Fig. 54 Benzotriazole-based polymer for photovoltaic cells

by polymer obtained from BBTA. Symmetrically substituted polymers obtained from BBTS also possessed excellent electrochemical properties and exhibited high transmittance changes in electrochromic switching studies [158].

Similarly, furan and thieno[3,2-*b*]thiophene end-capped benzotriazole-based donor–acceptor–donor (DAD) polymers (**139–140**) have been reported from their corresponding monomers. Furan- and thieno[3,2-*b*]thiophene-based monomers exhibited monomer oxidations at 1.15 and 1.25 V, respectively, which eased the formation of conducting polymer films without overoxidation. These polymers exhibited multicolored electrochromic behavior with reversible redox couples for PTTBT **139** and PFBT **140** at −2.3/−1.50 V and −1.99/−1.58 V, respectively (Fig. 54). This could be attributed to the stepwise oxidation of PTTBT and PFBT that permits the multicolored electrochromic states to these polymers. Both polymers also exhibited ambipolar characteristics with relatively low HOMO/LUMO energy levels [159].

3.5 Miscellaneous Uses of Benzotriazole Derivatives

Benzotriazole derivatives are also used as ink components in oil-based marking pen such as compounds **141** (Fig. 55) [160]. Water-soluble benzotriazole derivatives such as compounds **142, 143** were also employed as components for water-thinned ink compositions (Fig. 55) [160].

Only a few benzotriazole-based azo dyes were reported. For instance, benzotriazole-based azo dye **144** has been used as dye stuff for hair (Fig. 56) [161].

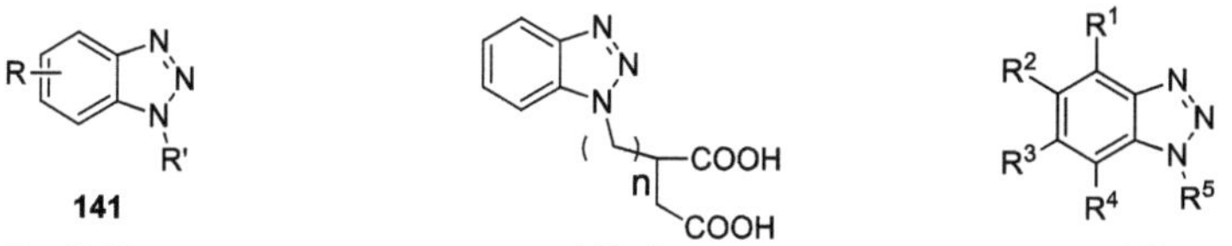

141

R = H, Me;
R' = $CH(CH_2COOH)COOH$, $CH_2CH(CH_2COOH)$ COOH, $CH_2NHC_8H_{17}$, $CH_2N(C_8H_{17})_2$, $CH_2N(CH_2Ch_2OH)_2$

142a-b

142a, n = 0
142b, n = 1

143

R^1, R^2, R^3, R^4, R^5 = H, $C_{1\text{-}4}$ substituted OH

Fig. 55 Benzotriazole derivatives used as ink components

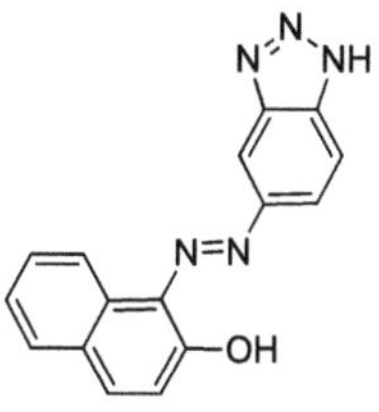

144

Fig. 56 Benzotriazole-based azo dye

Acknowledgments The authors are thankful to the Department of Science and Technology, New Delhi, India, for providing financial support (SB/FT/CS-140/2012).

References

1. Katritzky AR, Henderson SA, Yang B (1998) Applications of benzotriazole methodology in heterocycle ring synthesis and substituent introduction and modification. J Heterocycl Chem 35:1123–1159
2. Katritzky AR, Stanislaw R, Hitchings GJ (1991) Benzotriazole: a novel synthetic auxiliary. Tetrahedron 47:2683–2732
3. Katritzky AR, Linghong X, Toader D, Serdyuk L (1995) General and efficient carbon insertion route to one-carbon-homologated α-aryl, α-alkenyl, α-alkoxy, and α-phenylthio alkyl ketones. J Am Chem Soc 117:12015–12016
4. Branka Z, Zrinka RD, Ivan B (2012) Benzotriazole as a synthetic auxiliary. Croatica Chemica Acta 85:595–602
5. Panda SS, Dennis CH, Scriven E, Katritzky AR (2013) Aminoacyl benzotriazolides: versatile reagents for the preparation of peptides and their mimetics and conjugates. Aldrichim Acta 46:43
6. Kulagowski JJ, Moody CJ, Rees CW (1985) Generation and rearrangement of 4a H-carbazoles. J Chem Soc Perkin Trans 1 2725–2732
7. Murai H, Torres M, Strausz OP (1980) Electron spin resonance of iminocyclohexadienylidenes: photoinduced triplet geometrical isomerization. J Am Chem Soc 102:1421–1422
8. Monbaliu JCM, Katritzky AR (2012) Recent trends in Cys- and Ser/Thr-based synthetic strategies for the elaboration of peptide constructs. Chem Commun 48:11601–11622

9. Hagan DJ, Chan D, Schwalbe CH, Stevens MFG (1998) Antitumour polycyclic acridines. Part 3. 1 A two-step conversion of 9-azidoacridine to 7 H-pyrido [4, 3, 2-kl] acridines by Graebe–Ullmann thermolysis of substituted 9-(1, 2, 3-triazol-1-yl) acridines. J Chem Soc Perkin Trans 1 915–924
10. Katritzky AR, Stanislaw R (2010) Synthesis of heterocycles mediated by benzotriazole. 1. Monocyclic systems. Chem Rev 110:1564–1610
11. Katritzky AR, Stanislaw R (2011) Synthesis of heterocycles mediated by benzotriazole. 2. Bicyclic systems. Chem Rev 111:7063–7120
12. Katritzky AR, Rogovoy BV (2003) Benzotriazole: an ideal synthetic auxiliary. Chem Eur J 9:4586–4593
13. Sparatore F, La Rotonda MI, Ramundo E, Silipo C, Vittoria A (1978) Effect of benzotriazole derivatives on plant growth. II. Il Farmaco; edizione scientifica 33:924–944
14. Paglietti G, Sanna P, Carta A, Sparatore F, Vazzana I, Peana A, Satta M (1994) Choleretic activity of 3-[ring substituted benzotriazol-1(2)-yl]alkanoic and alkenoic acids. Farmaco (Societa chimica italiana:1989) 40:693–702
15. Ochal Z, Bretner, M, Wolinowska R, Tyski S (2013) Synthesis and in vitro antibacterial activity of 5-halogenomethylsulfonyl-benzimidazole and benzotriazole derivatives. Medicinal chemistry (Shariqah (United Arab Emirates)) 9:1129–1136
16. López-Vallejo F, Castillo R, Yépez-Mulia L, Medina-Franco JL (2011) Benzotriazoles and indazoles are scaffolds with biological activity against Entamoeba histolytica. J Biomol Screen 16:862–868
17. Borowski P, Deinert J, Schalinski S, Bretner M, Ginalski K, Kulikowski T, Shugar D (2003) Halogenated benzimidazoles and benzotriazoles as inhibitors of the NTPase/helicase activities of hepatitis C and related viruses. Eur J Biochem 270:1645–1653
18. Piccionello AP, Guarcello A (2010) Bioactive compounds containing benzoxadiazole, benzothiadiazole, benzotriazole. Curr Bioact Compd 6:266–283
19. Singh G, Kumar R, Swett J, Zajc B (2013) Modular synthesis of N-vinyl benzotriazoles. Org lett 15:4086–4089
20. Allam NK, Ashour EA, Nazeer AA, Ashour EA (2009) A review of the effects of benzotriazole on the corrosion of copper and copper alloys in clean and polluted environments. J Appl Electrochem 39:961–969
21. Mansfeld F, Smith T, Parry EP (1971) Benzotriazole as corrosion inhibitor for copper. Corrosion 27:289–294
22. Nagayoshi H, Kakimoto K, Takagi S, Konishi Y, Kajimura K, Matsuda T (2014) Benzotriazole ultraviolet stabilizers show potent activities as human aryl hydrocarbon receptor ligands. Environ Sci Technol 49:578–587
23. Carlini P, Bria E, Giannarelli D, Felici A, Papaldo P, Fabi A, Ruggeri EM, Milella M, Nuzzo C, Cecere FL (2007) Aromatase inhibitors in post-menopausal metastatic breast carcinoma. Expert Opin Investig Drugs 16:1023–1036
24. Kauste A, Tuominen M, Heikkinen H, Gordin A, Korttila K (1986) Droperidol, alizapride and metoclopramide in the prevention and treatment of post-operative emetic sequelae. Eur J Anaesthesiol 3:1–9
25. Semple G, Skinner PJ, Cherrier MC, Webb PJ, Sage CR, Tamura SY, Chen R, Richman JG, Connolly DT (2006) 1-Alkyl-benzotriazole-5-carboxylic acids are highly selective agonists of the human orphan G-protein-coupled receptor GPR109b. J Med Chem 49:1227–1230
26. Ren Y, Zhang L, Zhou CH, Geng RX (2014) Recent development of benzotriazole-based medicinal drugs. Med Chem (Shariqah (United Arab Emirates)) 4:640–662
27. Saour KY, Atto RA (2012) Synthesis of new levofloxacin derivatives and their biological activity. Pharmacie Globale 3:1–4
28. Zhang SL, Chang JJ, Damu GLV, Geng RX, Zhou CH (2013) Berberine azoles as antimicrobial agents: synthesis, biological evaluation and their interactions with human serum albumin. Med Chem Comm 4:839–846

29. Eustice DC, Feldman PA, Zajac I, Slee AM (1988) Mechanism of action of DuP 721: inhibition of an early event during initiation of protein synthesis. Antimicrob Agents Chemother 32:1218–1222
30. Dixit PP, Nair PS, Patil VJ, Jain S, Arora SK, Sinha N (2005) Synthesis and antibacterial activity of novel (un)substituted benzotriazolyl oxazolidinone derivatives. Bioorg Med Chem Lett 15:3002–3005
31. Dixit PP, Patil VJ, Nair PS, Jain S, Sinha N, Arora SK (2006) Synthesis of 1-[3-(4-benzotriazol-1/2-yl-3-fluoro-phenyl)-2-oxo-oxazolidin-5-ylmethyl]-3-substituted-thiourea derivatives as antituberculosis agents. Eur J Med Chem 41:423–428
32. Thomson I (2011) Narinder Pal Singh, Rajesh Kumar, Deo Nandan Prasad, Sarita Sharma and Om Silakari. Int J Biol Chem 5:193–199
33. Samadhiya P (2012) Synthesis and biological significance of some 2-azetidinone derivatives. J Sci I R Iran 23:139–146
34. Shukla DK, Sen JP, Jain AK (2012) Conventional and microwave assisted synthesis of new benzotriazole derivatives and their antimicrobial potential. Indian J Chem B 51:1776
35. Dubey A, Srivastava SK, Srivastava SD (2011) Conventional and microwave assisted synthesis of 4-oxo-2-substituted aryl-1, 3-thiazolidine derivatives of benzotriazole: a new class of biological compounds. Analele Universitatii Bucuresti 20:115–121
36. Gaikwad ND, Patil SV, Bobade VD (2012) Synthesis and biological evaluation of some novel thiazole substituted benzotriazole derivatives. Bioorg Med Chem Lett 22:3449–3454
37. Wan J, Lv PC, Tian NN, Zhu HL (2010) Facile synthesis of novel benzotriazole derivatives and their antibacterial activities. J Chem Sci 122:597–606
38. Shi Y, Zhou CH, Zhou XD, Geng RX, Ji QG (2011) Synthesis and antimicrobial evaluation of coumarin-based benzotriazoles and their synergistic effects with chloromycin and fluconazole. Acta Pharmaceutica Sinica 46:798–810
39. Bhardwaj B, Jain SC (2014) Synthesis and antimicrobial activities of novel aminoalkylated 2H-benzotriazoles. Am Chem Sci J 4:576–586
40. Ochal Z, Bretner M, Wolinowska R, Tyski S (2013) Synthesis and in vitro antibacterial activity of 5-halogenomethylsulfonyl-benzimidazole and benzotriazole derivatives. Med Chem 9:1129–1136
41. Visagaperumal D, Praveen V, Sudeepthi P, Prasanthi P, Sravani GG, Satyanarayana G (2010) Synthesis and antimicrobial evaluation for certain 1-(1H-benzotriazole-1-yl)-1-substituted phenyl methanamine. Int J Chem Sci 8:1315–1320
42. Ningegowda R, Kavitha CV, Priya BS, Gaonkar SL, Tejesvi MV, Rangappa KS, Swamy SN (2009) Microwave-assisted solvent-free synthesis of N-alkyl benzotriazole derivatives: antimicrobial studies. Lett Drug Des Discov 6:502–507
43. Muvvala SS, Ratnakaram VN (2014) Antibacterial activity of some newer 1, 2, 3–benzotriazole derivatives synthesized by ultrasonication in solvent–free conditions. Bulg Chem Commun 46:25–30
44. Patel JS, Garg CS, Sen DJ (2012) Synthesis of novel N-substituted 2-(1h-benzotriazol-1-yl)-acetohydrazide derivatives as antimicrobial agents. Int J Drug Dev Res 4:322–329
45. Hergert LY, Nieto MJ, Becerra MC, Albesa I, Mazzieri MR (2008) Synthesis of N-benzenesulfonylbenzotriazole derivatives, and evaluation of their antimicrobial activity. Lett Drug Des Discov 5:313–318
46. Augustynowicz-Kopec E, Zwolska Z, Orzeszko A, Kazimierczuk Z (2008) Synthesis and antimycobacterial activity of selected nitrobenzyloxylated benzotriazoles. Acta Pol Pharm Drug Res 65:435–439
47. Parekh DV, Desai PS (2012) Synthesis, characterization and antimicrobial activity studies of 5-(2-(5-benzoyl-1H-1, 2, 3-benzotriazole-1-yl) 2-oxoethylamino)-2-hydroxy benzoic acid and their transition metal chelates. Adv Appl Sci Res 3:1992–1996
48. Shukla SN, Gaur P, Kaur H, Srivastava RS (2011) Synthesis, spectroscopic characterization and antibacterial activity of some chloro dimethylsulphoxide/tetramethylenesulphoxide ruthenium (II)/(III) complexes with 1, 2, 3-benzotriazole. Acta Chim Slovenica 58:542–548

49. Cresnar B, Petric S, Trzaskos JM (2011) Cytochrome P450 enzymes in the fungal kingdom. Biochim Biophys Acta 1814:29–35
50. Kumari S, Kumar N, Drabu S, Sharma PK (2011) Advances in synthetic approach to and antifungal activity of triazoles. Beilstein J Org Chem 7:668–677
51. Zhang Y-Y M, Jia-Li ZC-H, Zhou X-D (2011) Synthesis of novel fluconazoliums and their evaluation for antibacterial and antifungal activities. Eur J Med Chem 46:4391–4402
52. Yu G-P, Xu L-Z, Yi X, Bi W-Z, Zhu Q, Zhai Z-W (2009) Synthesis and fungicidal evaluation of 2-Arylphenyl Ether-3-(1 H-1, 2, 4-triazol-1-yl) propan-2-ol derivatives. J Agric Food Chem 57:4854–4860
53. Wan K, Zhou C-H (2010) Synthesis of novel halobenzyloxy and alkoxy 1, 2, 4-triazoles and evaluation for their antifungal and antibacterial activities. Bull Korean Chem Soc 31:2003–2010
54. Rezaei Z, Khabnadideh S, Pakshir K, Hossaini Z, Amiri F, Assadpour E (2009) Design, synthesis, and antifungal activity of triazole and benzotriazole derivatives. Eur J Med Chem 44:3064–3067
55. Shivajirao P (2013) Synthesis and evaluation of antifungal activity of benzotriazole derivatives. Asian J Biomed Pharm Sci 3:29–34
56. Sudhir MS, Nadh RV, Radhika S (2013) Antifungal activities of novel 1, 2, 3-benzotriazole derivatives synthesized by ultrasonic and solvent-free conditions. Drug Invention Today 5:126–132
57. Swamy SN, Sarala G, Priya BS, Gaonkar SL, Prasad JS, Rangappa KS (2006) Microwave-assisted synthesis of N-alkylated benzotriazole derivatives: antimicrobial studies. Bioorg Med Chem Lett 16:999–1004
58. El-Zemity SR (2011) Antimicrobial activity of some 2-amino-5-subsituted pyridine derivatives. Arch Phytopathol Pflanzenschutz 44:381–389
59. Zhang SS, Jun W, X-m L, Li C-l XL-Z (2007) Syntheses, Structure and Biological Activities of 2-(1H-Benzo [1, 2, 3] triazole-1-yl)-1-(3-methyl-4-bromobiphenyl)-2-(1H-1, 2, 4-triazole-1-yl) ethanone. Chem Res Chin Univ 23:120–122
60. Saini R, Chaturvedi S, Kesari AN, Kushwaha S (2010) Synthesis of 2-(substituted)-5-(benzotriazomethyl)-1, 3, 4-oxadiazole for anti-fungal activity. Der Pharma Chem 2:297–302
61. Singh RJ (2009) Syntheses of some new 1, 2, 3-benzotriazoles as antimicrobial agents. Rasayan J Chem 2:598–601
62. Bapna M, Parashar B, Sharma VK, Chouhan LS (2012) Microwave-assisted synthesis of some novel and potent antibacterial and antifungal compounds with biological screening. Med Chem Res 21:1098–1106
63. Suma BV, Natesh NN, Venkataramana CHS, Jays J, Madhavan V (2012) Synthesis and antibacterial of some new 1, 2, 3 benzotriazoles derivatives containing pyrazolidinedione moieties. Int J Pharm Pharm Sci 4:16
64. Han X, An C, Zhang Z (2008) Two novel Ag (I) coordination polymers with triazoles derivatives: synthesis, crystal structures and biological activity. Appl Organomet Chem 22:565–572
65. Patel PK, Patel PD (2010) Synthesis, characterization, and metal complexation studies and biological screening of some newly synthesized metal complexes of 1-(4-carboxy-3-hydroxy-n-isopropyl amino methyl benzotriazole) with some transition metals. Int J Chem Tech 2(2):1147–1152
66. Becerra MC, Guiñazú N, Hergert LY, Pellegrini A, Mazzieri MR, Gea S, Albesa I (2012) In vitro activity of N-benzenesulfonylbenzotriazole on *Trypanosoma cruzi* epimastigote and trypomastigote forms. Exp Parasitol 131:57–62
67. Awasthi SK, Mishra N, Dixit SK, Singh A, Yadav M, Yadav SS, Rathaur S (2009) Antifilarial activity of 1, 3-diarylpropen-1-one: effect on glutathione-S-transferase, a phase II detoxification enzyme. Am J Trop Med Hyg 80:764–768
68. Kumar VM, Sanjay V (2008) Synthesis of nucleoside-based antiviral drugs in ionic liquids. Bioorg Med Chem Lett 18:5640–5642

69. Kang I-J, Wang L-W, Yeh T-K, Lee C, Lee Y-C, Hsu S-J, Wu Y-S, Wang J-C, Chao Y-S, Yueh A (2010) Synthesis, activity, and pharmacokinetic properties of a series of conformationally-restricted thiourea analogs as novel hepatitis C virus inhibitors. Bioorg Med Chem 18:6414–6421
70. Sakthi S, Periyasamy S, Sampath K, Clercq ED (2010) Synthesis, anti-viral and cytotoxicity studies of some 2-phenyl-3-substituted quinazolin-4-(3H)-ones. Int J Pharmacy Pharm 2:71–79
71. Tonelli M, Paglietti G, Boido V, Sparatore F, Marongiu F, Marongiu E, La Colla P, Loddo R (2008) Antiviral Activity of Benzimidazole Derivatives. I. Antiviral Activity of 1-Substituted-2-[(Benzotriazol-1/2-yl) methyl] benzimidazoles. Chem Biodivers 5:2386–2401
72. Carta A, Loriga M, Piras S, Paglietti G, Ferrone M, Fermeglia M, Pricl S, Colla PL, Collu G, Sanna T (2007) Synthesis and anti-picornaviridae in vitro activity of a new class of helicase inhibitors the N,N′-bis [4-(1H (2H)-benzotriazol-1 (2)-yl) phenyl] alkyldicarboxamides. Med Chem 3:520–532
73. Jemal A, Bray F, Center MM, Ferlay J, Ward E, Forman D (2011) Global cancer statistics. CA Cancer J Clin 61:69–90
74. Kattimani PP, Kamble RR, Kariduraganavar MY, Dorababu A, Hunnur RK (2013) Synthesis, characterization and in vitro anticancer evaluation of novel 1, 2, 4-triazolin-3-one derivatives. Eur J Med Chem 62:232–240
75. Tang Y, Zhang J, Zhang S, Geng R, Zhou C (2012) Synthesis and characterization of thiophene-derived amido bis-nitrogen mustard and its antimicrobial and anticancer activities. Chin J Chem 30:1831–1840
76. Zhou C, Gan L, Zhang Y, Zhang F, Wang G, Jin L, Geng R (2009) Review on supermolecules as chemical drugs. Sci China Ser B 52:415–458
77. Duncan JS, Gyenis L, Lenehan J, Bretner M, Graves LM, Haystead TA, Litchfield DW (2008) An unbiased evaluation of CK2 inhibitors by chemoproteomics characterization of inhibitor effects on CK2 and identification of novel inhibitor targets. Mol Cell Prot 7:1077–1088
78. Yuan J, Zhong Y, Li S, Zhao X, Luan G, Zhao Z, Huang J, Li H, Xu Y (2013) Triazole and benzotriazole derivatives as novel inhibitors for p90 ribosomal S6 protein kinase 2: synthesis, molecular docking and SAR analysis. Chin J Chem 31:1192–1198
79. Wang S, Jones KA (2006) CK2 controls the recruitment of Wnt regulators to target genes in vivo. Curr biol 16:2239–2244
80. Lin C-Y, Navarro S, Reddy S, Comai L (2006) CK2-mediated stimulation of Pol I transcription by stabilization of UBF–SL1 interaction. Nucl Acids Res 34:4752–4766
81. Najda-Bernatowicz A, Łebska M, Orzeszko A, Kopańska K, Krzywińska E, Muszyńska G, Bretner M (2009) Synthesis of new analogs of benzotriazole, benzimidazole and phthalimide—potential inhibitors of human protein kinase CK2. Bioorg Med Chem 17:1573–1578
82. Borowiecki P, Wawro AM, Wińska P, Wielechowska M, Bretner M (2014) Synthesis of novel chiral TBBt derivatives with hydroxyl moiety. Studies on inhibition of human protein kinase CK2α and cytotoxicity properties. Eur J Med Chem 84:364–374
83. Makowska M, Łukowska-C E, Wińska P, Kuś A, Bilińska-Chomik A, Bretner M (2011) Design and synthesis of CK2 inhibitors. Mol Cell Biochem 356:91–96
84. Świder R, Masłyk M, Martín-Santamaría S, Ramos A, de Pascual-Teresa B (2011) Multisite-directed inhibitors of protein kinase CK2: new challenges. Mol Cell Biochem 356:117–119
85. Kaminska B, Ellert-Miklaszewska A, Oberbek A, Wisniewski P, Kaza BMM, Bretner M, Kazimierczuk Z (2009) Efficacy and mechanism of anti-tumor action of new potential CK2 inhibitors toward glioblastoma cells. Int J Oncol 35:1091–1100
86. Xu Y, Zhao Z, Liu X, Huang J, Li H, Zhong Y, Lu W, Cao X, Liu K (2012) Synthesis of benzotriazol derivative as RSK2 inhibitor and application thereof. CN 102688233, A 20120926

87. Snegur LV, Nekrasov YS, Sergeeva NS, Zhilina ZV, Gumenyuk VV, Starikova ZA, Simenel AA, Morozova NB, Sviridova IK, Babin VN (2008) Ferrocenylalkyl azoles: bioactivity, synthesis, structure. Appl Organomet Chem 22:139–147
88. Ren Y, Wu Y (2014) Ferrocene benzotriazole derivative and its preparation method and application. CN 104031092, A 20140910
89. Zhang S, Luo Y, He L-Q, Liu Z-J, Jiang A-Q, Yang Y-H, Zhu H-L (2013) Synthesis, biological evaluation, and molecular docking studies of novel 1,3,4-oxadiazole derivatives possessing benzotriazole moiety as FAK inhibitors with anticancer activity. Bioorg Med Chem 21:3723–3729
90. Zhu H, Liu Z, Luo Yin, Zhang S (2014) Benzotriazole-oxadiazole derivatives as anticancer drugs and their preparation, pharmaceutical compositions and use in the treatment of cancer. CN 103848827, A 20140611
91. Zhang S-S, Zhang H-Q, Li D, Sun L-H, Ma C-P, Wang W, Wan J, Qu B (2008) A novel benzotriazole derivative inhibits proliferation of human hepatocarcinoma cells by increasing oxidative stress concomitant mitochondrial damage. Eur J Pharmacol 584:144–152
92. Wan J, Yan X, Ma C, Bi S, Zhu H-L (2010) Synthesis, structure characterization, and biological evaluation of some new 1, 2, 3-benzotriazole derivatives. Med Chem Res 19:970–983
93. Fu J, Yang Y, Zhang X-W, Mao W-J, Zhang Z-M, Zhu H-L (2010) Discovery of 1H-benzo [d][1, 2, 3] triazol-1-yl 3, 4, 5-trimethoxybenzoate as a potential antiproliferative agent by inhibiting histone deacetylase. Bioorg Med Chem 18:8457–8462
94. Li X, Fu J, Shi W, Luo Y, Zhang X, Zhu H, Yu H (2013) 3D-QSAR and molecular docking studies on benzotriazoles as antiproliferative agents and histone deacetylase inhibitors. Bull Korean Chem Soc 34:2387
95. Takagi K, Al-Amin M, Hoshiya N, Wouters J, Sugimoto H, Shiro Y, Fukuda H, Shuto S, Arisawa M (2014) Palladium-nanoparticle-catalyzed 1, 7-palladium migration involving C–H activation, followed by intramolecular amination: regioselective synthesis of N 1-arylbenzotriazoles and an evaluation of their inhibitory activity toward indoleamine 2, -3-dioxygenase. J Org Chem 79:6366–6371
96. Ratnam DV, Ankola DD, Bhardwaj V, Sahana DK, Kumar MNVR (2006) Role of antioxidants in prophylaxis and therapy: a pharmaceutical perspective. J Control Release 113:189–207
97. Šimunović M, Perković I, Zorc B, Ester K, Kralj M, Hadjipavlou-Litina D, Pontiki E (2009) Urea and carbamate derivatives of primaquine: synthesis, cytostatic and antioxidant activities. Bioorg Med Chem 17:5605–5613
98. Rajić Z, Hadjipavlou-Litina DPE, Balzarini J, Zorc B (2011) The novel amidocarbamate derivatives of ketoprofen: synthesis and biological activity. Med Chem Res 20:210–219
99. Perković I, Butula I, Kralj M, Martin-Kleiner I, Balzarini J, Hadjipavlou-Litina D, Katsori A-M, Zorc B (2012) Novel NSAID 1-acyl-4-cycloalkyl/arylsemicarbazides and 1-acyl-5-benzyloxy/hydroxy carbamoylcarbazides as potential anticancer agents and antioxidants. Eur J Med Chem 51:227–238
100. Jamkhandi CM, Disouza JI (2013) Evaluation of antioxidant activity for some benzotriazole substituted with n-phenyl acetamide and acetyl carbamic acid derivatives. Int J Pharm Pharmsci 5:249–253
101. Solanky S, Desai S, Singh R (2005) Antiozonant based on functionalized benzotriazole UV absorbers and the process thereof. US 20050143584, A1 20050630
102. Tomić M, Kundaković M, Butorović B, Janać B, Andrić D, Roglić G, ID, Kostić-Rajačić S (2004) Pharmacological evaluation of selected arylpiperazines with atypical antipsychotic potential. Bioorg Med Chem Lett 14:4263–4266
103. Caliendo G, Di CR, Meli R, Perissutti E, Santagada V, Silipo C, Vittoria A (1993) Synthesis and trazodone-like pharmacological profile of 1-and 2-[3-[4-(X)-1-piperazinyl]-propyl]-benzotriazoles. Eur J Med Chem 28:969–974

104. Caliendo G, Greco G, Grieco P, Novellino E, Perissutti E, Santagada V, Barbarulo D, Esposito E, De Blasi A (1996) Structure—affinity relationship studies on benzotriazole derivatives binding to 5-HT receptor subtypes. Eur J Med Chem 31:207–213
105. Boido A, Boido CC, Sparatore F (2001) Synthesis and pharmacological evaluation of aryl/heteroaryl piperazinyl alkyl benzotriazoles as ligands for some serotonin and dopamine receptor subtypes. Farmaco 56:263–275
106. Bovens S, Kaptur M, Elfringhoff ASc, Lehr M (2009) 1-(5-Carboxyindol-1-yl)propan-2-ones as inhibitors of human cytosolic phospholipase A2α: synthesis and properties of bioisosteric benzimidazole, benzotriazole and indazole analogues. Bioorg Med Chem Lett 19:2107–2111
107. Rajasekaran A, Rajagopal K (2009) Synthesis of some novel triazole derivatives as anti-nociceptive and anti-inflammatory agents. Acta pharmaceutica 59:355–364
108. Nakhare S, Garg SC (1991) Anthelmintic activity of the essential oil of artemisia pallens wall. Anc Sci Life 10:185
109. Sudhir MS, Nadh RV (2013) Evaluation of in vitro anthelmintic activities of novel 1, 2, 3-benzotriazole derivatives synthesized in ultrasonic and solvent free conditions. J Pharm Res 7:47–52
110. Pawar SS, Gorde PL, Kakde RB (2010) Synthesis of new N1-substituted benzotriazoles as anthelmintic agents. Arch Appl Sci Res 2:80–85
111. Patel D, Jain M, Shah SR, Bahekar R, Jadav P, Darji B, Siriki Y, Bandyopadhyay D, Joharapurkar A, Kshirsagar S (2011) Discovery of orally active, potent, and selective benzotriazole-based PTP1B inhibitors. ChemMedChem 6:1011–1016
112. Al-Kharafi FM, Ateya BG (2002) Effect of sulfides on the electrochemical impedance of copper in benzotriazole-inhibited media. J Electrochem Soc 149:B206–B210
113. Allam NK, Ashour EA (2008) Promoting effect of low concentration of benzotriazole on the corrosion of Cu10Ni alloy in sulfide polluted salt water. Appl Surface Sci 254:5007–5011
114. Allam NK, Hegazy HS, Ashour EA (2007) Inhibition of the sulfide induced pitting of copper nickel alloy using benzotriazole. Int J Electrochem Sci 2:549–562
115. Jayasree AC, Ravichandran R (2014) Anti-Corrosive effect of N-[(benzylidene hydrazino)-propyl]-benzotriazole and N-[(4-oxo-2-phenyl-1, 3-thiazolidineimino)-propyl]-benzotriazole for admiralty brass in marine environment. Int J Curr Microbiol App Sci 3:515–526
116. Chen Y, Pan Z, Ma C, Guo J, Xuan L (2014) Benzotriazole imidazoline and derivative and synthesis process and application thereof. CN 103539786 A 20140129
117. Gong Y, Li H, Gao , Yuan X (2013) Preparation of benzotriazole derivative as copper corrosion inhibitor. CN 103342684, A 20131009
118. Laguzzi G, Luvidi L (2010) Evaluation of the anticorrosive properties of benzotriazole alkyl derivatives on 6% Sn bronze alloy. Surf Coat Technol 204:2442–2446
119. Zhang J, Zhang X, Zhou H, Liu G, Kuang Y (2011) Synergetic inhibition effect of benzotriazole and its derivative on copper. Diandu Yu Tushi 30:46–49
120. Khaled KF, Fadl-Allah SA, Hammouti B (2009) Some benzotriazole derivatives as corrosion inhibitors for copper in acidic medium: experimental and quantum chemical molecular dynamics approach. Mater Chem Phys 117:148–155
121. Jia Z, Xia Y, Pang X, Hao J (2011) Tribological behaviors of different diamond-like carbon coatings on nitrided mild steel lubricated with benzotriazole-containing borate esters. Tribol Lett 41:247–256
122. Li J, Wu H, Ren T (2008) The tribological study of novel benzotriazolyl-containing anticorrosive polysulfides in rapeseed oil. J Tribol 131:011801
123. Formoshino SJ, Arnaut LG (1993) Excited-state proton transfer reactions. II. Intramolecular reactions. J Photochem Photobiol A Chem 75:21–48
124. Crawford JC (1999) 2(2-Hydroxyphenyl)2H-benzotriazole ultraviolet stabilizers. Prog Polym Sci 24:7–43
125. Farkas R, Lhiaubet-Vallet V, Corbera J, Törincsi M, Gorchs O, Trullas C, Jiménez O, Miranda MA, Novak L (2010) Synthesis of new 2-(2′-Hydroxyaryl)benzotriazoles and evaluation of their photochemical behavior as potential UV-filters. Molecules 15:6205

126. Mueller S, Giesinger J, Ehlis T, Herzog B, Koenig P, Kreyer G, Quass K (2006) Benzotriazole derivatives for photostabilization of cosmetic and pharmaceutical formulations. WO 2006114381, A1 20061102
127. Yamaguchi K, Motoya K (2011) Sunscreen cosmetics containing benzotriazole derivatives and spherical resin powders. JP 2011068566, A 20110407
128. Araki H, Abe K, Yajima I, Kanokogi H, Oguchi N (2007) Sunscreens containing 4-(1,1-dimethylethyl)-4′-methoxydibenzoylmethane and benzotriazole derivatives. JP 2007182389, A 20070719
129. Mendrok-Edinger C (2012) Sunscreen formulations comprising a dibenzoylmethane, a benzotriazole and an amino hydroxybenzophenone. WO 2012168275, A2 20121213
130. Friebel S, Wittenberg K, Hennecke S (2005) Use of benzotriazole derivatives for light-stabilizing wooden substrates. WO 2005058565, A1 20050630
131. Bara I (2009) Fragrance composition comprising a combination of a benzotriazole compound, a UVA sunscreening agent of the dibenzoylmethane type and a UVB sunscreening system. WO 2009059872, A1 20090514
132. Lonkar S P, Singh RP, Singh TN, Luc GJ, Florence J, Jacques L, Emmanuel B (2007) Mercaptobenzimidazole substituted benzotriazole derivatives, processes for preparing them, and their use as UV absorbers. WO 2007036948, A1 20070405
133. Shao Y, Zuo H (2007) Synthesis and characterization of light stabilizers of benzotriazole compound containing a hindered amine moiety. Jingxi Huagong 24:1163–1167
134. Chang C-H (2010) Ultraviolet light absorbing ketones of 2-(2-hydroxyphenyl) benzotriazole useful as protective agents and stabilizers. US 20100086500, A1 20100408
135. Cao H, Yuan X, Gong Y, Li H, Gao F (2013) Benzotriazole derivative containing o-nitrostilbene fragment. CN 103044344, A 20130417
136. Yamada K, Shimeno T, Kamisaka T, Ishitani T, Fujiwara T (2014) Benzotriazole derivatives, their polymers, organic fluorescent dyes containing them, and fluorescent polymer compositions and organic electroluminescent device substances containing the dyes. JP 2014144932, A 20140814
137. Kamisaka T, Ishitani T, Fujiwara T (2014) Benzotriazole derivatives, organic fluorescent dyes containing them, and fluorescent polymer compositions and organic electroluminescent device substances containing the dyes. JP 2014144931, A 20140814
138. Schlifke-Poschalko A, Mendrok-Edinger C, Westenfelder H (2011) Preparation of 2-phenyl-1,2,3-benzotriazoles for UV radiation absorbance. WO 2011086124, A1 20110721
139. Tanaka N, Fujiwara T (2014) Water-soluble benzotriazole derivatives having polyoxyethylene chain, their use as UV absorbers, and cosmetics containing them. JP 2014037352, A 20140227
140. Tanaka N, Yamagata M (2007) Blue laser-absorbing benzotriazole derivatives and organic pigments and ultraviolet absorbents containing them. JP 2007126385
141. Trullas C, Pelejero C, Corbera Arjona J, Jimenez Alonso O, Novak L (2008) New benzotriazole derivatives, substituted by silylated 1,3,5-benzenetriol groups as cosmetic skin UV photoprotectors and a process for preparation thereof. ES 2300199, A1 20080601
142. Kamisaka T, Tanaka N (2012) Benzotriazole derivatives and UV absorbers containing them. JP 2012025681, A 20120209
143. Xu J, Huang J, Deng A (2006) Process for preparation of 2-(2-hydroxyphenyl)benzotriazole derivatives as UV-absorbents. CN 1727338, A 20060201
144. Chang C-H (2012) Ultraviolet light absorbing compounds based on benzyl substituted 2-(2-hydroxyphenyl) benzotriazoles. A1 20121227
145. Perrier S (2004) Developments in polymer architecture and design applicable to surface coatings. Surf Coat Int Part B Coat Trans 87:235–239
146. Zhang G, Wang P, Xiang R, Fu X, Wei Y (2014) Preparation of benzotriazole-triazine-coumarin derivatives as thermostable fluorescent brightener and yellowing inhibitor. CN 103755690, A 20140430

147. Yu T, Zhao Y, Ding X, Fan D, Qian L, Dong W (2007) Synthesis, crystal structure and photoluminescent behaviors of 3-(1H-benzotriazol-1-yl)-4-methyl-benzo [7, 8] coumarin. J Photochem Photobiol A Chem 188:245–251
148. Yu T, Zhang P, Zhao Y, Zhang H, Meng J, Fan D, Chen L, Qiu Y (2010) Synthesis, crystal structure and photo-and electro-luminescence of the coumarin derivatives with benzotriazole moiety. Org Electron 11:41–49
149. Yu T, Zhao Y, Fan D (2006) Synthesis, crystal structure and photoluminescence of 3-(1-benzotriazole)-4-methyl-coumarin. J Mol Struct 791:18–22
150. Hansen P, Heppner A, Rillmann T, Schumann C (2009) Cosmetic sunscreen composition containing a derivative of benzotriazole and an AHR (aryl hydrocarbon) receptor antagonist. EP 2078521, A1 20090715
151. Satonaka K, Ito T, Tanaka Y (2009) Oil-in-water cosmetic emulsions containing methylene bis(benzotriazolylphenols) as UV absorbers. JP 2009120492, A 20090604
152. Bojinov VB, Grabchev IK (2005) Novel functionalized 2-(2-hydroxyphenyl)-benzotriazole–benzo [de] isoquinoline-1, 3-dione fluorescent UV absorbers: synthesis and photostabilizing efficiency. J Photochem Photobiol A Chem 172:308–315
153. Nakanishi S, Onoe H, Kuroki M (2014) Benzotriazole structure-containing chromophore and wavelength conversion luminescence medium for solar cells. JP 2014185286, A 20141002
154. Teixeira FC, Rangela CM, Teixeira APS (2013) New azaheterocyclic aromatic diphosphonates for hybrid materials for fuel cell applications. New J Chem 37:3084
155. Min J, Zhang Z-G, Zhang S, Zhang M, Zhang J, Li Y (2011) Synthesis and photovoltaic properties of D–A copolymers based on dithienosilole and benzotriazole. Macromol 44:7632–7638
156. Peng B, Najari A, Liu B, Berrouard P, Gendron D, He Y, Zhou K, Zou Y, Leclerc M (2010) A new dithienylbenzotriazole-based poly (2, 7-carbazole) for efficient photovoltaics. Macromol Chem Phys 211:2026–2033
157. Kim J-H, Kim HU, Song CE, Kang IN, Lee J-K, Shin WS, Hwang D-H (2013) Benzotriazole-based donor–acceptor type semiconducting polymers with different alkyl side chains for photovoltaic devices. Sol Energy Mater Sol Cells 108:113–125
158. Yigitsoy B, Karim SMA, Balan A, Baran D, Toppare L (2011) Multichromic polymers of benzotriazole derivatives: effect of benzyl substitution. Electrochim Acta 56:2263–2268
159. Akbaşoğlu N, Balan A, Baran D, Cirpan A, Toppare L (2010) Electrochemical and optical studies of furan and thieno [3, 2-b] thiophene end capped benzotriazole derivatives. J Polym Sci Part A Polym Chem 48:5603–5610
160. Shibata Y (2012) Oil-based marking pens using aluminum containers with good ink resistance. JP 2014114335, A 20140626
161. Giesa H, Gross WD, Kroos A (2010) Agent, useful for dyeing and optionally bleaching keratin fibers, preferably human hair, comprises at least one new or known (1H-benzotriazol-5-yl)-phenyl-diazene compound. DE 102009045575, A1 20100729

Index